Ajith Abraham, Rafael Falcón, and Rafael Bello (Eds.)

Rough Set Theory: A True Landmark in Data Analysis

Studies in Computational Intelligence, Volume 174

Editor-in-Chief
Prof. Janusz Kacprzyk
Systems Research Institute
Polish Academy of Sciences
ul. Newelska 6
01-447 Warsaw
Poland
E-mail: kacprzyk@ibspan.waw.pl

Further volumes of this series can be found on our homepage: springer.com

Vol. 151. Tomasz G. Smolinski, Mariofanna G. Milanova and Aboul-Ella Hassanien (Eds.)
Computational Intelligence in Biomedicine and Bioinformatics, 2008
ISBN 978-3-540-70776-9

Vol. 152. Jarosław Stepaniuk
Rough – Granular Computing in Knowledge Discovery and Data Mining, 2008
ISBN 978-3-540-70800-1

Vol. 153. Carlos Cotta and Jano van Hemert (Eds.)
Recent Advances in Evolutionary Computation for Combinatorial Optimization, 2008
ISBN 978-3-540-70806-3

Vol. 154. Oscar Castillo, Patricia Melin, Janusz Kacprzyk and Witold Pedrycz (Eds.)
Soft Computing for Hybrid Intelligent Systems, 2008
ISBN 978-3-540-70811-7

Vol. 155. Hamid R. Tizhoosh and M. Ventresca (Eds.)
Oppositional Concepts in Computational Intelligence, 2008
ISBN 978-3-540-70826-1

Vol. 156. Dawn E. Holmes and Lakhmi C. Jain (Eds.)
Innovations in Bayesian Networks, 2008
ISBN 978-3-540-85065-6

Vol. 157. Ying-ping Chen and Meng-Hiot Lim (Eds.)
Linkage in Evolutionary Computation, 2008
ISBN 978-3-540-85067-0

Vol. 158. Marina Gavrilova (Ed.)
Generalized Voronoi Diagram: A Geometry-Based Approach to Computational Intelligence, 2009
ISBN 978-3-540-85125-7

Vol. 159. Dimitri Plemenos and Georgios Miaoulis (Eds.)
Artificial Intelligence Techniques for Computer Graphics, 2009
ISBN 978-3-540-85127-1

Vol. 160. P. Rajasekaran and Vasantha Kalyani David
Pattern Recognition using Neural and Functional Networks, 2009
ISBN 978-3-540-85129-5

Vol. 161. Francisco Baptista Pereira and Jorge Tavares (Eds.)
Bio-inspired Algorithms for the Vehicle Routing Problem, 2009
ISBN 978-3-540-85151-6

Vol. 162. Costin Badica, Giuseppe Mangioni, Vincenza Carchiolo and Dumitru Dan Burdescu (Eds.)
Intelligent Distributed Computing, Systems and Applications, 2008
ISBN 978-3-540-85256-8

Vol. 163. Pawel Delimata, Mikhail Ju. Moshkov, Andrzej Skowron and Zbigniew Suraj
Inhibitory Rules in Data Analysis, 2009
ISBN 978-3-540-85637-5

Vol. 164. Nadia Nedjah, Luiza de Macedo Mourelle, Janusz Kacprzyk, Felipe M.G. França and Alberto Ferreira de Souza (Eds.)
Intelligent Text Categorization and Clustering, 2009
ISBN 978-3-540-85643-6

Vol. 165. Djamel A. Zighed, Shusaku Tsumoto, Zbigniew W. Ras and Hakim Hacid (Eds.)
Mining Complex Data, 2009
ISBN 978-3-540-88066-0

Vol. 166. Constantinos Koutsojannis and Spiros Sirmakessis (Eds.)
Tools and Applications with Artificial Intelligence, 2009
ISBN 978-3-540-88068-4

Vol. 167. Ngoc Thanh Nguyen and Lakhmi C. Jain (Eds.)
Intelligent Agents in the Evolution of Web and Applications, 2009
ISBN 978-3-540-88070-7

Vol. 168. Andreas Tolk and Lakhmi C. Jain (Eds.)
Complex Systems in Knowledge-based Environments: Theory, Models and Applications, 2009
ISBN 978-3-540-88074-5

Vol. 169. Nadia Nedjah, Luiza de Macedo Mourelle and Janusz Kacprzyk (Eds.)
Innovative Applications in Data Mining, 2009
ISBN 978-3-540-88044-8

Vol. 170. Lakhmi C. Jain and Ngoc Thanh Nguyen (Eds.)
Knowledge Processing and Decision Making in Agent-Based Systems, 2009
ISBN 978-3-540-88048-6

Vol. 171. Chi-Keong Goh, Yew-Soon Ong and Kay Chen Tan (Eds.)
Multi-Objective Memetic Algorithms, 2009
ISBN 978-3-540-88050-9

Vol. 172. I-Hsien Ting and Hui-Ju Wu (Eds.)
Web Mining Applications in E-Commerce and E-Services, 2009
ISBN 978-3-540-88080-6

Vol. 173. Tobias Grosche
Computational Intelligence in Integrated Airline Scheduling, 2009
ISBN 978-3-540-89886-3

Vol. 174. Ajith Abraham, Rafael Falcón and Rafael Bello (Eds.)
Rough Set Theory: A True Landmark in Data Analysis, 2009
ISBN 978-3-540-89886-3

Ajith Abraham
Rafael Falcón
Rafael Bello
(Eds.)

Rough Set Theory: A True Landmark in Data Analysis

Springer

Ajith Abraham
Centre of Excellence for Quantifiable Quality of Service
Norwegian University of Science and Technology
O.S. Bragstads plass 2E
N-7491 Trondheim
Norway
Email: ajith.abraham@ieee.org

Rafael Bello
Department of Computer Science
Central University of Las Villas
Carretera Camajuaní km 5 1/2
Santa Clara, Villa Clara 54830
CUBA
Email: rbellop@uclv.edu.cu

Rafael Falcón
School of Information Technology and Engineering (SITE)
University of Ottawa
800 King Edward Ave, STE 5012
Ottawa, ON K1N 6N5
Canada
Email: rfalcon@uottawa.ca

ISBN 978-3-642-10062-8 e-ISBN 978-3-540-89921-1

DOI 10.1007/978-3-540-89921-1

Studies in Computational Intelligence ISSN 1860949X

Typeset & Cover Design: Scientific Publishing Services Pvt. Ltd., Chennai, India.

Printed in acid-free paper

9 8 7 6 5 4 3 2 1

springer.com

To my loving parents

Ajith Abraham

To the Northey family,
for their continuous support
and encouragement

Rafael Falcón

To Marilyn and my daughters

Rafael Bello

This edited volume aims at continue stressing the benefits of applying rough sets in many real-life situations while still keeping an eye on topological aspects of the theory as well as strengthening its linkage with other reasoning paradigms.

Part I is comprised of four chapters which intend to outline some theoretical contributions. The original concept of rough set introduced by Z. Pawlak has been extended in many different directions shortly after its very inception. Some of these extensions have to do with loosening the requirement of the underlying indiscernibility relation to be an equivalence relation. Chapter 1 is devoted to present two different extensions of rough sets which rely upon fuzzy proximity and intuitionistic fuzzy proximity relations, respectively. The parallelism of their basic properties with those of the conventional rough set model becomes evident and several real life applications illustrating the practical use of these generalizations are considered along the chapter.

Categories arise in mathematics and appear frequently in computer science where algebraic and logical notions have powerful representations using categorical constructions. Chapter 2 proposes a generalization of the standard rough set operators, which are based on relations involving ordinary sets, to embrace partially-ordered monads from which a theory of rough monads can be derived. This turns rough set theory into a categorical methodology with a broad avenue of further developments both from the theoretical and application standpoints. The chapter illustrates the use of rough monads in management of drug interactions and medical diagnosis.

Chapter 3 is concerned with the interpretation of both rough set and knowledge space theories in light of granular computing. In order to arrive at a unified framework bridging these two methodologies, it becomes necessary to analyze their granular structures and approximations regardless the different types of elementary granules and granular structures each of them utilizes for information representation and processing. The result is a multilevel granular structure with a set-theoretic formulation of granular structures as its building blocks. The framework allows studying rough set analysis and knowledge spaces in a common setting with results being applied interchangeably.

Finally, a study on some topological aspects of rough sets and approximation of classifications is the subject of Chapter 4. The author sheds light on the different types of unions and intersections of rough sets in addition to the formulation of a new concept of rough equivalence which captures the approximate equality of sets at a higher level than the existing notion of rough equality. The approximation of classifications is also revisited and some observations about the properties of rules generated from information systems and their inherent structure are made.

The second part of the book has to do with some applications of rough sets in data mining activities. An interesting direction in rough set research is that of clustering algorithms. Chapter 5 elaborates on how to endow rough clustering approaches with mechanisms of partial supervision, thus far being solely conceived for fuzzy clustering. Partial supervision comes in the form of fuzzy membership grades and allocation of patterns to positive or negative regions of the sought knowledge structures (clusters). The chapter also provides some insight on the influence of the type and number of knowledge-based hints being furnished to the system.

Rough set theory has been regarded since its very inception as a powerful and feasible methodology for performing data mining and knowledge discovery activities. Prominent results, also embracing the financial and economic realms, can be witnessed in many research studies across literature. Although discriminant analysis continues to be the preferred and most popular approach for making predictions in these areas, the application of rough-set-based data analysis leads, most of the times, to high overall prediction accuracy and interpretable rule bases. Despite this fact, prediction still remains a challenging and difficult goal to achieve. In light of this, Chapter 6 presents a generic scheme for generating decision rules for stock market prediction. Numerical attributes are handled via rough set with Boolean reasoning (RSBR) approach and the rough confusion matrix is used to evaluate the performance of the predicted reducts and classes.

On the other hand, a framework for modeling uncertainty in Web caching scenarios has been crafted and formalized in Chapter 7. Web caching and prefetching have been widely acknowledged as effective schemes to relieve the service bottleneck and minimize the user access latency, both hot problems still encountered in today's web applications. The proposed framework comprises three modules, which exploit rough set capabilities to reduce the rules of log files and, at the same time, enhance the prediction performance of the user behavior. The induction engine module, for example, has an ID3-like learning algorithm based on the minimum entropy principle.

The last four chapters present some rough hybrid models tailored for classification and attribute reduction purposes. Hybridization among computational intelligence techniques has been regarded as both a successful paradigm and a well-settled tendency in decision-making and machine learning fields, thus giving rise to stronger, synergetic models, which can better cope with the huge amounts of information available nowadays and draw precise knowledge in a more interpretable way. Chapter 8 provides an overview of some rough hybrid approaches and a thorough comparative study on their performance versus prominent, conventional classifiers such as support vector machines and decision trees. Hybrid models include neuro-fuzzy decision trees, rough-neuro-fuzzy decision trees and fuzzy-rough classification trees in the context of software defect classification. Conclusions about the types of metric data that need to be collected and the readability of the generated rules are derived.

One of the most appealing properties of rough set theory lies in its ability to spot and remove redundant attributes which are present in the information systems we usually deal with. Given the exponential computational complexity of feature selection, the call for heuristic approaches becomes a must. In particular, evolutionary algorithms offer impressive optimization capabilities due to the multiple handling of potential problem solutions. Chapters 9 and 10 provide an overview of such nature-inspired heuristics in their quest for suitable feature subsets. Settled at the very core of the approaches, some concepts borrowed from rough set theory have become an integral part of the guiding principle ruling the optimization scheme. Chapter 9 moves from greedy to swarm intelligence approaches, also featuring a novel dynamic evolutionary technique and a search strategy that substantially reduces the time needed to find appropriate solutions (reducts). Chapter 10 emphasizes on particle swarm optimization and its application to functional magnetic resonance imaging.

The main workflow of rough set data analysis (RSDA) is applied in Chapter 11 to draw knowledge from a set of events captured by protection, control and monitoring devices in a substation network. Once again, the reduct computation phase is largely profited from the presence of evolutionary approaches (this time, Genetic Algorithms), which also make possible to dynamically add more events into the system owing to their amazing adaptive capabilities. The ensuing rule base identifies and isolates the most probable faulty section in the network, hence improving the outage response time. The case studies span from simple scenarios to complex distribution substations comprised of various types of relay models and even include time series data.

We want to thank the authors for submitting their contributions to this volume and the prominent experts in the field all over the world who were willing to devote a share of their valuable time in reviewing the articles and providing insightful remarks.

The kind assistance we got during the publication phase from Thomas Ditzinger and Heather King of Springer-Verlag is gratefully acknowledged.

Santa Clara, Cuba
September 2008

Ajith Abraham
Rafael Falcón
Rafael Bello

Contents

List of Contributors

Aboul Ella Hassanien
Information Tech. Dept., Cairo Univ.
5 Ahamed Zewal St.
Orman, Giza, Egypt
a.hassanien@fci-cu.edu.eg

Ajith Abraham
Center for Quantifiable QoS
Norwegian Univ. of Sci. & Tech.
O.S. Bragstads plass 2E NO-7491
Trondheim, Norway
ajith.abraham@ieee.org

Ann Nowe
CoMo Lab, Comp Science Dept.
Vrije Universiteit Brussel
Brussels, Belgium
asnowe@info.vub.ac.be

Balakrushna Tripathy
School of Computing Sciences
VIT University
Vellore, T.N. 632 014, India
tripathybk@rediffmail.com

Ching Lai Hor
CSM Renewable Energy Group
University of Exeter in Cornwall
Tremough Campus, Penryn
TR10 9EZ, United Kingdom
c.l.hor@exeter.ac.uk

Dean Millar
CSM Renewable Energy Group
University of Exeter in Cornwall
Tremough Campus, Penryn
TR10 9EZ, United Kingdom
dlmillar@csm.ex.ac.uk

Duoqian Miao
Dept. of Comp. Science & Technology
Tongji University
Shanghai, China 201804
miaoduoqian@163.com

Feifei Xu
Department of Computer Science
University of Regina
Regina, Saskatchewan
S4S 0A2, Canada
xufeifei1983@hotmail.com

Gwanggil Jeon
Dept. of Electronics & Comp. Eng.
Hanyang University
17 Haengdang-dong, Seongdong-gu
Seoul, Korea
windcap315@ece.hanyang.ac.kr

Hameed Al-Qaheri
Dept. of Quant. Methods & Inf. Syst.
College of Business Administration
Kuwait University, Kuwait
alqaheri@cba.edu.kw

Hongbo Liu
School of Comp. Science and Eng.
Dalian Maritime University
116026 Dalian, China
lhb@dlut.edu.cn

James F. Peters
Computational Intelligence Lab
Electrical & Comp. Eng. Department
University of Manitoba
Winnipeg, Manitoba
R3T 5V6 Canada
jfpeters@ee.umanitoba.ca

Jechang Jeong
Dept. of Electronics & Comp. Eng.
Hanyang University
17 Haengdang-dong, Seongdong-gu
Seoul, Korea
jjeong@ece.hanyang.ac.kr

Johan Karlsson
Department of Comp. Architecture
University of Málaga, Spain
johan@ac.uma.es

Maria A. Galán
Department of Applied Mathematics
University of Málaga, Spain
magalan@ctima.uma.es

Patrik Eklund
Department of Computing Science
Umeå University, Sweden
peklund@cs.umu.se

Peter Crossley
Electrical Energy and Power Systems
University of Manchester
P.O.Box 88,Manchester
M60 1QD, United Kingdom
p.crossley@manchester.ac.uk

Rafael Bello
Central Univ. of Las Villas (UCLV)
Carretera Camajuaní km 5.5
Santa Clara, Cuba 54830
rbellop@uclv.edu.cu

Rafael Falcón
School of Information Tech. & Eng.
800 King Edward Ave, U of Ottawa
Ottawa, ON K1N 6N5, Canada
rfalcon@uottawa.ca

Rajen Bhatt
Samsung India Software Center
Noida-201305, Uttar Pradesh, India
rajen.bhatt@gmail.com

Sarina Sulaiman
Soft Computing Research Group
Faculty of Comp. Sci. & Inf. System
Universiti Teknologi Malaysia
Johor, Malaysia
sarina@utm.my

Sheela Ramanna
Dept. of Applied Computer Science
University of Winnipeg
Winnipeg, Manitoba
R3B 2E9 Canada
s.ramanna@uwinnipeg.ca

Simon Watson
Dept. of Electronic & Electrical Eng.
University of Loughborough
Holywell Park, Leicestershire
LE11 3TU, United Kingdom
s.j.watson@lboro.ac.uk

Siti Mariyam Shamsuddin
Soft Computing Research Group
Faculty of Comp. Sci. & Inf. System
Universiti Teknologi Malaysia
Johor, Malaysia
mariyam@utm.my

Yailé Caballero
University of Camagüey
Carretera Circunvalación Norte km 5
Camagüey, Cuba 74650
yaile@inf.reduc.edu.cu

Yanheng Li
School of Comp. Science and Eng.
Dalian Maritime University
116026 Dalian, China

Yiyu Yao
Department of Computer Science
University of Regina
Regina, Saskatchewan
S4S 0A2, Canada
yyao@cs.uregina.ca

Yudel Gómez
Central Univ. of Las Villas (UCLV)
Carretera Camajuaní km 5.5
Santa Clara, Cuba 54830
ygomezd@uclv.edu.cu

Part I

Theoretical Contributions to Rough Set Theory

Rough Sets on Fuzzy Approximation Spaces and Intuitionistic Fuzzy Approximation Spaces

B.K. Tripathy

School of Computing Sciences, V.I.T. University, Vellore, T.N. 632 014, India
tripathybk@rediffmail.com

Summary. In the past few years the original concept of rough sets, as introduced by Pawlak [26] has been extended in many different directions. Some of these extensions are obtained by relaxing the requirement of the basic relations to be equivalence relations [14,15,31,32,34,35,37,38].That is by dropping the requirement of transitivity or symmetry. One such approach is to replace the equivalence relations by fuzzy proximity relations. The notions of rough sets thus generated are called rough sets defined upon fuzzy approximation spaces [14,15]. A generalization of this is obtained by taking intuitionistic fuzzy proximity relations instead of equivalence relations, called rough sets on intuitionistic fuzzy approximation spaces [37,38]. In this chapter we shall be concentrating on the study of these two notions of rough sets. It is our objective to define these types of rough sets along with related concepts and establish their properties which are parallel to those of basic rough sets. Several real life applications shall be considered in the sequel to illustrate the power and necessity of these generalized models of rough sets in the representation and study of imperfect knowledge.

1 Introduction

Most of our traditional tools for formal modeling, reasoning and computing are crisp, deterministic and precise in character. Real situations are very often not crisp and deterministic and they cannot be described precisely. For a complete description of a real system often one would require by far more detailed data than a human being could ever recognize simultaneously, process and understand. This observation leads to the extension of the concept of crisp sets so as to model imprecise data which can enhance the modeling power and applicability of Mathematics in different branches of knowledge. Philosophers, logicians and mathematicians have tackled the problem of imperfect knowledge for a long time. Recently, the problem also became a crucial issue for computer scientists, particularly in the area of Artificial Intelligence. There are many approaches for understanding and manipulating imperfect knowledge. The first and the most successful approach so far is, no doubt, Zadeh's fuzzy set theory [45].

In this chapter we shall be dealing with two generalization of basic rough sets introduced by Pawlak, namely rough sets on fuzzy approximation spaces and rough sets on intuitionistic fuzzy approximation spaces. There are several other generalizations of the basic rough set notion. However, our approach is

A. Abraham, R. Falcón, and R. Bello (Eds.): Rough Set Theory, SCI 174, pp. 3–44.
springerlink.com

different from the other extensions in the sense that from the base relations (fuzzy proximity relation in case of rough sets on fuzzy approximation spaces and intuitionistic fuzzy proximity relation in case of rough sets on intuitionistic fuzzy approximation spaces) we are deriving equivalence relations and basing upon these derived equivalence relations we define the notions of upper approximation and lower approximation of sets. So, must of the concepts on basic rough set theory can be mapped onto these new situations. The interesting part lies in the applications, which shall be illustrated through several examples in the chapter. Most importantly, the proximity and non-proximity parameters for attribute values on information spaces provide controls, which can be utilized as per the requirement of the situations.

1.1 Fuzzy Sets

Fuzzy set theory provides a strict mathematical framework in which vague conceptual phenomena can be preciously and rigorously studied. Fuzziness is particularly frequent, however in all areas in which human judgement, evaluation and decisions are important. The basic idea behind the introduction of fuzzy set is to allow graded membership of elements instead of dichotomous membership. There are two types of fuzziness with respect to their origin ; namely intrinsic fuzziness and informational fuzziness.

The meaning of a word might even be well defined, but when using the word as a label for a set, the boundaries within which objects do or do not belong to the set become fuzzy or vague. For example, in the terms such as "tall man" or "beautiful women" the meaning of the words "tall" or "beautiful" are fuzzy and context sensitive. A person who is considered to be tall in Africa may not be so in Asia and similarly a tall person in Asia may not be considered tall in Europe. Similarly, the word "beautiful" has varied meaning in different contexts. The above are examples of intrinsic fuzziness.

On the other hand, in the term "trustworthy customer" the word "trustworthy" can possibly be described completely and crisply if we use a large number of descriptors. All these descriptor values may not be known or may be known but difficult to handle simultaneously. So, such terms come under "subjective category" and become fuzzy. This is an example of informational fuzziness.

Fuzziness has a major role to play in the branch of Computer Science. The development of branches like fuzzy database systems [9,10], fuzzy pattern recognition [25], fuzzy image processing [4,5], fuzzy logic [47,48], fuzzy Boolean algebra [43] and soft computing are results of such applications and are only a few in a long list.

In fact, the notion of fuzzy set is defined through the concept of membership function, which generalizes the corresponding concept of characteristic function of crisp sets and is defined as follows [45].

Definition 1. A *fuzzy set* A over a universal set U is given by its *membership function* μ_A which is defined as

$$\mu_A : U \to [0, 1],$$

such that every $x \in U$ is associated with its membership value $\mu_A(x)$ which is a real number lying in [0,1].

As mentioned above, fuzzy sets have better modeling power than crisp sets. The graded membership value assigned to elements provides additional modeling power to fuzzy sets. Every element also has a non-membership value assigned to it, which is the one's complement of its membership value. Formally, for a fuzzy set 'A' with membership function μ_A, the non-membership function denoted by ν_A is defined as

$$\nu_A : U \to [0,1],$$

such that $\nu_A(x) = 1 - \mu_A(x)$, for every $x \in U$.

If we move back to crisp sets, then the membership function defines a set and the non-membership function defines its complement.

1.2 Intuitionistic Fuzzy Sets

It was observed by K.T.Attanasov [1,2] that the membership and non-membership values of an element with respect to a collection of elements from a universe may not add up to 1 in all possible cases. Let us consider some real life examples. In the exit polls in which we find that besides the options "yes" or "no" invariably there is a third option named "can't say". The last option expresses the hesitation or reluctance among the people participating in the poll. Using the information available with them, it is impossible for them to go for any of the first two options. Another example is the voting in parliament. Some members vote in favour of a resolution, some vote against it, while others abstain from voting. So, if we consider the collection of people voting for the resolution then the last group of people can neither be categorized as belonging to the collection nor not belonging to it. Similarly, let us consider the example of a "trustworthy customer". If a customer is trusted in 70 cases out of 100, we cannot conclude that he/she cannot be trusted in all the remaining 30 cases. There shall be obvious cases in which one does not have sufficient information to conclude in either way. The cases where an element can not be characterized as either belonging to or not belonging to a collection forms its hesitation part. Based upon the concept of the hesitation notion of intuitionistic fuzzy sets was introduced by Attanasov [2] as an extension of the notion of fuzzy sets. We formally define it as follows:

Definition 2. An *intuitionistic fuzzy set* A on a universe U is defined by the two functions; *membership function* μ_A and *non-membership function* ν_A such that

$$\mu_A, \nu_A : U \to [0,1],$$

where $0 \leq \mu_A(x) + \nu_A(x) \leq 1$, for all $x \in U$.

The hesitation function Π_A for an intuitionistic fuzzy set is given by

$$\Pi_A(x) = 1 - \mu_A(x) - \nu_A(x), \text{ for all } x \in A.$$

For a fuzzy set $\Pi_A(x) = 0$ for all $x \in A$. In the extreme case when $\mu_A(x) = 0$ and $\nu_A(x) = 1$, it is certain that the element x does not belong to A. Also, when $\mu_A(x) = 1$ and $\nu_A(x) = 0$, it is certain that the element x does not belong to A. Finally, when $\mu_A(x) = 0$ and $\nu_A(x) = 0$, we have $\Pi_A(x) = 1$ and hence it is completely undecidable whether the element is in A or not in A.

1.3 Rough Sets

Rough set theory, introduced by Pawlak [26] is a new mathematical approach to vagueness. The basic philosophy of rough sets is founded on the assumption that with every object of the universe of discourse, we associate some information (data, knowledge). Objects characterized by the same information are indiscernible in view of the available information about them. The indiscernibility relation generated in this way is the mathematical basis for rough set theory. Any set of all indiscernible objects is called an elementary set and forms a basic granule of knowledge about the universe. Any set of objects, being a union of some elementary sets, is referred to as crisp (precise);otherwise a set is rough (imprecise, vague). Consequently, each rough set has boundary line cases that are objects which can not be classified with certainty as members of a set or its complement. The idea of rough set consists of approximation of a set by a pair of sets, called the *lower* and *upper approximation* of the set.

In the beginning rough sets have been compared to fuzzy sets. These two theories were thought to be competitors in the study of imperfect knowledge. However, such a comparison is found to be incorrect [11]. Indiscernibility and vagueness are distinct facets of imperfect knowledge. Indiscernibility refers to the granularity of knowledge that affects the definition of universes of discourse. Vagueness is due to the fact that categories of natural language are often gradual notions and refer to sets with smooth boundaries. The basic assumption in rough set theory is that knowledge is deeply seated in the classificatory abilities of human beings and other species. Classification on more abstract levels seems to be a key issue in reasoning, learning and decision making. Thus knowledge in rough set approach is necessarily connected with the variety of classification patterns related to specific parts of the real or abstract world, which we call as universe. Knowledge consists of a family of classification patterns of a domain of interest, which provide explicit facts about reality. Together with the reasoning capacity we are able to deliver implicit facts derivable from explicit knowledge.

We know that equivalence relations induce classifications on a universal set and vice versa. So, for mathematical reasons equivalence relations are used instead of classifications to define a rough set.

Let $U(\neq \phi)$ be a finite set of objects, called the universe and R be an equivalence relation over U. By U/R we denote the family of all equivalence classes of R (or classification of U) referred to as categories or concepts of R and $[x]_R$ denotes a category in R containing an element $x \in U$. By a knowledge base, we understand a relation system $K = (U, \Re)$, where U is as above and $\Re$ is a family of equivalence relations over U. A knowledge base is also called an *approximation space* [14,15].

For any subset $P(\neq \phi) \subseteq \Re$, the intersection of all equivalence relations in P is denoted by $IND(P)$ and is called the *indiscernibility relation* over P. The equivalence classes of $IND(P)$ are called *P-basic knowledge categories* about U in K. For any $Q \in \Re$, Q is called a *Q-elementary knowledge* about U in K and equivalence classes of Q are called *Q-elementary concepts* of knowledge $\Re$. The family of P-basic categories for all $P(\neq \phi) \subseteq \Re$ is called the family of basic categories in knowledge base K. By $IND(K)$, we denote the family of all equivalence relations defined in K. Symbolically, $IND(K) = \{IND(P) : P(\neq \phi) \subseteq \Re\}$.

For any $X \subseteq U$ and an equivalence relation $R \in IND(K)$, we associate two subsets,$\underline{R}X = \cup\{Y \in U/R : Y \subseteq X\}$ and $\bar{R}X = \cup\{Y \in U/R : Y \cap X \neq \phi\}$, which are called the *R-lower* and *R-upper approximations* of X respectively. The *R-boundary* of X is denoted by $BN_R(X)$ and is given by $BN_R(X) = \bar{R}X - \underline{R}X$. The elements of $\underline{R}X$ are those elements of U which can be certainly classified as elements of X with the knowledge of R and $\bar{R}X$ is the set of elements of X which can be possibly classified as elements of X employing knowledge of R. The borderline region is that area of the universe which is not decidable. We say X is *rough* with respect to R if and only if $\underline{R}X \neq \bar{R}X$; equivalently $BN_R(X) \neq \phi$. X is said to be *R-definable* if and only if $\underline{R}X = \bar{R}X$, or $BN_R(X) = \phi$. So, a set is *rough* with respect to R if and only if it is not R-definable.

1.4 Motivation

The original rough set theory proposed by Pawlak [26] is based upon equivalence relations defined over a universe. It is the simplest formalization of indiscernibility. However, it cannot deal with some granularity problems we face in real information systems. This has lead to many meaningful and interesting extensions of the original concept. One such extension is to take tolerance or similarity relations [28,31,32,34,35] instead of equivalence relations. The rough sets thus generated have better generality and applicability than the basic ones. Another such approach is to take fuzzy similarity relations instead of equivalence relations as the basics, which can deal with data sets having both vagueness and fuzziness and the models of rough sets obtained are called fuzzy rough sets [18,19,20,21,23,44]. A covering of a universe is a generalization of the concept of partition of the universe. Rough sets based on coverings instead of partitions have been studied in several papers [6,7,8,22,29,30,36,49,50,51]. A complete completely distributive (CCD) lattice is selected as the mathematical foundation on which definitions of lower and upper approximations that form the basics concepts of rough set theory are proposed in [16]. These definitions result from the basic concepts of cover introduced on a CCD lattice and improve the approximations of the existing crisp generalizations of rough sets with respect to similarity relation and covers.

We would like to note that covers are also available in case of similarity relation based rough sets introduced by Slowinski and Vanderpooten [34,35]. So, similarity relation-based rough sets can be brought into the framework of covering rough set theory from the theoretical viewpoint. However, there are just some

Definition 14. Let R be an IF-proximity relation on U. If $(x, y) \in R_{\alpha,\beta}$ then we say that x and y are (α, β) *-similar* and write $'xR_{\alpha,\beta}y'$.

Definition 15. Let R be an IF-proximity relation on U. For any $(\alpha, \beta) \in J$ we say that two elements x and y are (α, β)*-identical* if $xR_{\alpha,\beta}y$ or x is *transitively* (α, β) *-similar* to y; that is, there exists elements $u_1, u_2, ..., u_n \in U$ such that $xR_{\alpha,\beta}u_1, u_1R_{\alpha,\beta}u_2, ..., u_nR_{\alpha,\beta}y$ and write $'xR(\alpha, \beta)y'$.

Note 1. It can be proved that the relation $R(\alpha, \beta)$ is an equivalence relation on U for each fixed $(\alpha, \beta) \in J$. For any $(\alpha, \beta) \in J$,

(i) $xR(\alpha, \beta)x$ for all x, as by definition of R, $\mu_R(x, x) = 1 \geq \alpha$ and $\upsilon_R(x, x) = 0 \leq \beta$. So, $R(\alpha, \beta)$ is reflexive.
(ii) Next, suppose $xR(\alpha, \beta)y$. Then two cases arise. In the first case, $xR_{\alpha,\beta}y$. So $\mu_R(x, y) \geq \alpha$ and $\upsilon_R(x, y) \leq \beta$.

But $\mu_R(x, y) \geq \alpha \Rightarrow \mu_R(y, x) \geq \alpha$ and $\upsilon_R(x, y) \leq \beta \Rightarrow \upsilon_R(y, x) \leq \beta$. So we get $yR(\alpha, \beta)x$. The proof in the other case is similar. Hence, $R(\alpha, \beta)$ is symmetric.
(iii) Finally, let $xR(\alpha, \beta)y$ and $yR(\alpha, \beta)z$. Then in the most general case, for some n and m there exists $u_1, u_2, ..., u_n, v_1, v_2, ..., v_m$ such that

$$xR_{\alpha,\beta}u_1, ..., u_nR_{\alpha,\beta}y, yR_{\alpha,\beta}v_1, ..., v_mR_{\alpha,\beta}z.$$

But, this in turn implies that $xR(\alpha, \beta)z$. So,$R(\alpha, \beta)$ is transitive.

For each fixed $(\alpha, \beta) \in J$, we shall denote the equivalence classes generated by the equivalence relation R (α, β) by $R^*_{\alpha,\beta}$.

1.7 Chapter Structure

In this sect. we have unwound the general background of imperfect or imprecise data and the basic models which capture much data. In the sequel the notions of fuzzy set, intuitionistic fuzzy set and rough sets are introduced. In sect. 1.4 we put the objective of the chapter in proper perspective with reference to current trends of research in this direction. The first major topic of discussion in this chapter is rough sets on fuzzy approximation space, which is based upon the concept of fuzzy proximity relation defined in sect. 1.5. Similarly the notion of intuitionistic fuzzy proximity relation is introduced in sect. 1.6 on which the major topic of study, "Rough sets on intuitionistic fuzzy approximation spaces" is based upon.

In sect. 2 we present rough sets on fuzzy approximation spaces. The first two subsections under this sect. deal with preliminaries leading to definition of the concept and elementary properties associated with it. Subsequently we shall study the problems of reduction of knowledge, relative reduct and relative core of knowledge, dependency of knowledge and partial dependency of knowledge in fuzzy approximation spaces. We conclude this sect. with an application of the concepts developed in it to the study of performance analysis of students and determine the important characteristics controlling it at a fixed level of proximity. This can be termed as an advanced case study as we have not considered real data. In sect. 3 the concept of intuitionistic fuzzy approximation space has been introduced and its general properties are studied. In sect. 4 the problem of knowledge reduction in IF-approximation spaces is studied. The topological property of kinds of rough sets has been extended to the context of both fuzzy approximation space and intuitionistic fuzzy approximation space. In sect. 5 this extended notion has been presented and properties of kinds of union and intersection of such rough sets have been presented. Section 6 deals with relative reduct and relative core of knowledge, sect. 7 deals with dependency of knowledge and sect. 8 deals with partial dependency of knowledge in IF-approximation spaces. An illustrative example which deals with the real life problem of selection of supporting books by a student has been tackled with the help of the concepts developed on IF-approximation spaces. Also, we have made comparison of the concepts with their counterparts on fuzzy approximation spaces. This follows with a conclusion and finally we end up with an extensive bibliography of materials referred for the compilation of the chapter.

2 Rough Sets on Fuzzy Approximation Spaces

In this sect., we shall discuss on an extension of basic rough sets, which is defined through less stringent relations than the equivalence relations. In fact, we consider fuzzy proximity relations defined in sect. 2.4, which are more general and abundant than equivalence relation as basic relations and define rough sets on the generated approximation space, which are called rough sets on fuzzy approximation spaces.

2.1 Preliminaries

Definition 16. For any set of fuzzy proximity relations $\Re$, the pair K = (U,$\Re$) is called a *fuzzy approximation space* [14,15].

For any fixed $\alpha \in [0,1]$, $\Re$ generates a set of equivalence relations $\Re(\alpha)$ and we call the associated space $\mathbf{K}(\alpha) = (U, \Re(\alpha))$ as the *generated approximation space* corresponding to $\mathbf{K}$ and α.

Definition 17. Let $\mathbf{K} = (U, \Re)$ be a fuzzy approximation space and $X \subseteq U$. Then for any fixed $\alpha \in [0,1]$, the rough set of X in the generated approximation space $\mathbf{K}(\alpha)$ and corresponding to the equivalence relation $R(\alpha) \in IND(\mathbf{K}(\alpha))$ is denoted by $(\underline{R}X_\alpha, \bar{R}X_\alpha)$, where

$$\underline{R}X_\alpha = \bigcup\{Y : Y \in R^*_\alpha Y \subseteq X\}. \tag{5}$$

and

$$\bar{R}X_\alpha = \bigcup\{Y : Y \in R^*_\alpha Y \cap X \neq \phi\}. \tag{6}$$

$\underline{R}X_\alpha$ and $\bar{R}X_\alpha$ are called the *α-lower approximation* and *α-upper approximation* of X respectively.

Definition 18. Let $X \subseteq U$, $\alpha \in [0,1]$ and $R(\alpha) \in IND(\mathbf{K}(\alpha))$. Then X is said to be *(R,α)-discernible* (that is *(R,α)-definable*) if and only if $\underline{R}X_\alpha = \bar{R}X_\alpha$.

For simplicity of notation, whenever the relation is clear from the context we write $(\underline{X}_\alpha, \bar{X}_\alpha)$ for $(\underline{R}X_\alpha, \bar{R}X_\alpha)$.

2.2 Properties

Several properties of rough sets on fuzzy approximation spaces have been established by De [14,15]. We first state these properties below, which are parallel to the corresponding properties in the basic case. The proofs of these properties are similar to those in the basic case and we omit them. In fact, some of the proofs in the generalised form shall be provided when we deal with rough sets on intuitionistic fuzzy approximation spaces. But the applications are noteworthy. We shall consider several examples to illustrate the applications. Also, we shall carry out the extension process further by the introduction of new parallel concepts and demonstrate their applicability.

Property 1. [14] Let R be a fuzzy relation from X to Y. Then for any two level values α_1 and $\alpha_2 \in [0,1]$ if $\alpha_1 \geq \alpha_2$ then

(i) $R_{\alpha_1} \subseteq R_{\alpha_2}$
(ii) $R(\alpha_1) \subseteq R(\alpha_2)$.
(iii) $R^*_{\alpha_1} \subseteq R^*_{\alpha_2}$.

Property 2. [14] If R and S are two fuzzy relations on $X \times Y$ then for all $\alpha \in [0,1]$,

(i) $(R \cup S)_\alpha = R_\alpha \cup S_\alpha$.
(ii) $(R \cap S)_\alpha = R_\alpha \cap S_\alpha$.
(iii) $(R \cup S)(\alpha) \subseteq R(\alpha) \cup S(\alpha)$
(iv) $(R \cap S)(\alpha) \supseteq R(\alpha) \cap S(\alpha)$.

Property 3. [14] If R and S are two fuzzy relations on U then for all $\alpha \in [0,1]$.

(i) $(R \cup S)^*_\alpha \subseteq R^*_\alpha \cup S^*_\alpha$.
(ii) $(R \cap S)^*_\alpha \supseteq R^*_\alpha \cap S^*_\alpha$.

Property 4. [14] For any fixed $\alpha \in [0,1]$ and $X \subseteq U$, we have

(i) $\underline{X}_\alpha \subseteq X \subseteq \bar{X}_\alpha$.
(ii) $\underline{\phi}_\alpha = \phi = \bar{\phi}_\alpha$ and $\underline{U}_\alpha = U = \bar{U}_\alpha$.
(iii) $\overline{(X \cup Y)}_\alpha = \bar{X}_\alpha \cup \bar{Y}_\alpha$.
(iv) $\underline{(X \cup Y)}_\alpha = \underline{X}_\alpha \cap \underline{Y}_\alpha$.
(v) $X \subseteq Y$ implies $\underline{X}_\alpha \subseteq \underline{Y}_\alpha$.
(vi) $X \subseteq Y$ implies $\bar{X}_\alpha \subseteq \bar{Y}_\alpha$.
(vii) $\underline{(X \cup Y)}_\alpha \supseteq \underline{X}_\alpha \cup \underline{Y}_\alpha$
(viii) $\overline{(X \cup Y)}_\alpha \subseteq \bar{X}_\alpha \cap \bar{Y}_\alpha$.

Property 5. [14] If $\alpha_1 \geq \alpha_2$ then

(i) $\underline{X}_{\alpha_1} \subseteq \underline{X}_{\alpha_2}$
(ii) $\bar{X}_{\alpha_1} \subseteq \bar{X}_{\alpha_2}$.

Property 6. [14] Let R and S be two fuzzy proximity relations on U and α be a chosen level threshold. Then

(i) $\underline{(R \cup S)}_{\alpha} X \subseteq \underline{R}_{\alpha} X \cup \underline{S}_{\alpha} X$.
(ii) $\overline{(R \cup S)}_{\alpha} X \supseteq \bar{R}_{\alpha} X \cup \bar{S}_{\alpha} X$.
(iii) $\underline{(R \cap S)}_{\alpha} X \supseteq \underline{R}_{\alpha} X \cap \underline{S}_{\alpha} X$.
(iv) $\overline{(R \cap S)}_{\alpha} X \subseteq \bar{R}_{\alpha} X \cup \bar{S}_{\alpha} X$.

2.3 Reduction of Knowledge in Fuzzy Approximation Spaces

It is always advantageous to know and find out whether the whole knowledge is necessary for defining some categories available in the knowledge considered or a portion of it is sufficient for the purpose. This topic is referred as reduction of knowledge and has importance in partial applications.

In this section we shall introduce the concept of α -dispensability and α -indispensability of relations along with α -dependency and α -independency of family or relations. The fundamental concepts of α -reduct and α -core are to be defined and their properties are to be studied, which are very important in the context of the topic of this section. The concept shall be illustrated through an example.

Definition 19. Let $\Re$ be a family of fuzzy proximity relations on U and $\alpha \in [0, 1]$. For any $R \in \Re$, we say that R is *α-dispensable* or *α-superfluous* in $\Re$ if and only if $IND(\Re(\alpha)) = IND(\Re(\alpha) - R(\alpha))$. Otherwise, R is *$\alpha$-indispensable* in $\Re$.

$\Re$ is said to be *α-independent* if each $R \in \Re$ is α -indispensable Otherwise, $\Re$ is *α-dependent.*

Definition 20. For a fixed $\alpha \in [0, 1]$, let $P(\neq \phi) \subseteq \Re$. Then $Q \subseteq P$ is a *α-reduct* of P if

$$Q \text{ is } \alpha - \text{independent}. \tag{7}$$

and

$$IND(Q(\alpha)) = IND(P(\alpha)). \tag{8}$$

It is denoted by α-RED(P). That is, α-RED(P) $= \{Q : Q \subseteq P, Q$ is α -independent and $IND(Q(\alpha)) = IND(P(\alpha))\}$. Hence, P may have many α-reducts, for each fixed $\alpha \in [0, 1]$.

Definition 21. Let $P(\neq \phi) \subseteq \Re$. Then, for a fixed $\alpha \in [0, 1]$, the set of all α *-indispensable* relations in P is called the *α-core* of P and it is denoted by α-CORE (P). That is,

α -CORE (P) $= \{R : R \in$ P and R is α-indispensable in P $\}$.

Theorem 1. For each fixed $\alpha \in [0, 1]$, if Q is α-independent and P $\subseteq$ Q then P is also α-independent.

Proof. Suppose on the contrary that, P $\subseteq$ Q and P is α-dependent.
Then there exists S $\subset$ P, such that $IND(S(\alpha)) = IND(P(\alpha))$.

Hence, $IND(S(\alpha) \cup (Q(\alpha) - P(\alpha))) = IND(P(\alpha) \cup (Q(\alpha) - (P(\alpha))) = IND(Q(\alpha))$, where $S \cup (Q - P) \subset Q$. Thus Q is α-independent, which is a contradiction.

The following result can be obtained from the Theorem 1.

Corollary 1. For a fixed $\alpha \in [0, 1]$, if $P \subset Q$ and P is α-dependent, then Q is α-dependent.

Theorem 2. For each fixed $\alpha \in [0, 1]$, α -CORE(P) $= \bigcap \alpha$ -RED(P), where α-RED(P) is the family of all α -reducts of P.

Proof. If Q is a α -reduct of P , then $IND(P(\alpha))) = IND(Q(\alpha))$.

Let $R \in P - Q$. Then $Q \subseteq P - \{R\} \subset P$ and hence $IND(Q(\alpha)) \subseteq IND(P(\alpha) - R(\alpha)) \subseteq IND(P(\alpha))$.

Thus R is α-superfluous in P, that is $R \notin \alpha$ -CORE(P).

Hence α-CORE $(P) \subseteq \{$ Q : Q is a α-reduct of P $\}$.

That is α-CORE $(P) \subseteq \bigcap \{$ Q : Q $\in \alpha$ -RED(P) $\}$.

Now, suppose R $\in \alpha$ -CORE(P). Then R is α-dispensable in P. Hence, $IND(P(\alpha) - \{R(\alpha)\}) = IND(P(\alpha))$.

So, there exists an α-independent subset $S \subseteq P - \{$ R $\}$ such that IND $(S(\alpha)) = IND(P(\alpha))$.

Obviously, S is a α-reduct of P and R $\notin$ S.

So, $\{$ S : S is a α-reduct of P $\} \subseteq \alpha$ -CORE(P).

That is, $\bigcap \{$ Q : Q α-RED(P) $\} \subseteq \alpha$ -CORE(P).

Note 2. As in the basic cases, from the above result, we conclude that the α-core can be interpreted as the most characteristic part of the knowledge to degree α, which can not be eliminated when reducing knowledge in a fuzzy approximation space. The smallest α-reduct of R is the α-core of R. So α-core can be used as a basis for computation of all α-reducts; that is, the α-core is included in every α-reduct. We illustrate the computation of α-core and α-reduct in fuzzy approximation spaces through the following example:

Example 1. Consider the universal set $U = \{x_1, x_2, x_3, x_4, x_5\}$. Suppose each object x_i is associated with the attributes a, b, c and d. Suppose each of the attributes defines a fuzzy proximity relation on U.

We define the fuzzy proximity relations P, Q, R and S over U corresponding to the attributes a, b, c and d respectively in the form of the following tables.

Table 1. Fuzzy proximity relation for attribute P

P	x_1	x_2	x_3	x_4	x_5
x_1	1	0.3	0.6	0.8	0.5
x_2	0.3	1	0.7	0.4	0.4
x_3	0.6	0.7	1	0.2	0.8
x_4	0.8	0.4	0.2	1	0.5
x_5	0.5	0.4	0.8	0.5	1

Table 2. Fuzzy proximity relation for attribute Q

Q	x_1	x_2	x_3	x_4	x_5
x_1	1	0.3	0.4	0.2	0.5
x_2	0.3	1	0.8	0.6	0.6
x_3	0.4	0.8	1	0.3	0.9
x_4	0.2	0.6	0.3	1	0.7
x_5	0.5	0.2	0.9	0.7	1

Table 3. Fuzzy proximity relation for attribute R

R	x_1	x_2	x_3	x_4	x_5
x_1	1	0.3	0.2	0.8	0.7
x_2	0.3	1	0.5	0.3	0.5
x_3	0.2	0.5	1	0.6	0.4
x_4	0.8	0.3	0.6	1	0.9
x_5	0.7	0.5	0.4	0.9	1

Table 4. Fuzzy proximity relation for attribute S

S	x_1	x_2	x_3	x_4	x_5
x_1	1	0.3	0.2	0.2	0.5
x_2	0.3	1	0.5	0.3	0.2
x_3	0.2	0.5	1	0.2	0.4
x_4	0.2	0.3	0.2	1	0.5
x_5	0.5	0.4	0.4	0.5	1

Table 5. Fuzzy proximity relation for $IND(\Re(\alpha))$

$IND(\Re(\alpha))$	x_1	x_2	x_3	x_4	x_5
x_1	1	0.3	0.2	0.2	0.5
x_2	0.3	1	0.3	0.3	0.2
x_3	0.2	0.3	1	0.2	0.4
x_4	0.2	0.3	0.2	1	0.4
x_5	0.5	0.2	0.4	0.4	1

Here $\Re = \{P, Q, R, S\}$ is a family of fuzzy proximity relations over U.

Suppose $\alpha = 0.6$. Then we get the following partitions based on fuzzy proximity relations.

$U/P(\alpha) = \{\{x_1, x_2, x_3, x_4, x_5\}\}$, $U/Q(\alpha) = \{\{x_1\}, \{x_2, x_3, x_4, x_5\}\}$, $U/R(\alpha) = \{\{x_1, x_3, x_4, x_5\}, \{x_2\}\}$, $U/S(\alpha) = \{\{x_1\}, \{x_2\}, \{x_3\}, \{x_4\}, \{x_5\}\}$.

The fuzzy proximity relation corresponding to $IND(\Re(\alpha))$ is given by Table 5.

So, $U/IND(\Re(\alpha)) = \{\{x_1\}, \{x_2\}, \{x_3\}, \{x_4\}, \{x_5\}\}$.

Now, $U/IND((\Re - P)(\alpha)) = \{\{x_1\}, \{x_2\}, \{x_3\}, \{x_4\}, \{x_5\}\} = U/IND(\Re(\alpha))$.
$U/IND((\Re - Q)(\alpha)) = \{\{x_1\}, \{x_2\}, \{x_3\}, \{x_4\}, \{x_5\}\} = U/IND(\Re(\alpha))$.
$U/IND((\Re - R)(\alpha)) = \{\{x_1\}, \{x_2\}, \{x_3\}, \{x_5\}, \{x_4\}\} = U/IND(\Re(\alpha))$.
$U/IND((\Re - S)(\alpha)) = \{\{x_1\}, \{x_2\}, \{x_3, x_4, x_5\}\} \neq U/IND(\Re(\alpha))$.

Here the relations P, Q and R are α-dispensable in $\Re$ whereas S is α-indispensable in $\Re$. So, { P, S }, { Q, S } and { S, R } are the α-reducts of $\Re$ and $\Re$ is α-dependent.

From above, we get α REDUCT ($\Re$) = { {P, S }, { Q, S }, { S, R } }.
So, α -CORE ($\Re$) = { P, S } $\cap$ { Q, S } $\cap$ { S, R } = S.

Similarly, we can get the different α-reducts and α-cores for different values of $\alpha \in [0, 1]$, which helps to find the core as per our level of requirement.

2.4 Relative Reducts and Relative Core of Knowledge in Fuzzy Approximation Spaces

In rough set philosophy, according to Pawlak [27], the discovery of inference rules in rough set context can be formulated as : how from a given knowledge another knowledge can be induced. In this section we generalize the concepts of α-core and α-reduct discussed in the previous section so that these generalized concepts are more suitable from applications point of view.

Definition 22. Let P and Q be two fuzzy proximity relations over the universe U. For every fixed $\alpha \in [0, 1]$, the *α-positive region* of P with respect to Q can be defined as

$$\alpha - POS_P Q = \bigcup_{X_\alpha \in U/Q} \underline{P} X_\alpha. \tag{9}$$

That means, (P, α) -positive region of Q is the set of all objects of the universe U which can be properly classified to the classes of U/Q(α) employing the knowledge expressed by the classification U/P(α).

Now, we provide the generalized concepts considered in Section 2.3.

Definition 23. Let **P** and **Q** be two families of fuzzy proximity relations on U. For every fixed $\alpha \in [0, 1]$, and $R \in \mathbf{P}$, R is $(\mathbf{Q}, \alpha)$-dispensable in **P** if

$$\alpha - POS_{IND(\mathbf{P})} IND(\mathbf{Q}) = \alpha - POS_{IND(\mathbf{P}-\{R\})} IND(\mathbf{Q}). \tag{10}$$

Otherwise R is $(\mathbf{Q}, \alpha)$-indispensable in **P**.

If every $R \in \mathbf{P}$ is $(\mathbf{Q}, \alpha)$ -indispensable in **P**, we say that **P** is $(\boldsymbol{Q}, \alpha)$-*independent.*

Definition 24. For every fixed $\alpha \in [0, 1]$, the family $\mathbf{S} \subseteq \mathbf{P}$ is a $(\mathbf{Q}, \alpha)$-reduct of **P** if and only of

$$\mathbf{S} \ is \ (\mathbf{Q}, \alpha) - independent. \tag{11}$$

and

$$\alpha - POS_{\mathbf{S}} \mathbf{Q} = \alpha - POS_{\mathbf{P}} \mathbf{Q}. \tag{12}$$

Definition 25. For a fixed $\alpha \in [0,1]$, The set of all $(\mathbf{Q},\alpha)$-indispensable elementary relations in **P** is called $(\mathbf{Q},\alpha)$-core of **P** and it is denoted by $\alpha - CORE_{\mathbf{P}}\mathbf{Q}$. When $Q = P$, we see that $\alpha - POS_P Q = U$ as $\underline{P}X_\alpha = X_\alpha$ for each $X_\alpha \in U/P$. So, $\alpha - CORE_P Q$ reduces to α -CORE defined earlier.

Proposition 1. For any $\alpha \in [0,1]$,
$\alpha - CORE_{\mathbf{Q}}\mathbf{P} = \cap\ \alpha - RED_{\mathbf{Q}}\mathbf{P}$, where $\alpha - RED_{\mathbf{Q}}\mathbf{P}$ is the family of all $(\mathbf{Q},\alpha)$-reducts of **P**.

Proof. Suppose a family of fuzzy proximity relations **S** be a $(\mathbf{Q},\alpha)$-reduct of **P**. Then **S** is $(\mathbf{Q},\alpha)$-independent of **P** and $\alpha - POS_{\mathbf{S}}\mathbf{Q} = \alpha - POS_{\mathbf{Q}}\mathbf{P}$.

Let R $\in$**P-S**. Then $\mathbf{S} \subseteq \mathbf{P}$- {R } $\subseteq$ **P**. But $\alpha - POS_{\mathbf{S}}\mathbf{Q} = \alpha - POS_{\mathbf{Q}}\mathbf{P}$. This gives $\alpha - POS_{\mathbf{S}}\mathbf{Q} = \alpha - POS_{(\mathbf{P}-\{R\})}\mathbf{Q} = \alpha - POS_{\mathbf{P}}\mathbf{Q}$.

Thus R is a $(\mathbf{Q},\alpha)$-dispensable relation in **P**, that is R $\in \alpha - CORE_{\mathbf{Q}}\mathbf{P}$.

Hence $\alpha - CORE_{\mathbf{Q}}\mathbf{P} \subseteq \mathbf{S}$, where **S** is a $(\mathbf{Q},\alpha)$-reducts of **P**.

Now, suppose R $\in \alpha - CORE_{\mathbf{Q}}\mathbf{P}$. Then R is $(\mathbf{Q},\alpha)$-indispensable in **P**. So, $\alpha - POS_{\mathbf{P}}\mathbf{Q} = POS_{(\mathbf{P}-\{R\})}\mathbf{Q})$.

Hence, there exists a subset $\mathbf{L} \subseteq \mathbf{P}$-{ R }, which is $(\mathbf{Q},\alpha$)-independent of **P**, such that $\alpha - POS_{\mathbf{L}}\mathbf{Q} = \alpha - POS_{\mathbf{P}}\mathbf{Q}$.

Which implies, $\mathbf{L} \subseteq \alpha - CORE_{\mathbf{Q}}\mathbf{P}$, where **L** is a $(\mathbf{Q},\alpha$)-reduct of **P**. So, $\cap$ { **L** : **L** is a $(\mathbf{Q},\alpha$)-reduct of **P** } $\subseteq \alpha - CORE_{\mathbf{Q}}\mathbf{P}$.

Hence $\alpha - CORE_{\mathbf{Q}}\mathbf{P} = \cap\alpha - RED_{\mathbf{Q}}\mathbf{P}$.

When the knowledge **P** has only one $(\mathbf{Q},\alpha$)-reduct, the knowledge **P** is deterministic, that is there is only one way to use the elementary α-categories of knowledge **P** when classifying the objects to elementary α -categories of knowledge **Q**.

Example 2. Consider the same information table, given in the Example 1. Now, we add another attribute e which generates a fuzzy proximity relation T as follows:

Let us find the relative reduct and the relative core, that is (T,α)-reduct and (T,α)-core of the family of fuzzy proximity relations $\mathcal{R} = \{P,Q,R,S\}$.

We take $\alpha = 0.6$. Then

$\mathrm{U/T}(\alpha) = \{\{x_1,x_4,x_5\},\{x_2,x_3\}\}$.

We have

$\mathrm{U/IND}\ \mathcal{R}(\alpha) = \{\{x_1\},\{x_2\},\{x_3\},\{x_4\},\{x_5\}\}$ and

$\alpha - POS_{\mathcal{R}_{(\alpha)}}T(\alpha) = \bigcup_{X\in U/T(\alpha)} \mathcal{R}\ X_\alpha = \{x_1,x_2,x_3,x_4,x_5\} = U.$

Now we will check whether $\mathcal{R}$ is $\mathrm{T}(\alpha)$-dependent or not.

Table 6. Fuzzy proximity relation for attribute T

T	x_1	x_2	x_3	x_4	x_5
x_1	1	0.4	0.5	0.6	0.8
x_2	0.4	1	0.9	0.3	0.5
x_3	0.5	0.9	1	0.2	0.4
x_4	0.6	0.3	0.2	1	0.5
x_5	0.8	0.5	0.4	0.5	1

Here $\alpha - POS_{(\mathcal{R}-\{P\})}T = \alpha - POS_{\mathcal{R}}T$,
$\alpha - POS_{(\mathcal{R}-\{Q\})}T = \{x_1, x_2, x_3, x_4, x_5\} = \alpha - POS_{\mathcal{R}}T$,
$\alpha - POS_{(\mathcal{R}-\{R\})}T = \{x_1, x_2, x_4\} \neq \alpha - POS_{\mathcal{R}}T$,
$\alpha - POS_{(\mathcal{R}-\{S\})}T = \{x_1, x_2, x_3, x_4, x_5\} = \alpha - POS_{\mathcal{R}}T$.
This gives that P,Q and S are (T, α)-dispensable, and R is (T, α)-dispensable, that is R is (T, α)-independent. So the (T, α)-reduct of $\mathcal{R}$ is $\{R\}$ and the (T, α)-core is also $\{R\}$.

2.5 Dependency of Knowledge in Fuzzy Approximation Spaces

In this section, we shall discuss how a knowledge in a fuzzy approximation space can be induced from a given knowledge in the fuzzy approximation space.

In the following definitions, we have $\alpha \in [0,1], \mathcal{K}(\alpha) = (U, \mathcal{R}(\alpha))$ is a fuzzy approximation space and $\mathbf{P}, \mathbf{Q} \subseteq \mathcal{R}$.

Definition 26. We say knowledge $\mathbf{Q}$ is $\alpha -$ *derivable* from knowledge $\mathbf{P}$ if the elementary α-categories of $\mathbf{Q}$ can be defined in terms of some elementary α-categories of knowledge $\mathbf{P}$.

Definition 27. If $\mathbf{Q}$ is α-derivable from $\mathbf{P}$, we say that knowledge $\mathbf{Q}$ α*-depends* on knowledge $\mathbf{P}$ and we denote it by $\mathbf{P} \overset{\alpha}{\Rightarrow} \mathbf{Q}$. So, $\mathbf{P} \overset{\alpha}{\Rightarrow} \mathbf{Q}$ if and only if IND$\mathbf{P}(\alpha) \subseteq$ IND$\mathbf{Q}(\alpha)$.

Definition 28. Knowledge $\mathbf{P}$ and $\mathbf{Q}$ are said to be $\alpha -$ *equivalent*, denoted by $\mathbf{P} \overset{\alpha}{\equiv} \mathbf{Q}$ if and only if $\mathbf{P} \overset{\alpha}{\Rightarrow} \mathbf{Q}$ and $\mathbf{Q} \overset{\alpha}{\Rightarrow} \mathbf{P}$.

So, $\mathbf{P} \overset{\alpha}{\equiv} \mathbf{Q}$ if and only if IND$\mathbf{P}(\alpha) =$ IND$\mathbf{Q}(\alpha)$.

Definition 29. Knowledge $\mathbf{P}$ and knowledge $\mathbf{Q}$ are $\alpha -$ *independent* if and only if neither $\mathbf{P} \overset{\alpha}{\Rightarrow} \mathbf{Q}$ nor $\mathbf{Q} \overset{\alpha}{\Rightarrow} \mathbf{P}$ holds.

We state below some properties on α-dependency of knowledges in fuzzy approximation spaces which can be proved easily like their base cases.

Proposition 2. For fixed $\alpha \in [0,1]$, the following conditions are equivalent.

(i) $\mathbf{P} \overset{\alpha}{\Rightarrow} \mathbf{Q}$
(ii) IND$((\mathbf{P} \cup \mathbf{Q})(\alpha)) =$ IND$\mathbf{P}(\alpha)$.
(iii) $\alpha - POS_{\mathbf{P}}\mathbf{Q} = U$.
(iv) $\underline{P}X_\alpha = X_\alpha$ for all $X \in U/\mathbf{Q}(\alpha)$, where $\underline{P}X_\alpha$ is referred as $\underline{\text{IND}\mathbf{P}}X_\alpha$.

Proposition 3. Let $\alpha \in [0,1]$. If $\mathbf{P}$ is a α-reduct of $\mathbf{Q}$, then
$\mathbf{P} \overset{\alpha}{\Rightarrow} \mathbf{Q} - \mathbf{P}$ and IND$\mathbf{P}(\alpha) =$ IND$\mathbf{Q}(\alpha)$.

Proposition 4. Let $\alpha \in [0,1]$.

(i) If $\mathbf{P}$ is α-dependent, then there exists a subset $\mathbf{Q} \subseteq \mathbf{P}$ such that $\mathbf{Q}$ is a α-reduct of $\mathbf{P}$.
(ii) If $\mathbf{P} \subseteq \mathbf{Q}$ and $\mathbf{P}$ is α-independent, then all basic relations in $\mathbf{P}$ are pair-wise α-independent.
(iii) If $\mathbf{P} \subseteq \mathbf{Q}$ and $\mathbf{P}$ is α-independent, the every subset $\mathbf{R}$ of $\mathbf{P}$ is α-independent.

Proposition 5. Let $\alpha \in [0,1]$.

(i) If $\mathbf{P} \overset{\alpha}{\Rightarrow} \mathbf{Q}$ and $\mathbf{P'} \supset \mathbf{P}$ then $\mathbf{P'} \overset{\alpha}{\Rightarrow} \mathbf{Q}$.
(ii) If $\mathbf{P} \overset{\alpha}{\Rightarrow} \mathbf{Q}$ and $\mathbf{Q'} \subset \mathbf{Q}$ then $\mathbf{P} \overset{\alpha}{\Rightarrow} \mathbf{Q'}$.
(iii) $\mathbf{P} \overset{\alpha}{\Rightarrow} \mathbf{R}$ and $\mathbf{Q} \overset{\alpha}{\Rightarrow} \mathbf{R}$ imply $\mathbf{P} \bigcup \mathbf{Q} \overset{\alpha}{\Rightarrow} \mathbf{R}$.

2.6 Partial Dependency of Knowledge in Fuzzy Approximation Spaces

It may happen that the α -derivation of one knowledge P from another knowledge Q can be partial. That is only a part of knowledge P can be α -derivable from P. We define and derive properties of such partial α -dependencies of knowledge in this section.

Let $\alpha \in [0,1]$. Suppose $\mathcal{K}(\alpha) = (U, \mathcal{R}(\alpha))$ is a fuzzy approximation space and, $\mathrm{P}, \mathrm{Q} \subseteq \mathcal{R}$. Then, we say that knowledge Q $\alpha -$ *depends in a degree* $\mathrm{k}(\alpha), 0 \leq \mathrm{k}(\alpha) \leq 1$, from the knowledge P, denoted by $\mathrm{P} \overset{\alpha}{\Longrightarrow}_{\mathrm{k}} \mathrm{Q}$ if $\mathrm{k}(\alpha) = \gamma_{\mathcal{P}(\alpha)}(\mathrm{Q}(\alpha)) = |\mathrm{POS}_{\mathcal{P}(\alpha)}(\mathrm{Q}(\alpha))|/|\mathrm{U}|$.

If $\mathrm{k}(\alpha) = 1$, we will say that Q is $totally\ \alpha - dependent\ from$ P. If $0 < \mathrm{k}(\alpha) < 1$, we say that Q $roughly(partially)\ \alpha - depends$ from P. If $\mathrm{k}(\alpha) = 0$, we say Q is $totally\ \alpha - independent\ from$ P.

If $\mathrm{P} \overset{\alpha}{\Longrightarrow}_{1} \mathrm{Q}$ we simply write $\mathrm{P} \overset{\alpha}{\Rightarrow} \mathrm{Q}$. In this case, all the elements of the universe can be classified to elementary categories $\mathrm{U}/\mathrm{Q}(\alpha)$ by using the knowledge P.

When $\mathrm{k}(\alpha) \neq 1$, only those elements of the universe which belong to the positive region can be classified to the categories of knowledge Q employing knowledge P. When $\mathrm{k}(\alpha) = 0$, none of the elements of the universe can be classified using knowledge P to elementary categories of knowledge Q.

Hence the co-efficient $\gamma_{\mathcal{P}(\alpha)}(Q(\alpha))$ can be understood as a degree of α-dependency of the knowledge Q relative to the knowledge P.

We establish the following propositions:

Proposition 6. For fixed $\alpha \in [0,1]$,

(i) If $\mathrm{R} \overset{\alpha}{\Longrightarrow}_{\mathrm{k}} \mathrm{P}$ and $\mathrm{Q} \overset{\alpha}{\Longrightarrow}_{\mathrm{l}} \mathrm{P}$, then $\mathrm{R} \cup \mathrm{Q} \overset{\alpha}{\Longrightarrow}_{\mathrm{m}} \mathrm{P}$ for some $m \geq \max\{k, l\}$.
(ii) If $\mathrm{R} \cup \mathrm{P} \overset{\alpha}{\Longrightarrow}_{\mathrm{k}} \mathrm{Q}$, then, $\mathrm{R} \overset{\alpha}{\Longrightarrow}_{\mathrm{l}} \mathrm{Q}$ and $\mathrm{P} \overset{\alpha}{\Longrightarrow}_{\mathrm{m}} \mathrm{Q}$ for some $l, m \leq k$.
(iii) If $\mathrm{R} \overset{\alpha}{\Longrightarrow}_{\mathrm{k}} \mathrm{Q}$ and $\mathrm{R} \overset{\alpha}{\Longrightarrow}_{\mathrm{l}} \mathrm{P}$ then $\mathrm{R} \overset{\alpha}{\Longrightarrow}_{\mathrm{m}} \mathrm{Q} \cup \mathrm{P}$, for some $m \leq \min\{k, l\}$.
(iv) If $\mathrm{R} \overset{\alpha}{\Longrightarrow}_{\mathrm{k}} \mathrm{Q} \cup \mathrm{P}$, then $\mathrm{R} \overset{\alpha}{\Longrightarrow}_{\mathrm{l}} \mathrm{P}$ and $\mathrm{R} \overset{\alpha}{\Longrightarrow}_{\mathrm{m}} \mathrm{P}$, for some $l, m \geq k$.
(v) If $\mathrm{R} \overset{\alpha}{\Longrightarrow}_{\mathrm{k}} \mathrm{P}$ and $\mathrm{P} \overset{\alpha}{\Longrightarrow}_{\mathrm{l}} \mathrm{Q}$, then $\mathrm{R} \overset{\alpha}{\Longrightarrow}_{\mathrm{m}} \mathrm{Q}$, for some $m \geq l + k - 1$.
Here we have represented $l(\alpha), m(\alpha)$ and $\mathrm{k}(\alpha)$ by l, m and k respectively.

Proof
(i) It is known that, $\underline{\mathrm{P}}\mathrm{X}_{\alpha} = \underline{\mathrm{INDP}}\mathrm{X}_{\alpha}$.
So, $\mathrm{P} \cup \mathrm{Q} \supseteq \mathrm{P} \Rightarrow \mathrm{IND}(\mathrm{P}(\alpha) \cup \mathrm{Q}(\alpha)) \subseteq \mathrm{INDP}(\alpha)$, which is meaningless.
So, $\mathrm{IND}(\mathrm{P}(\alpha) \cup \mathrm{Q}(\alpha)) \supseteq \mathrm{INDP}(\alpha)$.
This gives, card $\underset{X \in U/P(\alpha)}{\cup} \underline{R \cup Q}\mathrm{X}_{\alpha} \geq$ card $\underset{\mathrm{X} \in \mathrm{U}/\mathrm{P}(\alpha)}{\cup} \underline{\mathrm{R}}\mathrm{X}_{\alpha}$.
$\Rightarrow \dfrac{\text{card } \underset{X \in U/P(\alpha)}{\cup} \underline{(R \cup Q)}\mathrm{X}_{\alpha}}{\text{card U}} \geq \dfrac{\text{card } \underset{X \in U/P(\alpha)}{\cup} \underline{R}\mathrm{X}_{\alpha}}{\text{card } U}$.
$\Rightarrow \gamma_{R(\alpha) \cup Q(\alpha)}(P(\alpha)) \geq \gamma_{R(\alpha)}(P(\alpha)) \Rightarrow m \geq k$.

Similarly we can show that $m \geq l$. Hence $m \geq \max\{k, l\}$.

(ii) We know that,

$$\bigcup_{X \in U/Q(\alpha)} \underline{R(\alpha)X} \subseteq \bigcup_{X \in U/Q(\alpha)} \underline{R(\alpha) \cup Q(\alpha)X}.$$

$\Rightarrow \frac{card\ [POS_{R(\alpha)}Q(\alpha)]}{card\ U} \leq \frac{card\ [POS_{(R \cup P)(\alpha)}Q(\alpha)]}{card\ U}$.
$\Rightarrow l \leq k$.
Similarly we can prove $m \leq k$.
In a similar manner we can prove statements (iii) and (iv) of the proposition.
(v) Given $P \overset{\alpha}{\Longrightarrow}_l Q$ and $R \overset{\alpha}{\Longrightarrow}_k P$.
This gives, $INDP(\alpha) \subseteq_l INDQ(\alpha)$ and $\mathrm{INDR}(\alpha) \subseteq_k \mathrm{INDP}(\alpha)$.
Since, $INDP(\alpha) \subseteq INDQ(\alpha)$ and $INDR(\alpha) \subseteq INDP(\alpha)$, we get $P \subseteq Q$ and $R \subseteq P$.
Now, $\mathrm{POS}_{P(\alpha)}Q(\alpha) = \bigcup_{X \in U/Q(\alpha)} \underline{P}X_\alpha$ and $\mathrm{POS}_{R(\alpha)}P(\alpha) = \bigcup_{X \in U/P(\alpha)} \underline{R}X_\alpha$.

Hence, $\mathrm{POS}_{P(\alpha)}Q(\alpha) \cup \mathrm{POS}_{R(\alpha)}P(\alpha) = \bigcup_{X \in U/Q(\alpha)} \underline{P}X_\alpha \cup \bigcup_{X \in U/P(\alpha)} \underline{R}X_\alpha$
$\subseteq \bigcup_{X \in U/P(\alpha)} \underline{P}X_\alpha \cup \underline{R}X_\alpha$ (as $P \subseteq Q$).
$\subseteq \bigcup_{X \in U/P(\alpha)} \underline{R}X_\alpha$ (as $R \subseteq P$)
$\subseteq \mathrm{POS}_{R(\alpha)}Q(\alpha)$.
So, $\frac{\mathrm{card\ POS}_{P(\alpha)}Q(\alpha) + \mathrm{card}\ \alpha - \mathrm{POS}_{P(\alpha)}Q(\alpha)}{\mathrm{card\ U}} < \frac{\mathrm{card\ POS}_{P(\alpha)}Q(\alpha)}{\mathrm{card\ U}}$
Hence, $\gamma_{P(\alpha)}Q(\alpha) + \gamma_{R(\alpha)}P(\alpha) < \gamma_{R(\alpha)}Q(\alpha)$. This gives, $m > k + l$ which implies $m \geq k + l + 1$.

Interpretation: The intuitive interpretation of the properties established in Proposition 6 are as follows:

(i) This means that the degree of partial α -dependency of a knowledge P on two knowledge Q and R taken together is at least equal to the larger of the degrees of partial α -dependency of P on Q and R taken individually.
(ii) This means that partial α -dependency of a knowledge P on two knowledge Q and R taken together implies individual partial α -dependency of P on Q and R and the later degree of dependency is less than the earlier.
(iii) The partial α -dependency of two knowledge P and Q taken together on a knowledge R follows from the individual α -dependencies of P and Q on R and the combined α -dependency is almost the smaller of the individual α -dependencies.
(iv) The partial α -dependency of two knowledge P and Q taken together on a knowledge R implies individual α -dependencies of P and Q on R to degree not less than the degree of combined α -dependency.
(v)A variant of transitive property holds for α-dependencies of knowledge in the sense that the degree of dependency in the conclusion satisfies a constraint involving degree of dependencies in the hypothesis part.

Note 3. For $\alpha, \beta \in [0,1], R^*_\alpha \supset R^*_\beta$ for $\alpha > \beta$. That is, $R(\alpha) \Rightarrow R(\beta)$ for $\alpha > \beta, R \in \mathcal{R}$. Hence every equivalence class of $R(\alpha)$ is contained in some equivalence class of $R(\beta)$. Also, we find that, if $P(\alpha)$ is a α-reduct of $\mathcal{R}(\alpha)$, then

Table 7. Fuzzy proximity relation R

R	x_1	x_2	x_3	x_4	x_5	x_6
x_1	1	0.8	0.6	0.3	0.5	0.2
x_2	0.8	1	0.7	0.6	0.4	0.3
x_3	0.6	0.7	1	0.9	0.6	0.5
x_4	0.3	0.6	0.9	1	0.7	0.6
x_5	0.5	0.4	0.6	0.7	1	0.9
x_6	0.2	0.3	0.5	0.6	0.9	1

Table 8. Fuzzy proximity relation S

S	x_1	x_2	x_3	x_4	x_5	x_6
x_1	1	0.8	0.2	0.5	0.6	0.7
x_2	0.8	1	0.4	0.7	0.5	0.2
x_3	0.2	0.4	1	0.5	0.3	0.6
x_4	0.5	0.7	0.5	1	0.9	0.7
x_5	0.6	0.5	0.3	0.9	1	0.6
x_6	0.7	0.2	0.6	0.7	0.6	1

$P(\beta)$ is also a β-reduct of $\mathcal{R}(\beta)$, for some $\alpha < \beta$. We state the next proposition without proof, which is similar to the base case.

Proposition 7. For fixed $\alpha \in [0, 1]$, let $\mathcal{R}$ be a family of fuzzy proximity relation on U.

(i) For any $R \in \mathcal{R}$ which is α-indispensable in $\mathcal{R}$, R is also β-indispensable in $\mathcal{R}$ for $\alpha < \beta$.
(ii) If R is α-indispensable in $\mathcal{R}$, then R is also β-indispensable in $\mathcal{R}$ for $\alpha > \beta$.
(iii) $R(\alpha) \Rightarrow R(\beta)$ for $\alpha > \beta$, $R \in \mathcal{R}$.
(iv) $P(\alpha) \Rightarrow P(\beta)$ for $\alpha > \beta$, when $P(\alpha) \subseteq R(\alpha)$ and $P(\beta) \subseteq R(\beta)$.
(v) If P is a α-reduct of R, then P is β-reduct of R for $\alpha < \beta$.

Example 3. We illustrate the concepts defined in this section through the following example :

Let $U = \{x_1, x_2, x_3, x_4, x_5, x_6\}$ be a universe. R, S and T be fuzzy proximity relations on U defined in Tables 7-9:

Here $\mathcal{R} = \{R, S, T\}$ is a set of proximity relations on U. Let us determine the different equivalence classes of R, S and T for different values of $\alpha, \alpha \in [0, 1]$.

(i) $\alpha = 0.7$
From the relation table of R, R_α

$$= \{(x_1, x_1), (x_1, x_2), (x_2, x_1), (x_2, x_2), (x_2, x_3), (x_3, x_2), (x_3, x_3), (x_3, x_4), (x_4, x_3), (x_4, x_4), (x_4, x_5), (x_5, x_4), (x_5, x_5), (x_5, x_6), (x_6, x_6)\}$$

Thus $R(\alpha) = \{x_1, x_2, x_3, x_4, x_5, x_6\}$

So, $R^*_\alpha = U/R(\alpha) = \{\{x_1, x_2, x_3, x_4, x_5, x_6\}\}$.

Table 9. Fuzzy proximity relation T

T	x_1	x_2	x_3	x_4	x_5	x_6
x_1	1	0.9	0.5	0.6	0.7	0.2
x_2	0.9	1	0.8	0.2	0.2	0.8
x_3	0.5	0.8	1	0.3	0.4	0.2
x_4	0.6	0.2	0.3	1	0.8	0.3
x_5	0.7	0.2	0.4	0.8	1	0.4
x_6	0.2	0.8	0.2	0.3	0.4	1

From the relation table of S, $S_\alpha == \{(x_1,x_1),(x_1,x_2),(x_1,x_6),(x_2,x_1),(x_2,x_2),(x_2,x_4),$
$(x_3,x_3),(x_4,x_2),(x_4,x_4),(x_4,x_5),(x_5,x_4),(x_5,x_5),(x_5,x_6),(x_6,x_1),(x_6,x_4),(x_6,x_6)\}$
So, $S^*_\alpha = U/S(\alpha) = \{\{x_1,x_2,x_4,x_5,x_6\},\{x_3\}\}$.
From the relation table of T, $T_\alpha = \{(x_1,x_1),(x_1,x_2),(x_1,x_5),(x_2,x_1),(x_2,x_2),(x_2,x_3),(x_2,x_6),$
$(x_3,x_2),(x_3,x_3),(x_4,x_4),(x_4,x_5),(x_5,x_1),(x_5,x_4),(x_5,x_5),(x_6,x_2),(x_6,x_6)\}$
So, $T^*_\alpha = U/T(\alpha) = \{\{x_1,x_2,x_3,x_4,x_5,x_6\}\}$.

(ii) $\alpha = \mathbf{0.8}$
As in case (i), we have
$R^*_\alpha = U/R(\alpha) = \{\{x_1,x_2\},\{x_3,x_4\},\{x_5,x_6\}\}$.
$S^*_\alpha = U/S(\alpha) = \{\{x_1,x_2\},\{x_3\},\{x_4,x_5\},\{x_6\}\}$.
$T^*_\alpha = U/T(\alpha) = \{\{x_1,x_2,x_3,x_6\},\{x_4,x_5\}\}$.
(iii) $\alpha = \mathbf{0.9}$
As in case (i), we have
$R^*_\alpha = U/R(\alpha) = \{\{x_1\},\{x_2\},\{x_3,x_4\},\{x_5,x_6\}\}$.
$S^*_\alpha = U/S(\alpha) = \{\{x_1\},\{x_2\},\{x_3\},\{x_4,x_5\},\{x_6\}\}$.
$T^*_\alpha = U/T(\alpha) = \{\{x_1,x_2\},\{x_3\},\{x_4\},\{x_5\},\{x_6\}\}$.

2.7 An Application to Performance Analysis

The concepts and results of this section have wider and more realistic applications than the basic rough set approach. We shall illustrate this through an example in which the performance of a student is studied in a group of students of any institute. In the table below, we enumerate the possible facts which control the performance of a student in the examinations of a semester; their possible range of values and the fuzzy proximity relations which characterize the relationship between attribute values. The students devoting more hours for study having better regularity in attending the classes, having higher IQ, using more books for reference being involved in more group discussions and attending more seminars shall definitely score more marks in the examinations. However, this being an ideal case, is rare in practice. So, a student need not excel in all the characteristics noted in order to perform well. However, out of these, some characteristics may have higher influence on the scoring than the others. For different levels of proximity values α these important characteristics may be different.

Table 10. Attribute characteristics

Sl No.	Facts	Attribute	Domains	Relationship Function
1	No. of hours of study	A	A non−negative Number in [0,24]	$1 - \frac{\lvert x-y \rvert}{24}$
2	Regularity in the class /Percentage of attendance	B	An integral value in [0,100]	$1 - \frac{\lvert x-y \rvert}{100}$
3	IQ	C	An integral value in [0,100]	$1 - \frac{\lvert x-y \rvert}{100}$
4	No. of reference books used	D	*A non−negative number*	$1 - \frac{\lvert x-y \rvert}{2(x+y)}$
5	Attending in seminars /Group discussions	E	{F, AA, A BA, NA}	From Similarity Relation Table

Table 11. Students' information table

Student	A	B	C	D	E
x_1	8	90	60	0	Full
x_2	10	85	75	4	Below Average
x_3	10	60	70	6	Full
x_4	7	60	90	3	Average
x_5	6	55	85	5	Average
x_6	2	80	70	0	Average
x_7	5	55	40	8	Above Average
x_8	8	40	60	12	Above Average

Table 12. Proximity relation for attribute A

R_1	x_1	x_2	x_3	x_4	x_5	x_6	x_7	x_8
x_1	1	.92	.92	.96	.92	.75	.87	1
x_2	.92	1	1	.87	.83	.67	.79	.92
x_3	.92	1	1	.87	.83	.67	.79	.92
x_4	.96	.87	.87	1	.96	.79	.92	.87
x_5	.92	.83	.83	.96	1	.83	.96	.92
x_6	.75	.67	.67	.79	.83	1	.87	.75
x_7	.87	.79	.79	.92	.96	.87	1	.87
x_8	1	.92	.92	.87	.92	.75	.87	1

Characteristics of Attributes :
For attribute 'E' we use the following abbreviations:
Full(F)
Above Average(AA)
Average(A)
Below Average(BA)
Not at All(NA)

We consider a small universe of 8 students $\{x_1, x_2, x_3, x_4, x_5, x_6, x_7, x_8\}$ and the information about them is presented in the table below. It may be noted that if

Table 13. Proximity relation for attribute B

R_2	x_1	x_2	x_3	x_4	x_5	x_6	x_7	x_8
x_1	1	.95	.70	.70	.65	.90	.65	.50
x_2	.95	1	.75	.75	.70	.95	.70	.55
x_3	.70	.75	1	1	.95	.80	.95	.80
x_4	.70	.75	1	1	.95	.80	.95	.80
x_5	.65	.70	.95	.95	1	.75	1	.85
x_6	.90	.95	.80	.80	.75	1	.75	.60
x_7	.65	.70	.95	.95	1	.75	1	.85
x_8	.50	.55	.80	.80	.85	.60	.85	1

Table 14. Proximity relation for attribute C

R_3	x_1	x_2	x_3	x_4	x_5	x_6	x_7	x_8
x_1	1	.85	.90	.70	.75	.90	.80	1
x_2	.85	1	.95	.85	.90	.95	.65	.85
x_3	.90	.95	1	.80	.85	1	.70	.90
x_4	.70	.85	.80	1	.95	.80	.50	.70
x_5	.75	.90	.85	.95	1	.85	.55	.75
x_6	.90	.95	1	.80	.85	1	.70	.90
x_7	.80	.65	.70	.50	.55	.70	1	.80
x_8	1	.85	.90	.70	.75	.90	.80	1

Table 15. Proximity relation for attribute D

R_4	x_1	x_2	x_3	x_4	x_5	x_6	x_7	x_8
x_1	1	.50	.50	.50	.50	1	.50	.50
x_2	.50	1	.90	.93	.93	.50	.83	.75
x_3	.50	.90	1	.83	.87	.50	.93	.83
x_4	.50	.93	.83	1	.87	.50	.81	.70
x_5	.50	.93	.87	.87	1	.50	.88	.79
x_6	1	.50	.50	.50	.50	1	.50	.50
x_7	.50	.83	.93	.81	.88	.50	1	.87
x_8	.50	.75	.93	.70	.79	.50	.87	1

Table 16. Proximity relation for attribute E

SA	F	AA	A	BA	NA
F	1	.95	.90	.80	.45
AA	.95	1	.94	.70	.55
A	.90	.94	1	.80	.70
BA	.80	.70	.80	1	.78
NA	.40	.55	.70	.78	1

the number of students increases then it only requires additional computations and the structure of analysis remains same.

Proximity relations corresponding to attributes A–E are given in Tables 12–16.

We take here $R = \{R_1, R_2, R_3, R_4, R_5\}$. We take a reasonably high level of proximity value for α as 0.88. The equivalence classes of different relations corresponding to this value of α are given by,

$U/R_1(\alpha) = \{\{x_1, x_2, x_3, x_4, x_5, x_7, x_8\}, \{x_6\}\}$,
$U/R_2(\alpha) = \{\{x_1, x_2, x_6\}, \{x_3, x_4, x_5, x_7\}, \{x_8\}\}$,
$U/R_3(\alpha) = \{\{x_1, x_2, x_3, x_6, x_8\}, \{x_4, x_5\}, \{x_7\}\}$,
$U/R_4(\alpha) = \{\{x_1, x_6\}, \{x_2, x_3, x_4, x_5, x_7, x_8\}\}$,
$U/R_5(\alpha) = \{\{x_1, x_3, x_4, x_5, x_6, x_7, x_8\}, \{x_2\}\}$ and
$U/R(\alpha) = \{\{x_1\}, \{x_2\}, \{x_3\}, \{x_4, x_5\},, \{x_6\}, \{x_7\}, \{x_8\}\}$.
Now,
$U/(R - \{R_1(\alpha)\}) = \{\{x_1, x_6\}, \{x_2\}, \{x_3\}, \{x_4\}, \{x_5\}, \{x_7\}, \{x_8\}\} \neq U/R(\alpha)$.
$U/(R - \{R_2(\alpha)\}) = \{\{x_1\}, \{x_6\}, \{x_2\}, \{x_3, x_8\}, \{x_4, x_5\}, \{x_7\}\} \neq U/R(\alpha)$.
$U/(R - \{R_3(\alpha)\}) = \{\{x_1\}, \{x_6\}, \{x_2\}, \{x_3, x_4, x_5, x_7\}, \{x_8\}\} \neq U/R(\alpha)$.
$U/(R - \{R_4(\alpha)\}) = \{\{x_1\}, \{x_6\}, \{x_2\}, \{x_3\}, \{x_4, x_5\}, \{x_7\}, \{x_8\}\} = U/R(\alpha)$.
and
$U/(R - \{R_5(\alpha)\}) = \{\{x_1\}, \{x_6\}, \{x_2\}, \{x_3\}, \{x_4, x_5\}, \{x_7\}, \{x_8\}\} = U/R(\alpha)$.

So, R_1, R_2 and R_3 are α-indispensable relations in R, whereas R_4 and R_5 are α-dispensable relations in R.

The α-reducts are $\{R_1, R_2, R_3, R_4\}, \{R_1, R_2, R_3, R_5\}$ and the α-core of $R = \{R_1, R_2, R_3\}$.

Basing upon the above facts, we conclude that attribute 'A' (number of hour of study), attribute 'B' (regularity in classes) and attribute 'C' (IQ of a student) are the controlling factors for the performance of a student in his/her examinations at 0.88 level of proximity.

The conclusion will be different for different levels of proximity values. For example, if we take $\alpha = 1$ then all the attributes become indispensable. But this is a case when we adher to complete matching of attribute values and the flexibility is lost. Similarly, if we reduce the value of α to smaller values then the number of controlling factors decreases and obviously at $\alpha = 0$ there is no controlling factor. So, one has to select the value of α suitably and carefully.

It is obvious that the number of attributes varies according to different circumstances and facilities available to students. Here, we have considered some common attributes which control the performance of a student. However, one can add attributes like availability of study materials, quality of faculties, laboratory facilities and financial status of a student for performance evaluation. On the other way, some attributes can be removed.

3 Rough Sets in Intuitionistic Fuzzy Approximation Spaces

We continue from section 2.5 with the definitions and notations.

Definition 30. Let $\mathcal{R}$ be a set of intuitionistic fuzzy proximity relations on U. Then the pair $\mathcal{K} = (U, \mathcal{R})$ is called an *intuitionistic fuzzy approximation space* (IF-approximation space).

For each pair of values of $(\alpha, \beta) \in J$, an IF-approximation space $(U, \mathcal{R})$ generates usual approximation space $\mathcal{K}(\alpha, \beta) = (U, \mathcal{R}(\alpha, \beta))$.

Definition 31. Let $\mathcal{K} = (U, \mathcal{R})$ be an intuitionistic fuzzy approximation space and $X \subseteq U$. Then for any fixed $(\alpha, \beta) \in J$, the rough set of X in the generalized approximation space $\mathcal{K}(\alpha, \beta)$ and corresponding to the equivalence relation $R(\alpha, \beta) \in IND(\mathcal{K}(\alpha, \beta))$ is denoted by $(\underline{R}_{\alpha,\beta}X, \bar{R}_{\alpha,\beta}X)$ or in short $(\underline{X}_{\alpha,\beta}, \bar{X}_{\alpha,\beta})$, when the relation is clear from the context and is given by

$$\underline{R}_{\alpha,\beta}X = \underline{X}_{\alpha,\beta} = \cup\{Y : Y \in R^*_{\alpha,\beta} \text{ and } Y \subseteq X\}. \tag{13}$$

$$\bar{R}_{\alpha,\beta}X = \bar{X}_{\alpha,\beta} = \cup\{Y : Y \in R^*_{\alpha,\beta} \text{ and } Y \cap X \neq \phi\}. \tag{14}$$

where $R^*_{\alpha,\beta}$ is the family of equivalence classes of $R(\alpha, \beta)$.

Definition 32. Let X be a rough set in the generalized approximation space $(U, \mathcal{R}(\alpha, \beta))$. Then we define

$$BNR_{\alpha,\beta}(X) = \bar{X}_{\alpha,\beta} - \underline{X}_{\alpha,\beta}. \tag{15}$$

called the (α, β)-*boundary* of X.

Definition 33. Let X be a rough set in the generalized approximation space $(U, \mathcal{R}(\alpha, \beta))$. Then we say

$$X \text{ is } (\alpha, \beta) - \textit{discernible} \text{ if and only if } \underline{X}_{\alpha,\beta} = \bar{X}_{\alpha,\beta}. \tag{16}$$

and

$$X \text{ is } (\alpha, \beta) - \textit{rough} \text{ if and only if } \underline{X}_{\alpha,\beta} \neq \bar{X}_{\alpha,\beta}. \tag{17}$$

Next, we shall establish some properties of (α, β) -cuts of IF proximity relations.

Theorem 3. If $\alpha_1 \geq \alpha_2$ *and* $\beta_1 \leq \beta_2$ then

(i) $R_{\alpha_1,\beta_1} \subseteq R_{\alpha_2,\beta_2}$,
(ii) $R(\alpha_1, \beta_1) \subseteq R(\alpha_2, \beta_2)$ and
(iii) $R^*_{\alpha_1,\beta_1} \subseteq R^*_{\alpha_2,\beta_2}$,
in the sense that every equivalence class in $R^*_{\alpha_1,\beta_1}$ is contained in some equivalence class in $R^*_{\alpha_2,\beta_2}$.

Proof. We have by hypothesis,

$$(x, y) \in R_{\alpha_1,\beta_1} \Rightarrow \mu_R(x, y) \geq \alpha_1 \textit{ and } \nu_R(x, y) \leq \beta_1$$
$$\Rightarrow \mu_R(x, y) \geq \alpha_2 \textit{ and } \nu_R(x, y) \leq \beta_2$$
$$\Rightarrow (x, y) \in R_{\alpha_2,\beta_2}, \text{ which proves (i).}$$

Again $(x, y) \in R(\alpha_1, \beta_1) \Rightarrow xR_{\alpha_1,\beta_1}y$ *or* $xR_{\alpha_1,\beta_1}u_1, u_1R_{\alpha_1,\beta_1}u_2, \ldots, u_nR_{\alpha_1,\beta_1}y$. In the first case by (i) $xR_{\alpha_2,\beta_2}y$. So, $(x, y) \in R(\alpha_2, \beta_2)$. In the second case $xR_{\alpha_2,\beta_2}u_1, u_1R_{\alpha_2,\beta_2}u_2, \ldots, u_nR_{\alpha_2,\beta_2}y$. So, $(x, y) \in R(\alpha_2, \beta_2)$. This proves (ii).

Next, let $[x]_{\alpha_1,\beta_1} \in R^*_{\alpha_1,\beta_1}$. Then for any $y \in [x]_{\alpha_1,\beta_1}$, $yR_{\alpha_1,\beta_1}x$. So by (i) $yR_{\alpha_2,\beta_2}x$. Hence $y \in [x]_{\alpha_2,\beta_2}$. Thus we get $[x]_{\alpha_1,\beta_1} \subseteq [x]_{\alpha_2,\beta_2}$. This proves (iii).

Theorem 4. Let R and S be two IF-proximity relations on U. Then for any pair $(\alpha, \beta) \in J$,

(i) $(R \cup S)(\alpha, \beta) \subseteq R(\alpha, \beta) \cup S(\alpha, \beta)$

and

(ii) $(R \cap S)(\alpha, \beta) \supseteq R(\alpha, \beta) \cap S(\alpha, \beta)$.

Proof. Let $(x, y) \in (R \cup S)(\alpha, \beta)$. Then we have $x(R \cup S)(\alpha, \beta)y$. This implies $x(R \cup S)_{\alpha,\beta}y$ or there exists a sequence of elements $u_1, u_2, \ldots, u_n$ such that

$$x(R \cup S)_{\alpha,\beta}u_1, u_1(R \cup S)_{\alpha,\beta}u_2, ..., u_n(R \cup S)_{\alpha,\beta}y.$$

The second case being similar, we consider only the first case. We have,

$$x(R \cup S)_{\alpha,\beta}y \Rightarrow \mu_{R\cup S}(x, y) \geq \alpha \textit{ and } \nu_{R\cup S}(x, y) \leq \beta$$
$$\Rightarrow \max\{\mu_R(x, y), \mu_S(x, y)\} \geq \alpha \textit{ and } \max\{\nu_R(x, y), \nu_S(x, y)\} \leq \beta$$
$$\Rightarrow \mu_R(x, y) \geq \alpha \textit{ or } \mu_S(x, y) \geq \alpha \textit{ and } \nu_R(x, y), \nu_S(x, y) \leq \beta$$
$$\Rightarrow \{\mu_R(x, y) \geq \alpha \textit{ and } \nu_R(x, y) \leq \beta\} \textit{ or } \{\mu_S(x, y) \geq \alpha \textit{ and } \nu_S(x, y) \leq \beta\}$$
$$\Rightarrow xR(\alpha, \beta)y \textit{ or } xS(\alpha, \beta)y$$
$$\Rightarrow xR(\alpha, \beta) \cup S(\alpha, \beta)y$$
$$\Rightarrow (x, y) \in R(\alpha, \beta) \cup S(\alpha, \beta)$$

This proves (i).

Theorem 5. If R and S are two IF-proximity relations on U then for any pair $(\alpha, \beta) \in J$,

(i) $(R \cup S)^*_{\alpha,\beta} \subseteq R^*_{\alpha,\beta} \cup S^*_{\alpha,\beta}$

and

(ii) $(R \cap S)^*_{\alpha,\beta} \supseteq R^*_{\alpha,\beta} \cap S^*_{\alpha,\beta}$.

Proof. Let $[x] \in (R \cup S)^*_{\alpha,\beta}$. Then by Theorem 4(i), for any y,

$$y \in [x] \Rightarrow (x, y) \in (R \cup S)(\alpha, \beta)$$

$$\Rightarrow (x, y) \in R(\alpha, \beta) \cup S(\alpha, \beta)$$

$$\Rightarrow (x, y) \in R(\alpha, \beta) \textit{ or } (x, y) \in S(\alpha, \beta)$$

$$\Rightarrow [x] \in R^*_{\alpha,\beta} \textit{ or } [x] \in S^*_{\alpha,\beta}$$

$$\Rightarrow [x] \in R^*_{\alpha,\beta} \cup S^*_{\alpha,\beta}.$$

This proves (i). Proof of (ii) is similar.

Table 17. IF-proximity relation R

R	x_1	x_2	x_3	x_4	x_5
x_1	(1,0)	(.8,.1)	(.6,.3)	(.5,.3)	(.2,.6)
x_2	(.8,.1)	(1,0)	(.7,.2)	(.6,.3)	(.4,.5)
x_3	(.6,.3)	(.7,.2)	(1,0)	(.9,.1)	(.6,.3)
x_4	(.5,.3)	(.6,.3)	(.9,.1)	(1,0)	(.4,.5)
x_5	(.2,.6)	(.4,.5)	(.6,.3)	(.4,.5)	(1,0)

In the next example, we illustrate the computation of (α,β) -lower and (α,β)-upper approximations of sets and deduce whether the sets are (α,β) -rough or (α,β) - discernible.

Example 4. Let $U = \{x_1, x_2, x_3, x_4, x_5\}$. We define an IF-proximity relation R on U as shown in Table 17.
Taking $\alpha = 0.7$ and $\beta = 0.2$ we find that

$$R^*_{\alpha,\beta} = \{\{x_1, x_2\}, \{x_3\, x_4\}, \{x_5\}\}.$$

Again for $\alpha = 0.9$ and $\beta = 0.1$, we see that

$$R^*_{\alpha,\beta} = \{\{x_1\}, \{x_2\}, \{x_3\}, \{x_4\}, \{x_5\}\}.$$

Let us consider two subsets $X_1 = \{x_1, x_2\}$ and $X_2 = \{x_1, x_4, x_5\}$ of U. We see that for $\alpha = 0.7$ and $\beta = 0.1$

$$\underline{X}_{1,\alpha,\beta} = \phi \text{ and } \bar{X}_{1,\alpha,\beta} = \{x_1, x_2, x_3, x_4\}.$$

So that X_1 is (α,β)-rough. On the other hand for $\alpha = 0.9$ and $\beta = 0.1$,

$$\underline{X}_{2,\alpha,\beta} = \{x_2, x_4, x_5\} \text{ and } \bar{X}_{2,\alpha,\beta} = \{x_2, x_4, x_5\}.$$

So that X_2 is (α,β)-discernible.

The following theorem establishes properties of (α,β)-lower approximation and (α,β)-upper approximations of rough sets. We shall provide proofs only of two properties. The proofs of the other properties are similar.

Theorem 6. Let X and Y be rough sets in the generalised approximation space $(U, R(\alpha,\beta)), (\alpha,\beta) \in J$. Then
(i) $\underline{X}_{\alpha,\beta} \subseteq X \subseteq \bar{X}_{\alpha,\beta}$.
(ii) $\underline{\phi}_{\alpha,\beta} = \bar{\phi}_{\alpha,\beta} = \phi, \quad \underline{U}_{\alpha,\beta} = \bar{U}_{\alpha,\beta} = U.$
(iii) $\overline{(X \cup Y)}_{\alpha,\beta} = \bar{X}_{\alpha,\beta} \cup \bar{Y}_{\alpha,\beta}$.
(iv) $\underline{(X \cap Y)}_{\alpha,\beta} = \underline{X}_{\alpha,\beta} \cap \underline{Y}_{\alpha,\beta}$.
(v) $X \subseteq Y \Rightarrow \underline{X}_{\alpha,\beta} \subseteq \underline{Y}_{\alpha,\beta}$.

(vi) $X \subseteq Y \Rightarrow \bar{X}_{\alpha,\beta} \subseteq \bar{Y}_{\alpha,\beta}$.
(vii) $\underline{(X \cup Y)}_{\alpha,\beta} \supseteq \underline{X}_{\alpha,\beta} \cup \underline{Y}_{\alpha,\beta}$.
(viii) $\overline{(X \cap Y)}_{\alpha,\beta} \subseteq \bar{X}_{\alpha,\beta} \cap \bar{Y}_{\alpha,\beta}$.

Proof. We have, $x \in \underline{X}_{\alpha,\beta} \Rightarrow x \in [y] \in R^*_{\alpha,\beta}$ and $[y] \subseteq X$. So, $x \in X$.
Again, $x \in X \Rightarrow xR(\alpha,\beta)x$ for any $(\alpha,\beta) \in J$.
So, $\Rightarrow x \in [x] \in R^*_{\alpha,\beta}$, such that $[x] \cap X \neq \phi$.
Hence $x \in \bar{X}_{\alpha,\beta}$. This proves (i).
Next, suppose $x \in \overline{(X \cup Y)}_{\alpha,\beta}$.
Thus there exists an equivalence class $[z]$ with respect to $R(\alpha,\beta)$ such that $x \in [z]$ and $[z] \cap (X \cup Y) \neq \phi$.
So, $x \in [z]$ and $[z] \cap X \neq \phi$ or $[z] \cap Y \neq \phi$.
This implies $x \in \bar{X}_{\alpha,\beta}$ or $x \in \bar{Y}_{\alpha,\beta}$. Hence,

$$x \in \bar{X}_{\alpha,\beta} \cup \bar{Y}_{\alpha,\beta}. \tag{18}$$

Also, $x \in \bar{X}_{\alpha,\beta} \cup \bar{Y}_{\alpha,\beta} \Rightarrow x \in \bar{X}_{\alpha,\beta}$ or $x \in \bar{Y}_{\alpha,\beta}$. So, $\exists[y]$ such that $x \in [y]$ and $[y] \cap X \neq \phi$ or $\exists[z]$ such that $x \in [z]$ and $[z] \cap Y \neq \phi$.

Hence, in any case there exists an equivalence class containing x which has nonempty intersection with $X \cup Y$. That is,

$$x \in \overline{X \cup Y}_{\alpha,\beta}. \tag{19}$$

We get (iii) from (18) and (19).

Theorem 7. If $\alpha_1 \geq \alpha_2$ and $\beta_1 \leq \beta_2$, then
(i) $\underline{X}_{\alpha_2,\beta_2} \subseteq \underline{X}_{\alpha_1,\beta_1}$
(ii) $\bar{X}_{\alpha_2,\beta_2} \subseteq \bar{X}_{\alpha_1,\beta_1}$

Proof. Let $x \in \underline{X}_{\alpha_2,\beta_2}$. Then $[x]_{\alpha_2,\beta_2} \subseteq X$. Now,
$y \in [x]_{\alpha_1,\beta_1} \Rightarrow \mu_R(x,y) \geq \alpha_1$ and $\nu_R(x,y) \leq \beta_1$
$\Rightarrow \mu_R(x,y) \geq \alpha_2$ and $\nu_R(x,y) \leq \beta_2$
$\Rightarrow y \in [x]_{\alpha_2,\beta_2}$.
So, $[x]_{\alpha_1,\beta_1} \subseteq X$. Hence, $x \in \underline{X}_{\alpha_1,\beta_1}$. This proves (i).
Again, $x \in \bar{X}_{\alpha_1,\beta_1} \Rightarrow [x]_{\alpha_1,\beta_1} \cap X \neq \phi$
$\Rightarrow [x]_{\alpha_2,\beta_2} \cap X \neq \phi$, as above.
That is, $x \in \bar{X}_{\alpha_2,\beta_2}$.
This completes the proof.

Theorem 8. Let R and S be two IF-proximity relations on U and α,β be chosen threshold values with $(\alpha,\beta) \in J$. Then
(i) $\underline{(R \cup S)}_{\alpha,\beta}X \subseteq \underline{R}_{\alpha,\beta}X \cup \underline{S}_{\alpha,\beta}X$,
(ii) $\overline{(R \cup S)}_{\alpha,\beta}X \supseteq \bar{R}_{\alpha,\beta}X \cup \bar{S}_{\alpha,\beta}X$,
(iii) $\underline{(R \cap S)}_{\alpha,\beta}X \supseteq \underline{R}_{\alpha,\beta}X \cap \underline{S}_{\alpha,\beta}X$,
and
(iv) $\overline{(R \cap S)}_{\alpha,\beta}X \subseteq \bar{R}_{\alpha,\beta}X \cap \bar{S}_{\alpha,\beta}X$,

Proof. We prove only (i). Rest of the proofs are similar.

$\underline{(R \cup S)}_{\alpha,\beta} X \subseteq \underline{R}_{\alpha,\beta} X \cup \underline{S}_{\alpha,\beta} X.$

$$\underline{(R \cup S)}_{\alpha,\beta} X = \bigcup\{Y \in (R \cup S)^*_{\alpha,\beta} : Y \subseteq X\}$$
$$\subseteq \bigcup\{Y \in (R^*_{\alpha,\beta} \cup S^*_{\alpha,\beta}) : Y \subseteq X\}$$
$$\subseteq [\bigcup\{Y \in R^*_{\alpha,\beta} : Y \subseteq X\}] \bigcup [\bigcup\{Y \in S^*_{\alpha,\beta} : Y \subseteq X\}]$$
$$= \underline{R}_{\alpha,\beta} X \cup \underline{S}_{\alpha,\beta} X.$$

This completes the proof.

4 Knowledge Reduction in IF-Approximation Spaces

Definition 34. Let **R** be a family of IF-proximity relations on U and $(\alpha, \beta) \in J$. For any R $\in$ **R**, we say that R is (α, β)-dispensable in **R** if IND $\mathbf{R}(\alpha, \beta) =$ IND($\mathbf{R}(\alpha, \beta)$ $-$ R(α, β)), otherwise R is $(\alpha, \beta) -$ *indispensable* in **R**. **R** is $(\alpha, \beta) -$ *independent* if each R $\in$ **R** is (α, β)-indispensable. Otherwise $\mathbf{R}(\alpha, \beta)$ is $(\alpha, \beta) -$ *dependent*.

Definition 35. For a fixed $(\alpha, \beta) \in J$, let $(\mathbf{P} \neq \phi) \subseteq \mathbf{R}$. Then $\mathbf{Q} \subseteq \mathbf{P}$ is a $(\alpha, \beta) -$ *reduct* of **P** if (denoted as (α, β) -RED(**P**)) if

$$\mathbf{Q} \;\; is \;\; (\alpha, \beta) - independent, \tag{20}$$

and

$$\mathrm{IND}(\mathbf{Q}(\alpha, \beta)) = \mathrm{IND}(\mathbf{P}(\alpha, \beta)). \tag{21}$$

It may be noted that **P** may have many (α, β)-reducts, for $(\alpha, \beta) \in J$.

Definition 36. Let $\mathbf{P}(\neq \phi) \subseteq \mathbf{R}$. For a fixed $(\alpha, \beta) \in J$, the set of all (α, β)-indispensable relations in **P** is called (α, β)-*core* of **P** and it is denoted by $(\alpha, \beta)-$CORE(**P**).

We shall give the following theorems and propositions which can be proved as their counterparts in section 3.

Theorem 9. For fixed $(\alpha, \beta) \in J$, if **Q** is (α, β)-independent and $\mathbf{P} \subseteq \mathbf{Q}$, then **P** is also (α, β)-independent.

The following result can be obtained from Theorem 9.

Corollary 2. For fixed $(\alpha, \beta) \in J$, if $\mathbf{P} \subseteq \mathbf{Q}$ and **P** is (α, β)-dependent, then **Q** is (α, β)-dependent.

Theorem 10. For each fixed $(\alpha, \beta) \in J$,

$$(\alpha, \beta) - CORE(\mathbf{P}) = \bigcap (\alpha, \beta) - RED(\mathbf{P}),$$

where (α, β)-RED(**P**) is the family of all (α, β)-reducts of **P**.

Note 4. As in case of α -core, (α, β)-core can be interpreted as the set of the most characteristic part of the knowledge, which can not be eliminated when reducing the knowledge. The smallest (α, β)-reduct of **R** is the (α, β)-core of **R**.

Table 18. Another information system

Object	a	b	c	d
x_1	a_1	b_2	c_3	d_1
x_2	a_2	b_2	c_2	d_2
x_3	a_1	b_1	c_1	d_2
x_4	a_3	b_3	c_1	d_1
x_5	a_2	b_1	c_3	d_2

Table 19. IF-proximity relation P

P	x_1	x_2	x_3	x_4	x_5
x_1	(1,0)	(0.3,0.5)	(0.6,0.2)	(0.8,0.1)	(0.5,0.2)
x_2	(0.3,0.5)	(1,0)	(0.7,0.2)	(0.4,0.4)	(0.4,0.3)
x_3	(0.6,0.2)	(0.7,0.2)	(1,0)	(0.2,0.5)	(0.8,0.1)
x_4	(0.8,0.1)	(0.4,0.4)	(0.2,0.5)	(1,0)	(0.5,0.3)
x_5	(0.5,0.2)	(0.4,0.3)	(0.8,0.1)	(0.5,0.3)	(1,0)

Table 20. IF-proximity relation Q

Q	x_1	x_2	x_3	x_4	x_5
x_1	(1,0)	(0.3,0.5)	(0.4,0.2)	(0.2,0.4)	(0.5,0.2)
x_2	(0.3,0.5)	(1,0)	(0.8,0.2)	(0.6,0.4)	(0.2,0.5)
x_3	(0.4,0.2)	(0.8,0.2)	(1,0)	(0.3,0.4)	(0.9,0.1)
x_4	(0.2,0.4)	(0.6,0.4)	(0.3,0.4)	(1,0)	(0.7,0.1)
x_5	(0.5,0.2)	(0.3,0.4)	(0.9,0.1)	(0.7,0.1)	(1,0)

Table 21. IF-proximity relation R

R	x_1	x_2	x_3	x_4	x_5
x_1	(1,0)	(0.3,0.6)	(0.2,0.5)	(0.8,0.1)	(0.7,0.2)
x_2	(0.3,0.6)	(1,0)	(0.5,0.3)	(0.3,0.6)	(0.5,0.2)
x_3	(0.2,0.5)	(0.5,0.3)	(1,0)	(0.6,0.3)	(0.4,0.4)
x_4	(0.8,0.1)	(0.3,0.6)	(0.6,0.3)	(1,0)	(0.9,0.1)
x_5	(0.7,0.2)	(0.5,0.2)	(0.4,0.4)	(0.9,0.1)	(1,0)

So, (α, β)-core can be used as a basis for computation of all (α, β)-reducts, that is, the (α, β)-core is included in every (α, β)-reduct.

Example 5. Consider the universal set $U = \{x_1, x_2, x_3, x_4, x_5\}$. Suppose each object $x_i, i = 1, 2, 3, 4, 5$ is associated with the attributes a, b, c and d, as represented in the following table:

We have assumed here that the attributes a , b, c and d have the domains $\{a_1, a_2, a_3\}$, $\{b_1, b_2, b_3\}$, $\{c_1, c_2, c_3\}$ and $\{d_1, d_2\}$ respectively.

Table 22. IF-proximity relation S

S	x_1	x_2	x_3	x_4	x_5
x_1	(1,0)	(0.3,0.6)	(0.2,0.5)	(0.2,0.6)	(0.5,0.3)
x_2	(0.3,0.6)	(1,0)	(0.5,0.2)	(0.3,0.4)	(0.2,0.6)
x_3	(0.2,0.5)	(0.5,0.2)	(1,0)	(0.2,0.6)	(0.4,0.4)
x_4	(0.2,0.6)	(0.3,0.4)	(0.2,0.6)	(1,0)	(0.5,0.3)
x_5	(0.5,0.3)	(0.2,0.6)	(0.4,0.4)	(0.5,0.3)	(1,0)

We define the IF-proximity relations P, Q, R and S over U corresponding to the attributes a, b, c and d respectively according to Tables 19–22.

Taking $\Re = \{P, Q, R, S\}$ as a family of IF-proximity relations over U and $\alpha = 0.6, \beta = 0.2$, we get the following partitions based on IF-proximity relations.

$U/P(\alpha, \beta) = \{x_1, x_2, x_3, x_4, x_5\}$
$U/Q(\alpha, \beta) = \{\{x_1\}, \{x_2, x_3, x_4, x_5\}\}$
$U/R(\alpha, \beta) = \{\{x_1, x_4, x_5\}, \{x_2\}, \{x_3\}\}$
$U/S(\alpha, \beta) = \{\{x_1\}, \{x_2\}, \{x_3\}, \{x_4\}, \{x_5\}\}$

Now, $U/IND\ \Re(\alpha, \beta) = \{\{x_1\}, \{x_2\}, \{x_3\}, \{x_4\}, \{x_5\}\}$

Now

$U/IND(\Re - P)(\alpha, \beta) = \{\{x_1, x_2, x_3, x_5\}, \{x_4\}\} \neq U/INDR(\alpha, \beta)$.
$U/IND(\Re - Q)(\alpha, \beta) = \{\{x_1, x_3\}, \{x_2, x_4, x_5\}\} \neq U/INDR(\alpha, \beta)$.
$U/IND(\Re - R)(\alpha, \beta) = \{\{x_1\}, \{x_3\}, \{x_2\}, \{x_5\}, \{x_4\}\} = U/IND\Re(\alpha, \beta)$.
$U/IND(\Re - S)(\alpha, \beta) = \{\{x_1\}, \{x_3\}, \{x_2\}, \{x_5\}, \{x_4\}\} = U/IND\Re(\alpha, \beta)$.

Hence, the relations P and Q are (α, β)-indispensable in $\Re$, whereas R and S are (α, β)-dispensable in $\Re$.

So, $\{P, Q, R\}$ and $\{P, Q, S\}$ are the (α, β)-reducts of R, and
(α, β)-CORE$(R) = \{P, Q, R\} \cap \{P, Q, S\} = \{P, Q\}$.

When the values of α and β are changed, the (α, β) -reducts and (α, β) - cores change. According to one's requirement the values of α, β can be adjusted and for this pair of values, the most important knowledge can be obtained. We emphasize that the parameters α and β reflect the minimum proximity of values and maximum non proximity of values for the attributes under consideration.

5 Kinds of Rough Sets in IF-Approximation Spaces

Four kinds of rough sets were defined by Pawlak [27] depending upon whether the lower approximation set is empty or nonempty and whether the upper approximation set is the universal or not. Also, Pawlak has provided physical interpretations for these kinds of rough sets. Similar kinds of rough sets on fuzzy approximation spaces are introduced by Tripathy [40] and their physical interpretations have been provided. One interesting observation in Pawlak's classification is to determine the positions of union or intersection of two elements of the same kind. The answers to these questions are provided by Tripathy and Mitra [39].However, one observes that the union or intersection in most of the cases falls beyond the range of a particular kind. He has also extended these results to the setting of rough sets on fuzzy approximation spaces [40]. The fundamental reason behind the extension of rough sets to the general setting is the abundance of fuzzy proximity relations and their applicability.

Following is the classification of a rough set on intuitionistic fuzzy approximation spaces [37,38]:

Definition 37

(i) Type 1. If $\underline{X}_{\alpha,\beta} \neq \phi$, $\overline{X}_{\alpha,\beta} \neq U$ then we say that X is *roughly* $R_{\alpha,\beta}$ -*definable*(Kind 1).

(ii) Type 2. If $\underline{X}_{\alpha,\beta} = \phi$, $\overline{X}_{\alpha,\beta} \neq U$ then we say that X is *internally* $R_{\alpha,\beta}$ -*undefinable* (Kind 2).

(iii) Type 3. If $\underline{X}_{\alpha,\beta} \neq \phi$, $\overline{X}_{\alpha,\beta} = U$ then we say that X is *externally* $R_{\alpha,\beta}$ -*undefinable* (Kind 3).

(iv) Type 4. If $\underline{X}_{\alpha,\beta} = \phi$, $\overline{X}_{\alpha,\beta} = U$ then we say that X is *totally* $R_{\alpha,\beta}$ -*undefinable* (Kind 4).

Physical Interpretation

(i) A set X is roughly $R_{\alpha,\beta}$-definable means that we are able to decide for some elements of U whether they are (α, β)- similar or transitively (α, β)-similar to some elements of X or $-X$ with respect to R.

(ii) A set X is internally $R_{\alpha,\beta}$-undefinable means that we are able to decide whether some elements of U are $R_{\alpha,\beta}$-similar or transitively $R_{\alpha,\beta}$-similar to some elements of $-X$ but we are unable to indicate this property for any element of X with respect to R.

(iii) A set X is externally $R_{\alpha,\beta}$-undefinable means that we are able to decide whether some elements of U are $R_{\alpha,\beta}$-similar or transitively $R_{\alpha,\beta}$-similar to some elements of X but we are unable to indicate this property for any element of $-X$ with respect to R.

(iv) A set X is totally$R_{\alpha,\beta}$-undefinable means that we are unable to decide for any element of U whether it is (α, β)-similar or transitively (α, β)-similar to some element of X or $-X$ with respect to R.

The following theorem characterizes the position of intersection or union of two elements of one kind of rough sets.

Theorem 11

(i) If X and Y are internally $R_{\alpha,\beta}$-undefinable then $X \cap Y$ is internally $R_{\alpha,\beta}$-undefinable.

(ii) If X and Y are internally $R_{\alpha,\beta}$-undefinable then $X \cup Y$ can be in any one of the four classes.

(iii) If X and Y are roughly $R_{\alpha,\beta}$-definable then $X \cap Y$ can be roughly $R_{\alpha,\beta}$-definable or internally $R_{\alpha,\beta}$-undefinable.

(iv) If X and Y are roughly $R_{\alpha,\beta}$-definable then $X \cup Y$ may be roughly $R_{\alpha,\beta}$-definable or externally $R_{\alpha,\beta}$-undefinable.

(v) If X and Y are externally $R_{\alpha,\beta}$-definable then $X \cap Y$ can be in any one of the four classes.

(vi) If X and Y are externally $R_{\alpha,\beta}$-undefinable, then $X \cup Y$ is externally $R_{\alpha,\beta}$-undefinable.
(vii) If X and Y are totally $R_{\alpha,\beta}$-undefinable, then $X \cap Y$ can be internally $R_{\alpha,\beta}$-undefinable or totally $R_{\alpha,\beta}$-undefinable.
(viii) If X and Y are totally $R_{\alpha,\beta}$-undefinable, then $X \cup Y$ can be externally $R_{\alpha,\beta}$-undefinable or totally $R_{\alpha,\beta}$-undefinable.

6 Relative Reduct and Relative Core of Knowledge in IF-Approximation Spaces

In this section, continuing with the concept in section 2.4, we develop concepts which are more general and applicable than (α,β) -core and (α,β) -reduct introduced in the previous section. Basically these concepts shall be useful in discovering inference rules.

Definition 38. Let $\mathbf{P}$ and $\mathbf{Q}$ be two IF-proximity relations over the universe U. For a fixed $(\alpha,\beta) \in J$, the (α,β)-*positive region of* $\boldsymbol{P}$ *with respect to* $\boldsymbol{Q}$ is defined as

$$(\alpha,\beta) - POS_{\mathbf{P}}\mathbf{Q} = \bigcup_{X_{(\alpha,\beta)} \in U/\mathbf{Q}(\alpha,\beta)} \underline{P}X_{\alpha,\beta}. \tag{22}$$

That is, the (α,β)-positive region of $\mathbf{P}$ with respect to $\mathbf{Q}$ is the set of all objects of the universe U which can be properly classified to the classes of $U/\mathbf{Q}(\alpha,\beta)$ employing the knowledge expressed by the classification $U/\mathbf{P}(\alpha,\beta)$.

Now, we provide the generalized concepts considered in Section 4.

Definition 39. Let $\mathbf{P}$ and $\mathbf{Q}$ be two families of IF-proximity relations on U. For a fixed $(\alpha,\beta) \in J$, suppose $\mathbf{R} \in P$, then R is $\boldsymbol{Q}(\alpha,\beta)$-*dispensable in* $\boldsymbol{P}$ if

$$(\alpha,\beta) - POS_{IND(\mathbf{P})}IND(\mathbf{Q}) = (\alpha,\beta) - POS_{IND(\mathbf{P}-\{R\})}IND(\mathbf{Q}). \tag{23}$$

Otherwise, R is $\mathbf{Q}(\alpha,\beta)$-indispensable in $\mathbf{P}$.

If every $R \in \mathbf{P}$ is $\mathbf{Q}(\alpha,\beta)$-indispensable, we can say that $\mathbf{P}$ is $\mathbf{Q}(\alpha,\beta)$-independent or $\mathbf{P}$ is (α,β)-*independent* with respect to $\mathbf{Q}$.

Definition 40. For fixed $(\alpha,\beta) \in J$, the sub-family $\mathbf{S} \subseteq \mathbf{P}$ *is a* $\boldsymbol{Q}(\alpha,\beta)$-*reduct of* $\boldsymbol{P}$ if and only if

$$\mathbf{S}\ is\ \mathbf{Q}(\alpha,\beta) - independent\ of\ \mathbf{P}. \tag{24}$$

$$(\alpha,\beta) - POS_{\mathbf{S}}\mathbf{Q} = (\alpha,\beta) - POS_{\mathbf{P}}\mathbf{Q}. \tag{25}$$

Here $\mathbf{P}$ means $IND\mathbf{P}(\alpha,\beta)$ and similarly for others.

Obviously, $\mathbf{P}$ may have many $\mathbf{Q}(\alpha,\beta)$-reducts for different values of $(\alpha,\beta) \in J$.

Definition 41. For a fixed $(\alpha,\beta) \in J$, the set of all $\mathbf{Q}(\alpha,\beta)$-indispensable elementary relations in $\mathbf{P}$ is called $\mathbf{Q}(\alpha,\beta)$-*core* of $\mathbf{P}$ and it is denoted by $(\alpha,\beta) - CORE_{\mathbf{Q}}\mathbf{P}$, where $\mathbf{P},\mathbf{Q} \subseteq \mathbf{R}$.

If $\mathbf{P} = \mathbf{Q}$, it reduces to the general concept of (α,β)-core.

Table 23. Modified information system, now adding attribute e

object	a	b	c	d	e
x_1	a_1	b_2	c_3	d_1	e_2
x_2	a_2	b_2	c_2	d_2	e_3
x_3	a_1	b_1	c_1	d_2	e_1
x_4	a_3	b_3	c_1	d_1	e_3
x_5	a_2	b_1	c_3	d_2	e_2

Table 24. IF-proximity relation T

T	x_1	x_2	x_3	x_4	x_5
x_1	(1,0)	(0.7,0.3)	(0.6,0.3)	(0.7,0.3)	(1,0)
x_2	(0.7,0.3)	(1,0)	(0.4,0.50)	(1,0)	(0.7,0.3)
x_3	(0.6,0.3)	(0.4,0.5)	(1,0)	(0.4,0.5)	(0.6,0.3)
x_4	(0.7,0.3)	(1,0)	(0.4,0.5)	(1,0)	(0.7,0.3)
x_5	(1,0)	(0.7,0.3)	(0.6,0.3)	(0.7,0.3)	(1,0)

Proposition 8. For each fixed $(\alpha, \beta) \in J$,

$$(\alpha, \beta) - CORE_{\mathbf{Q}}\mathbf{P} = \cap(\alpha, \beta) - RED_{\mathbf{Q}}\mathbf{P},$$

where $(\alpha, \beta) - RED_{\mathbf{Q}}\mathbf{P}$ is the family of all $\mathbf{Q}(\alpha, \beta)$-reducts of $\mathbf{P}$.

When the knowledge $\mathbf{P}$ has only one $\mathbf{Q}(\alpha, \beta)$-reduct, the knowledge $\mathbf{P}$ is deterministic, that is there is only one way to use the elementary categories of knowledge $\mathbf{P}$ when classifying the object to elementary categories of knowledge $\mathbf{Q}$.

In the following example we illustrate the computation of relative (α, β)-core and relative (α, β) -reduct in a intuitionistic fuzzy knowledge base.

Example 6. Consider the same information table given in the above Example 5. Now, we add another atribute e to the objects whose domain is $\{e_1, e_2, e_3\}$, which generates a IF-proximity relation T on e. The modified information table is given below.

Here we find the relative reduct and the relative core, $T(\alpha, \beta)$-reduct and $T(\alpha, \beta)$-core of the family of IF-proximity relations $\mathcal{R} = \{P, Q, R, S\}$

Let $\alpha = 0.6,\ \beta = 0.2$. Then $U/T(\alpha, \beta) = \{\{x_1, x_5\}, \{x_2, x_4\}, \{x_3\}\}$ and $U/IND\mathcal{R}(\alpha, \beta) = \{\{x_1, x_5\}, \{x_2, x_4\}, \{x_3\}\}$.

So,

$$(\alpha, \beta) - POS_{R}T = \bigcup_{X \in U/T(\alpha, \beta)} \underline{\mathcal{R}}X_{\alpha, \beta} = \{x_3\}.$$

Now we will check whether $\mathcal{R}$ is $T(\alpha, \beta)$-dependent or not.

Here $(\alpha, \beta) - POS_{\mathcal{R}-\{P\}}T = \{\{x_1, x_2\}, \{x_2, x_5\}, \{x_4\}\} \neq (\alpha, \beta) - POS_{\mathcal{R}}T$

$$(\alpha,\beta)-POS_{\mathcal{R}-\{Q\}}T=\{\{x_1,x_2\},\{x_2,x_5\},\{x_4\}\}\neq(\alpha,\beta)-POS_{\mathcal{R}}T$$

$$(\alpha,\beta)-POS_{(\mathcal{R}-\{R\})}T=\{\{x_1,x_2\},\{x_2,x_5\},\{x_4\}\}\neq(\alpha,\beta)-POS_{\mathcal{R}}T$$

$$(\alpha,\beta)-POS_{(\mathcal{R}-\{S\})}T=\{\{x_1,x_2\},\{x_2,x_5\},\{x_4\}\}\neq(\alpha,\beta)-POS_{\mathcal{R}}T$$

This gives that P, Q, R and S are $T(\alpha,\beta)$-indispensable. That is $\mathcal{R}$ is $T(\alpha,\beta)$-independent. So the $T(\alpha,\beta)$-reduct of $\mathcal{R}$ is $\{P,Q,R,S\}$ and the $T(\alpha,\beta)$-core is also $\{P,Q,R,S\}$.

But, with different pair of values of (α,β) the situation may be different.

7 Dependency of Knowledge in IF - Approximation Space

In this section we discuss, how another knowledge base can be induced from a given knowledge base.

Definition 42. Suppose $\mathcal{K}=(U,\mathcal{R})$ is an IF-approximation space and let $\mathbf{P},\mathbf{Q}\subseteq\mathcal{R}$. For a fixed $(\alpha,\beta)\in J$, knowledge $\mathbf{Q}$ is (α,β)-*derivable* from knowledge $\mathbf{P}$ if all (α,β) categories of $\mathbf{Q}$ can be defined in terms of some elementary (α,β)-categories of knowledge $\mathbf{P}$.

Definition 43. Let $\mathcal{R}$ be a family of IF-proximity relations on the IF-approximation space $\mathcal{K}=(U,\mathcal{R})$ and $\mathbf{P},\mathbf{Q}\subset\mathcal{R}$. Now $\mathbf{Q}$ is (α,β) derivable from $\mathbf{P}$, we say that knowledge $\mathbf{Q}(\alpha,\beta)$ dependent on knowledge $\mathbf{P}$ and we denote it by $\mathbf{P}\overset{(\alpha,\beta)}{\Rightarrow}\mathbf{Q}$ if and only if $IND\ \mathbf{P}(\alpha,\beta)\subseteq IND\ \mathbf{Q}(\alpha,\beta)$.

Definition 44. Knowledge $\mathbf{P}$ and $\mathbf{Q}$ are said to be (α,β)-*equivalent*, denoted by $\mathbf{P}\overset{(\alpha,\beta)}{\equiv}\mathbf{Q}$ if and only if $\mathbf{P}\overset{(\alpha,\beta)}{\Rightarrow}\mathbf{Q}$ and $\mathbf{Q}\overset{(\alpha,\beta)}{\Rightarrow}\mathbf{P}$.

Obviously, $\mathbf{P}\overset{(\alpha,\beta)}{\equiv}\mathbf{Q}$ if and only if $IND\mathbf{P}\overset{(\alpha,\beta)}{=}IND\ \mathbf{Q}(\alpha,\beta)$

Knowledge $\mathbf{P}$ is (α,β)-independent to knowledge $\mathbf{Q}$ if and only if neither $\mathbf{P}\overset{(\alpha,\beta)}{\Rightarrow}\mathbf{Q}$ nor $Q\overset{(\alpha,\beta)}{\Rightarrow}\mathbf{P}$ holds.

Propositions extending Propositions 2, 3, 4 and 5 can be established in this general setting.

8 Partial Dependency of Knowledge in IF-Approximation Spaces

Extending the concepts developed in section 2.6, we shall introduce the notion of partial (α,β) -dependency of one knowledge base upon another.

Definition 45. Let $\mathcal{K}=(U,\mathcal{R})$ be a knowledge base. For $(\alpha,\beta)\in J$, let $\mathbf{P},\ \mathbf{Q}\in\mathcal{R}$. Then we say that knowledge $\mathbf{Q},(\alpha,\beta)$-depends in a degree $k(\alpha,\beta), 0\leq k(\alpha,\beta)\leq 1$ from the knowledge $\mathbf{P}$ denoted by $\mathbf{P}\overset{(\alpha,\beta)}{\Rightarrow}{}_k\mathbf{Q}$ if

$$k(\alpha,\beta)=\gamma_{\mathbf{P}(\alpha,\beta)}(\mathbf{Q}(\alpha,\beta))=|POS_{\mathbf{P}(\alpha,\beta)}(\mathbf{Q}(\alpha,\beta))|/|U|. \tag{26}$$

(i) If $k(\alpha, \beta) = 1$, we will say that $\mathbf{Q}$ is totally (α, β)-dependent on $\mathbf{P}$.
(ii) If $0 < k(\alpha, \beta) < 1$, we say that $\mathbf{Q}$ is *roughly or partially* (α, β)*-dependent* on $\mathbf{P}$ and if $k(\alpha, \beta) = 0, \mathbf{Q}$ is *totally* (α, β)*-independent* from $\mathbf{P}$.

Hence $\mathbf{P} \overset{(\alpha,\beta)}{\Rightarrow}{}_1 \mathbf{Q}$ means $\mathbf{P} \overset{(\alpha,\beta)}{\Rightarrow} \mathbf{Q}$, that is when $k(\alpha, \beta) = 1$, all the elements of the universe can be classified into elementary categories $U/\mathbf{Q}(\alpha, \beta)$ by using the knowledge $\mathbf{P}$. When $k(\alpha, \beta) \neq 1$, only those elements of the universe which belong to the positive region can be classified into the categories of knowledge $\mathbf{Q}$ employing knowledge $\mathbf{P}$. When $k(\alpha, \beta) = 0$, none of the elements of the universe can be classified using knowledge $\mathbf{P}(\alpha, \beta)$ to elementary categories of knowledge $\mathbf{Q}(\alpha, \beta)$.

Hence the coefficient can be understood as the degree of dependency of knowledge $\mathbf{Q}(\alpha, \beta))$ related to the knowledge $\mathbf{P}(\alpha, \beta))$.

Note 5
(i) Propositions extending Proposition 6 and Proposition 7 can be established in this general setting.
(ii) We have $(\alpha, \beta), (\alpha_1, \beta) \in J$, $R^*(\alpha, \beta) \supset R^*(\alpha_1, \beta)$ for $\alpha > \alpha_1$. That is R $(\alpha, \beta) \Rightarrow$ R $(\alpha_1, \beta)$for $\alpha > \alpha_1$, R $\in R$. Hence every equivalence class of R (α, β) is contained in some equivalence class of R (α_1, β). Also, we find that, if $P(\alpha, \beta)$ is a (α, β)-reduct of $(\alpha, \beta), (\alpha_1, \beta)$ is (α_1, β)-reduct of $R(\alpha_1, \beta)$ for some $\alpha < \alpha_1$. So, we have the result.
(iii)A relation (IF-proximity relation) can be indispensable in Pawlak's sense, but it may be dispensable in the generalized sense. So that, by reducing the accuracy we can find a smaller core.

9 An Application

Let us consider a situation of selection of reference books by a student in a particular subject depending upon different factors of suitability of the books.

Suppose $U = \{b_1, b_2, b_3, b_4, b_5\}$ be the set of available books for the given syllabus of a particular subject. Let $X = \{b_1, b_3\}$ be the books prescribed for the syllabus. The student is interested to purchase some additional books for the enhancement of his/her knowledge according to his/her requirement based on the following criteria.

(i) based on matching of chapters
(ii) based on matching of chapters and more examples in matching chapters.
(iii) based on matching of chapters and more exercises in matching chapters.
(iv) based on economy to purchase a book.

Based on different observations and calculations, we formed the following relations on the above four criteria. Formulas used in generating membership and non-membership values in R_1, R_2, R_3 and R_4 are follows.

For R_1**:**

$$\mu(\mathrm{A,B}) = \frac{|\text{Completely matching chapters of A\&B}|}{|\text{Chapters of chapters of A\&B}| - |\text{Completely matching chapters of A\&B}|}$$

$$\nu(\mathrm{A,B}) = \frac{|\text{Completely unmatching chapters of A\&B}|}{|\text{Chapters of A\&B}| - |\text{Completely matching chapters of A\&B}|}$$

Table 25. $\mathbf{R}_1$: Matching of chapters

R_1	b_1	b_2	b_3	b_4	b_5
b_1	**(1,0)**	(0.33,0.40)	(0.83,0.17)	(0.55,0.33)	(0.50,0.33)
b_2	(0.33,0.40)	**(1,0)**	(0.37,0.50)	(0.57,0.33)	(0.57,0.27)
b_3	(0.83,0.17)	(0.37,0.50)	**(1,0)**	(0.62,0.12)	(0.42,0.41)
b_4	(0.55,0.33)	(0.5,0.33)	(0.62,0.12)	**(1,0)**	(0.42,0.43)
b_5	(0.50,0.33)	(0.57,0.27)	(0.42,0.43)	(0.42,0.43)	**(1,0)**

Table 26. $\mathbf{R}_2$: Matching of chapters & more examples in matching chapters

R_2	b_1	b_2	b_3	b_4	b_5
b_1	**(1,0)**	(0.35,0.19)	(0.57,0.20)	(0.15,0.55)	(0.75,0.25)
b_2	(0.35,0.19)	**(1,0)**	(0.61,0.06)	(0.53,0.03)	(0.46,0.50)
b_3	(0.57,0.20)	(0.61,0.06)	**(1,0)**	(0.10,0.27)	(0.08,0.58)
b_4	(0.15,0.55)	(0.53,0.03)	(0.10,0.27)	**(1,0)**	(0.28,0.42)
b_5	(0.75,0.25)	(0.46,0.50)	(0.08,0.58)	(0.28,0.42)	**(1,0)**

Table 27. $\mathbf{R}_3$: Matching of chapters & more examples in matching chapters

R_3	b_1	b_2	b_3	b_4	b_5
b_1	**(1,0)**	(0.47,0.29)	(0.50,0.25)	(0.11,0.67)	(0.47,0.27)
b_2	(0.47,0.29)	**(1,0)**	(0.52,0.39)	(0.37,0.45)	(0.59,0.41)
b_3	(0.50,0.25)	(0.52,0.39)	**(1,0)**	(0.39,0.46)	(0.61,0.39)
b_4	(0.11,0.67)	(0.37,0.45)	(0.39,0.46)	**(1,0)**	(0.03,0.42)
b_5	(0.47,0.27)	(0.59,0.41)	(0.61,0.39)	(0.03,0.42)	**(1,0)**

For R_2:

$$\mu(\mathrm{A},\mathrm{B}) = \frac{|\text{Matching examples in matching chap. of A\&B}|}{|\text{Examp. in matching chap. of A\&B}|-|\text{Matching examp. in matching chap. of A\&B}|}$$

$$\nu(\mathrm{A},\mathrm{B}) = \frac{|\text{Unmatching examples of matching chap. of A\&B}|}{|\text{Examp. in matching chap. of A\&B}|-|\text{Matching examp. in matching chap. of A\&B}|}$$

For R_3:

$$\mu(\mathrm{A},\mathrm{B}) = \frac{|\text{Complete matching Exercises of matching chap. of A\&B}|}{|\text{Exercises of matching chap. of A\&B}|-|\text{Matching exercises of matching chap. of A\&B}|}$$

$$\nu(\mathrm{A},\mathrm{B}) = \frac{|\text{Complete unmatching Exercises of matching chap. of A\&B}|}{|\text{Exercise of matching chap. of A\&B}|-|\text{Matching exercises of matching chap. of A\&B}|}$$

For R_4:

$$\mu(\mathrm{A},\mathrm{B}) = \frac{|\text{Completely matching chap. of A\&B}|\times(\text{Price of A}+\text{Price of B})}{|\text{Completely chap. of A}|\times\text{Price of B}+|\text{Chap. of B}|\times\text{Price of A}}$$

$$\nu(\mathrm{A},\mathrm{B}) = \frac{|\text{Unmatching chap. of A\&B}|\times(\text{Price of A}+\text{Price of B})}{|\text{Chap. of A}|\times\text{Price of B}+|\text{Chap. of B}|\times\text{Price of A}}$$

1. Let us find the upper approximation of the books to be acquired by a student as per his/her necessity.
The IF-approximation space generated from $\mathcal{R} = \{R_1, R_2, R_3, R_4\}$ is given in Table 29.

If a student gives importance to all criteria, that is, having more common chapters with the text, more examples and exercises and also economic for

Table 28. **R**$_4$: Price factor of books

R_4	b_1	b_2	b_3	b_4	b_5
b_1	**(1,0)**	(0.44,0.23)	(0.10,0.10)	(0.72,0.14)	(0.67,0.20)
b_2	(0.44,0.23)	**(1,0)**	(0.37,0.61)	(0.61,0.32)	(0.73,0.21)
b_3	(0.10,0.10)	(0.37,0.61)	**(1,0)**	(0.81,0.11)	(0.50,0.50)
b_4	(0.72,0.14)	(0.61,0.32)	(0.81,0.11)	**(1,0)**	(0.50,0.50)
b_5	(0.67,0.20)	(0.73,0.21)	(0.50,0.50)	(0.50,0.50)	**(1,0)**

Table 29. IF-approximation space for $\mathcal{R}$

IND($\mathcal{R}$)	b_1	b_2	b_3	b_4	b_5
b_1	**(1,0)**	(0.33,0.40)	(0.10,0.25)	(0.11,0.67)	(0.47,0.33)
b_2	(0.10,0.25)	**(1,0)**	(0.37,0.61)	(0.37,0.45)	(0.46,0.50)
b_3	(0.10,0.25)	(0.37,0.61)	**(1,0)**	(0.10,0.46)	(0.08,0.58)
b_4	(0.11,0.67)	(0.37,0.45)	(0.10,0.46)	**(1,0)**	(0.03,0.50)
b_5	(0.47,0.33)	(0.08,0.58)	(0.08,0.58)	(0.03,0.50)	**(1,0)**

purchase up to particular levels of membership and non-membership then he/she has to consider the common cases by taking the intersection of all the above relations.

Case 1. Suppose $\alpha = 0.45$ and $\beta = 0.4$. $R(\alpha, \beta) = \{\{b_1, b_5\}, \{b_2\}, \{b_3\}, \{b_4\}\}$ and $\overline{X}_{\alpha,\beta} = \{b_1, b_3, b_5\}$.

Conclusion: The student is required to purchase the book b_5 for his/her use in addition to the textbooks b_1 and b_3 if importance is given to all the four characteristics above but with low positive value of 0.45 and high negative value of 0.4.

Case 2. Suppose $\alpha = 0.8$ and $\beta = 0.2$. $R(\alpha, \beta) = \{\{b_1\}, \{b_2\}, \{b_3\}, \{b_4\}, \{b_5\}\}$ and $\overline{X}_{\alpha,\beta} = \{b_1, b_3\}$.

Conclusion: The student is not required to purchase any book for his/her use besides the textbooks b_1 and b_2 if importance is given to all the four characteristic above with high positive value of 0.8 and low negative value of 0.2.

On the other hand, if some criteria are not important for the student; like he/she has no problem for price and does not insist for more number of exercises then he/she has to consider the intersection of R_1 and R_2.

2. Next we shall determine the prime factor to select a book(s) by a student using the reduct and core of the relation $R(\alpha, \beta)$.
Let $\alpha = 0.55$ and $\beta = 0.33$. Then

$$U/R_1(\alpha,\beta) = \{(b_1,b_1),(b_1,b_3),(b_2,b_2),(b_2,b_5),(b_3,b_1),(b_3,b_3),(b_3,b_4),(b_4,b_3),(b_4,b_4),(b_5,b_2),(b_5,b_5)\}.$$
$$= \{\{b_1,b_3,b_4\},\{b_2,b_5\}\}$$

$$U/R_2(\alpha,\beta) = \{(b_1,b_1),(b_1,b_3),(b_2,b_2),(b_2,b_3),(b_2,b_4),(b_3,b_1),(b_3,b_2),(b_3,b_3),(b_4,b_2),(b_4,b_4),(b_5,b_1),(b_5,b_5)\}.$$
$$= \{\{b_1,b_2,b_3,b_4,b_5\}\}$$
$$U/R_3(\alpha,\beta) = \{(b_1,b_1),(b_2,b_2),(b_3,b_3),(b_3,b_5),(b_4,b_4),(b_5,b_3),(b_5,b_5)\}.$$
$$= \{\{b_1\},\{b_2\},\{b_3,b_5\},\{b_4\}\}$$
$$U/R_4(\alpha,\beta) = \{(b_1,b_1),(b_1,b_4),(b_1,b_5),(b_2,b_2),(b_2,b_5),(b_3,b_3),(b_3,b_4),(b_4,b_1),(b_4,b_3),(b_4,b_4),(b_5,b_1),(b_5,b_2),(b_5,b_5)\}.$$
$$= \{\{b_1,b_2,b_3,b_4,b_5\}\}$$
$$U/(R_1\cap R_2\cap R_3\cap R_4)(\alpha,\beta) = U/IND\mathcal{R}(\alpha,\beta)$$
$$= \{\{b_1\},\{b_2\},\{b_3\},\{b_4\},\{b_4\},\{b_5\}\}.$$

Now,

$$U/IND(\mathcal{R}-R_1)(\alpha,\beta) = \{\{b_1\},\{b_2\},\{b_3,b_5\},\{b_4\}\} \neq U/IND\mathcal{R}(\alpha,\beta)$$

$$U/IND(\mathcal{R}-R_2)(\alpha,\beta) = \{\{b_1\},\{b_2\},\{b_3\},\{b_4\},\{b_5\}\} = U/IND\mathcal{R}(\alpha,\beta)$$

$$U/IND(\mathcal{R}-R_3)(\alpha,\beta) = \{\{b_1,b_3,b_4\},\{b_2,b_5\}\} \neq U/IND\mathcal{R}(\alpha,\beta)$$

$$U/IND(\mathcal{R}-R_4)(\alpha,\beta) = \{\{b_1\},\{b_2\},\{b_3\},\{b_4\},\{b_5\}\} = U/IND\mathcal{R}(\alpha,\beta)$$

Here R_1 and R_3 are (α,β)-indispensable, while R_2 and R_4 are (α,β)-dispensable in $\mathcal{R}$.

The (α,β)-reducts of $\mathcal{R}$ are

$\{R_1,R_2,R_3\}$ and $\{R_1,R_3,R_4\}$.

$$\text{CORE}(R(\alpha,\beta)) = \{R_1,R_2,R_3\}\cap\{R_1,R_3,R_4\}$$
$$= \{R_1,R_3\}.$$

So, the number of matching chapters and number of matching examples in matching chapters are the primary factors for selecting a book by students.

3. Comparison of result on Fuzzy Approximation Spaces

(a) let us compare the IF-approximation space approach with the Fuzzy approximation Space approach for this particular example.

We drop the non-membership values from the above example, leading to fuzzy approximation spaces corresponding to the fuzzy proximity relations R_1, R_2, R_3 and R_4.

Case 1. Suppose $\alpha = 0.45$. Then
$\mathcal{R}(\alpha) = \{\{b_1,b_2,b_5\},\{b_3\},\{b_4\}\}$ and $\overline{X}_\alpha = \{b_1,b_2,b_3,b_5\}$.

Conclusion: The student is required to purchase two new books b_2 and b_5 for his/her use besides the textbooks b_1 and b_3.

Comparing this inference with IF-approximation spaces (Case 1. of 1 above) the student requires purchasing only one additional book b_2, we see that in the fuzzy approximation space we don't have any control over the non-membership value, which is automatically set to 0.55. Because of this high non-membership value one has to go for purchasing the additional book.

Case 2. suppose $\alpha = 0.8$. Then
$\mathcal{R}(\alpha) = \{\{b_1\}, \{b_2\}, \{b_3\}, \{b_4\}, \{b_5\}\}$ and $\overline{X}_\alpha = \{b_1, b_3\}$.

Conclusion: So, a student is not required to purchase any book other than b_1 and b_3. Here the non-membership value is automatically set to 0.2 and so, there is no change.
(b) Next, we find the $\alpha-$core and the $\alpha-$reduct of the relations IND$\mathcal{R}(\alpha)$.

Let $\alpha = 0.55$. Then
$U/R_1(\alpha) = \{\{b_1, b_3, b_4\}, \{b_2, b_5\}\}$.
$U/R_2(\alpha) = \{\{b_1, b_2, b_3, b_5\}, \{b_4\}\}$.
$U/R_3(\alpha) = \{\{b_1\}, \{b_2, b_3, b_5\}, \{b_4\}\}$.
$U/R_4(\alpha) = \{b_1, b_2, b_3, b_4, b_5\}$.
$U/IND\mathcal{R}(\alpha) = \{\{b_1\}, \{b_2\}, \{b_3\}, \{b_4\}, \{b_5\}\}$.
Now,
$U/IND(\mathcal{R} - R_1)(\alpha) = \{\{b_1\}, \{b_2, b_3, b_5\}, \{b_4\}\} \neq U/IND\mathcal{R}(\alpha)$
$U/IND(\mathcal{R} - R_2)(\alpha) = \{\{b_1\}, \{b_2, b_5\}, \{b_3\}, \{b_4\}\} \neq U/IND\mathcal{R}(\alpha)$
$U/IND(\mathcal{R} - R_3)(\alpha) = \{\{b_1, b_3\}, \{b_2, b_5\}, \{b_4\}\} \neq U/IND\mathcal{R}(\alpha)$
$U/IND(\mathcal{R} - R_4)(\alpha) = \{\{b_1\}, \{b_2, b_5\}, \{b_3\}, \{b_4\}\} \neq U/IND\mathcal{R}(\alpha)$

Here, all the relations R_1, R_2,R_3 and R_4 are α-indispensable. Hence, all the factors are important to select a book for the study. This analysis fails to distinguish the significance of the four relations properly. However in Case 2, we find that with the control of non-membership value β if we reduce its value then only two characteristic are indispensable for the selection of reference books. So, by considering the intuitionistic fuzzy approximation space approach we can control both proximity and non proximity parameters which are to get better results. The values of these parameters are to be selected judiciously depending upon the situations and requirements.

10 Conclusion

The introduction of the concept of Rough set by Z. Pawlak was mainly intended to capture impreciseness. It was primarily dependent on the notion of equivalence relations. However equivalence relations are rare in nature. So, many real life situations could not be covered by this basic version of rough sets. Extensions have been made in many directions. Mainly these extensions are by the relaxation of the restriction of the basic relations to be equivalence relations [32,33,34,35,37,38,40,41,42]. In this chapter we have discussed some such new versions of the notion of rough sets; namely rough sets on fuzzy approximation spaces [14,15,40,42] and rough sets on intuitionistic fuzzy approximation spaces [37,38,41]. By the introduction of the fuzzy concept more reality could be brought into the notion of rough sets and also its modelling power and applicability have been improved. The definition of extended version are dependent upon fuzzy proximity relation and intutionistic fuzzy proximity relation, which are generalized versions of equivalence relation. These new versions depend upon the initial version for their development but with separate identities and with better modelling power.

References

1. Attanasov, K.T.: Intuitionistic Fuzzy sets. In: VII ITKR's Session, Deposed in central Sci. Techn. Library of Bulg. Acad. of Sci., Sofia, pp. 1684–1697 (1983)
2. Attanasov, K.T.: Intuitionistic Fuzzy sets. Fuzzy Sets and Systems 20, 87–96 (1986)
3. Attanasov, K.T.: Intuitionistic Fuzzy relations, First Scientific Session of the Mathematical Foundations of Artificial Intelliginence, Sofia, IM-MFAIS, 1–3 (1989)
4. Bezdek, J.C.: Pattern Recognition with Fuzzy Objective Function Algorithms. Plenum Press, New York (1981)
5. Bezdek, J.C., Pal, S.K. (eds.): Fuzzy Models for Pattern Recognition, Method that Search for patterns in data. IEEE Press, New York (1992)
6. Bonikowski, Z.: Algebraic Structures of Rough Sets. Rough Sets, Fuzzy Sets and Knowledge Discovery, 243–247 (1994)
7. Bonikowski, Z., Bryniarski, E., Wybraniec, U.: Extensions and intentions in the rough set theory. Information Sciences 107, 149–167 (1998)
8. Bryniaski, E.: A Calculus of Rough Sets of the First Order. Bull. Pol. Acad. Sci. 36(16), 71–77 (1989)
9. Buckles, B.P., Petry, F.E.: A Fuzzy Representation of data from relation databases. Fuzzy Sets and Systems 7(3), 213–226 (1982)
10. Buckles, B.P., Petry, F.E.: Information -theoretical Characterization of fuzzy relational data bases. IEEE Trans. on Systems, Man and Cybernatics 13(1), 74–77 (1983)
11. Burillo, P., Bustince, H.: Intutionistic Fuzzy relations (Part-I). Mathware Soft Computing 2, 5–38 (1995)
12. Burillo, P., Bustince, H.: Intutionistic Fuzzy relations (Part-II). Mathware Soft Computing 2, 117–148 (1995)
13. Chakraborty, M.K., Das, M.: Studies of fuzzy relations over fuzzy subsets. Fuzzy sets and sys. 9, 79–89 (1983)
14. De, S.K.: Some aspects of Fuzzy Sets, Rough Sets and Intuitionistic Fuzzy Sets, Ph.D. Thesis, I.I.T. Kharagpur (1999)
15. De, S.K., Biswas, R., Roy, A.R.: Rough sets on fuzzy approximation space. The Journal Fuzzy Maths. 11, 1–15 (2003)
16. Chen, D., Wenxice, Z., Yeung, D., Tsang, E.C.C.: Rough Approximations on a Complete Completely Distributive Lattice with Applications to Generalised Rough Sets. Information Science 176, 1829–1848 (2006)
17. Chen, D., Changzhong, W., Qinghua, H.: A new approach to attribute reduction of consistent and inconsistent covering decision systems with covering rough sets. Information Science 177, 3500–3518 (2007)
18. Dubois, D., Prade, H.: Fuzzy sets and systems: Theory and applications. Academic press, New York (1980)

19. Dubois, D., Prade, H.: Rough Fuzzy Sets and Fuzzy Rough Sets. Fuzzy Sets and Systems 23, 3–18 (1987)
20. Dubois, D., Prade, H.: Rough Fuzzy Sets and Fuzzy Rough Sets. International Journal of General Systems 17, 191–209 (1990)
21. Mi, J.S., Zhang, W.X.: An Axiomatic Characterization of a Fuzzy Generalization of Rough Sets. Information Sciences 160(1-4), 235–249 (2004)
22. Mordeson, J.: Rough Set Theory Applied to (Fuzzy) Ideal Theory. Fuzzy Sets and system 121, 315–324 (2001)
23. Morsi, N.N., Yakout, M.M.: Axiomatics for Fuzzy Rough Sets. Fuzzy Sets and Systems 100, 327–342 (1998)
24. Nemitz, W.C.: Fuzzy Relations and Fuzzy Functions. Fuzzy Sets and Systems 19, 177–191 (1986)
25. Pal, S.K., Majumdar, D.K.D.: Fuzzy Mathematical Approach to Pattern Recognition. John Wiley, New York (1986)
26. Pawlak, Z.: Rough Sets. Int. Jour. Inf. Comp. Sc. II, 341–356 (1982)
27. Pawlak, Z.: Rough Sets: Theoretical Aspects of Reasoning about Data. Kluwer Academic Publishers, Dordrecht (1991)
28. Peters, J., Skowron, A., Stepaniuk, J.: Nearness in Approximation Spaces. In: Lindemann, G., Schlilngloff, H., et al. (eds.) Proc. Concurrency, Specification and Programming (CS and P 2006), Humboltd-Universitat zu Berlin, informatik-Berichte, vol. 206, pp. 435–445 (2006)
29. Pomykala, J.A.: Approximation Operations in Approximation Spaces. Bull. Pol. Acad. Sci. 35(9-10), 653–662 (1987)
30. Pomykala, J.A.: Approximation Similarity and Rough Constructions, IILC Publication Series, University of Amaterdam CT-93-07 (1993)
31. Skowron, A., Stepaniuk, J.: Generalized Approximation Spaces. In: Lin, T.Y., Wildberger, A.M. (eds.) Soft Computing: Rough Sets, Fuzzy Logic, Neural Networks, Uncertainty Management, pp. 18–21. San Diago Simulation Councils, Inc. (1995)
32. Skowron, A., Stepaniuk, J.: Tolerance Approximation Spaces. Fundamenta Informaticae 27(2-3), 245–253 (1996)
33. Skowron, A., Peters, J., Stepanink, J., Swiniarski: Calculi of Approximation Spaces. Fundamenta Informaticae 72, 363–378 (2006)
34. Slowinski, R., Vanderpooten, D.: Similarity Relation as a Basis for Rough Approximations. In: Wang, P.P. (ed.) Advances in Machine Intelligence & Soft-Computing, vol. IV, pp. 17–33. Duke University Press, Durham (1997)
35. Slowinski, R., Vanderpooten, D.: A Generalized Definition of Rough Approximations Based On Similarity. IEEE Transactions On Data and Knowledge Engineering 12, 331–336 (2000)
36. Tsang, E., Cheng, D., Lee, J., Yeung, D.: On the Upper Approximation of Covering Generalized Rough Sets. In: Proc. 3rd Interntional Conf. Machine Learning and Cybernetics, pp. 4200–4203 (2004)
37. Tripathy, B.K.: Rough sets on Intuitionistic Fuzzy Approximation Spaces. In: Proceedings of 3rd International IEEE Conference on Intelligent Systems, IS 2006, London, September 4-6, pp. 776–779 (2006)
38. Tripathy, B.K.: Rough sets on intituionistic fuzzy approximation spaces. NIFS 12(1), 45–54 (2006)
39. Tripathy, B.K., Mitra, A.: Some Topological Properties of Rough Sets and their Applications, Communicated to International Journal of Granular Computing, Rough Sets and Intelligent Systems

40. Tripathy, B.K., Acharjya, D.P.: Rough Sets on Fuzzy Approximation Spaces and Application to Distributed Knowledge Systems. International Journal of Artificial Intelligence and Soft Computing 1(1), 1–15 (2008)
41. Tripathy, B.K., Gantayat, S.S.: Rough sets on Intuitionistic Fuzzy Approximation Spaces and Knowledge Representation. In: Proceedings of National Conference on Soft Computing Techniques in Engineering Applications (SCT 2006), NIT Rourkela, March 24-26, pp. 319–328 (2006)
42. Tripathy, B.K., Gantayat, S.S., Mohanty, D.: Properties of Rough Sets on Fuzzy Approximation Spaces and Knowledge Representation. In: Proceeding of the National Conference on Recent Trends in Intelligent Computing (RTIC 2006), Kalyani Govt. Engg. College (W.B), November 17-19, pp. 3–8 (2006)
43. Tripathy, B.K., Choudhury, P.K.: Fuzzy lattices and Fuzzy Boolean Algebra. Journal of Fuzzy Mathematics 16(1), 185–198 (2008)
44. Wu, W.Z., Mi, J.S., Zhang, W.X.: Generalized Fuzzy Rough Sets. Information Sciences 151(6), 263–282 (2003)
45. Zadeh, L.A.: Fuzzy Sets. Information and Control 8, 338–353 (1965)
46. Zadeh, L.A.: Similarity Relations and Fuzzy Ordering. Information sciences 3(2), 177–200 (1970)
47. Zadeh, L.A.: Fuzzy logic and approximate reasoning. Synthese 30(1), 407–428 (1975)
48. Zadeh. L. A.: The concept of linguistic variable and its application to approximate reasoning I,II,III, Information Sciences 8, 199–251, 301–357, 9, 43–80 (1975)
49. Zakowski, W.: Approximations in the spaces (μ, π). Demonstratio mathematica 16, 761–769 (1983)
50. Zhu, F.: On Covering generalised Rough Sets, Master thesis - The University of Arizona, Tuscon, Arizona, USA (May 2002)
51. Zhu, W., Wang, F.Y.: Reduction and Axiomatization of Covering Generalized Rough Sets. Information Sciences 152, 217–230 (2003)
52. Wilium, Z.: Topological Approaches to Covering Rough Sets. Information Sciences 177, 1499–1508 (2007)

Categorical Innovations for Rough Sets

P. Eklund[1], M.A. Galán[2,*], and J. Karlsson[3]

[1] Department of Computing Science, Umeå University, Sweden
peklund@cs.umu.se
[2] Department of Applied Mathematics, University of Málaga, Spain
magalan@ctima.uma.es
[3] Department of Computer Architecture, University of Málaga, Spain
johan@ac.uma.es

Summary. Categories arise in mathematics and appear frequently in computer science where algebraic and logical notions have powerful representations using categorical constructions. In this chapter we lean towards the functorial view involving natural transformations and monads. Functors extendable to monads, further incorporating order structure related to the underlying functor, turn out to be very useful when presenting rough sets beyond relational structures in the usual sense. Relations can be generalized with rough set operators largely maintaining power and properties. In this chapter we set forward our required categorical tools and we show how rough sets and indeed a theory of rough monads can be developed. These rough monads reveal some canonic structures, and are further shown to be useful in real applications as well. Information within pharmacological treatment can be structured by rough set approaches. In particular, situations involving management of drug interactions and medical diagnosis can be described and formalized using rough monads.

1 Introduction

Monads are useful e.g. for generalized substitutions as we have extended the classical concept of a term to a many-valued set of terms [21]. This builds essentially upon composing various set functors, as extendable to monads, with the term functor and its corresponding monad. The most trivial set functor is the powerset functor for which a substitution morphism in the corresponding Kleisli category is precisely a relation. Thus relations are seen as connected to a powerset functor that can be extended to a monad. Further, whenever general powerset monads can be extended to partially ordered monads, this structure is sufficient for the provision of rough set operations in a category theory setting. This categorical presentation of rough sets will establish connections to other categorical structures with the objective to enrich the theory. Key in these constructions is the first observation that relations are morphisms in the Kleisli category of the monad extended from the powerset functor.

Fuzzy sets, closely related to rough sets, are founded on the notion of many-valued membership, and is considered as a gradual property for fuzzy sets. Fuzzy

* Partially supported by Spanish projects P06-FQM-02049 and TIN2006-15455-C03-01.

A. Abraham, R. Falcón, and R. Bello (Eds.): Rough Set Theory, SCI 174, pp. 45–69.
springerlink.com

set theory offers a more expressive mathematical language and has many applications in a very wide variety of fields. Fuzzy sets, originally introduced by Zadeh [39] in 1965, increase the expressiveness of classical mathematics to deal with information systems that are incomplete, uncertain and imprecise. In 1967, Goguen, [22], extended the idea of fuzzy sets to L-fuzzy sets, considering more general order structures L beyond the unit interval. The extended notion of L-fuzzy sets also represent the extension of the crisp powerset monad. Thus the powerset monads is the categorical way to represent fuzzy sets. Beyond fuzzy sets, and introducing partial order into monads, using so called *partially ordered monads*, we can then also represent rough sets.

The outline of the chapter is as follows. Section 2 gives a historical background to the categorical apparatus used in the chapter. In sect. 3 we provide the categorical preliminaries and notations making the chapter easier to read and comprehend. Section 4 includes important examples of monads, one of the underlying categorical tools used in this chapter. Section 5 then describes partially ordered monads upon which rough monads are built in Sect. 7. In sect. 6, rough sets are conceptually embedded into the categorical machinery. Section 8 outlines applications related to management of drug interactions and cognitive disorder diagnosis, respectively. Section 9 concludes the chapter.

2 Historical Remarks and Related Work

Monads were initiated by Godement around 1958. Huber shows in 1961 that adjoint pairs give rise to monads. Kleisli [28] and also Eilenberg and Moore [4] proves the converse in 1965. Kleisli categories were explicitely constructed in those contributions. Partially ordered monads are monads [32], where the underlying endofunctor is equipped with an order structure, which makes them useful for various generalized topologies and convergence structures [18, 20]. They are indeed derived from studies on convergence, initiated by [30]. Partially ordered monads were initially studied in [18, 19]. Topology and convergence were forerunners in the development of partially ordered monads, but these monads contain structure also for modelling rough sets [33] in a generalized setting with set functors. Partially ordered monads contribute to providing a generalised notion of powerset Kleene algebras [5]. This generalisation builds upon a more general powerset functor setting far beyond just strings [27] and relational algebra [37]. These structures are typical representatives of Kleene algebras, which are widely used e.g. in formal languages [36] and analysis of algorithms [1]. Rough sets and their purely algebraic properties are studied e.g. within shadowed sets [3]. There is further an interesting interaction between monads and algebras, which is well-known. The tutorial example is the isomorphism between the Kleisli category of the powerset monad and the category of 'sets and relations'. The Eilenberg-Moore category of the powerset monad is isomorphic to the category of complete lattices and join-preserving mappings. The Kleisli category of the term monad coincides with its Eilenberg-Moore category and is isomorphic to the category of Ω-algebras. A rather intrepid example, although still folklore, is the isomorphism between the Eilenberg-Moore category of the ultrafilter monad and the category of compact Hausdorff spaces. Here is where "algebra and topology meet".

3 Categorical Preliminaries

A major advantage of category theory is its 'power of abstraction' in the sense that many mathematical structures can be characterized in terms of relatively few categorical ones. This fact enables to pursue a more general study towards generalizations of the structures. Category theory has been successfully applied in different areas such as topology, algebra, geometry or functional analysis. In recent years, category theory has also contributed to the development of computer science: the abstraction of this theory has brought the recognition of some of the constructions as categories. This growing interest towards categorical aspects can be found in, for instance, term rewriting systems, game semantics and concurrency. In a gross manner one can say a *category* is given by a class of *object* and a class of *morphisms* between the objects under certain mathematical conditions. Examples of categories come not only from mathematics (the category of groups and group homomorphisms, the category of topological spaces and continuous functions, etc.) but also from computer science. Deductive systems is a category where the objects are formulas and morphisms are proofs. Partially ordered sets form a category where objects are partially ordered sets and morphisms are monotone mappings. A particular partially ordered set also forms a category where objects are its elements and there is exactly one morphism from an element x to an element y if and only if $x \leq y$. We can go beyond categories and wonder if there is a category of categories. The answer is yes (provided the underlying selected set theory is properly respected). In this category of categories the objects are categories and the morphisms are certain structure-preserving mappings between categories, called functors. Examples of functors are for instance the list functor, the powerset functor and the term functor. The concept of naturality is important in many of the applications of category theory. Natural transformations are certain structure-preserving mappings from one functor to another. It might seem abstract to consider morphisms between morphisms of categories, but natural transformations appear in a natural way very frequently both in mathematics as well as in computer science. Natural transformations are cornerstones in the concept of monads.

3.1 Categories

A *category* $\mathtt{C}$ consists of *objects*, and for each pair (A, B) of objects we have *morphisms* f from A to B, denoted by $f : A \longrightarrow B$ or $A \xrightarrow{f} B$. Further there is an (A-)identity morphism $A \xrightarrow{id_A} A$ and a composition $\circ$ among morphisms that composes $A \xrightarrow{f} B$ and $B \xrightarrow{g} C$ to $A \xrightarrow{g \circ f} C$ in order to always guarantee $h \circ (g \circ f) = (h \circ g) \circ f$, and also $id_B \circ f = f \circ id_A = f$ for any $A \xrightarrow{f} B$. The set of $\mathtt{C}$-morphisms from A to B is written as $Hom_{\mathtt{C}}(A, B)$ or $Hom(A, B)$.

Example 1. The category of sets, `Set`, consists of sets as objects and functions as morphisms together with the ordinary composition and identity.

Example 2. The category of partially ordered sets, `Poset`, consists of partially ordered sets as objects and order-preserving functions as morphisms. The category of boolean algebras, `Boo`, consists of boolean algebras as objects and boolean homomorphisms as morphisms. The category of groups, `Grp`, consists of groups as objects and group homomorphisms as morphisms.

Example 3. A poset (partially ordered set) forms also a category where objects are its elements and there is exactly one morphism from an element x to an element y if and only if $x \leq y$. Composition is forced by transitivity. Identity is forced by reflexivity.

Example 4. The category of Ω-algebras, $\mathtt{Alg}(\Omega)$, consists of Ω-algebras as objects and Ω-homomorphisms as morphisms between them. Recall that a Ω-algebra is a pair $(X, (\omega_i)_{i \in I})$ where X is a set and $\omega_i : X^{n_i} \longrightarrow X$ are the n_i-ary operations on X. An Ω-homomorphism $f : (X, (\omega_i)_{i \in I}) \longrightarrow (\hat{X}, (\hat{\omega}_i)_{i \in I})$ consists of a function $f : X \longrightarrow \hat{X}$ such that the diagram

$$\begin{array}{ccc} X^{n_i} & \xrightarrow{f^{n_i}} & \hat{X}^{n_i} \\ {\scriptstyle \omega_i}\downarrow & & \downarrow{\scriptstyle \hat{\omega}_i} \\ X & \xrightarrow[f]{} & \hat{X} \end{array}$$

commutes, i.e. $f(\omega_i(x_1, \ldots, x_{n_i})) = \hat{\omega}_i(f(x_1), \ldots, f(x_{n_i}))$.

Further examples are the category of groups and group homomorphisms, the category of vector spaces and linear transformations, and the category of topological spaces with continuous functions as morphisms. A useful category in computer science is the following.

3.2 Functors

To relate category theory with typed functional programming, we identify the objects of the category with types and the morphisms between the objects with functions. In a given category, the set of morphisms $Mor(\mathcal{C})$ is useful to establish connections between the different objects in the category. But also it is needed to define the notion of a transformation of a category into another one. This kind of transformation is called a *functor*. In the previous context, functors are not only transformations between types, but also between morphisms, so, at the end, they will be mappings between categories. Let us see an example. Given a type, for instance, *Int*, we can consider the linear finite list type of elements of this type, *integer lists*. Let us denote by *List(S)* to indicate the lists of elements with type S. Let us see how *List* actuates not only over types, but also over functions between types. Given a function $f : S \longrightarrow T$ we want to define a function $List(f)$,

$$List(f) : List(S) \longrightarrow List(T).$$

Note that here we are using the same name for two operations, one over objects and the other one over functions. This is the standard when using functors. To understand how to define *List* over functions, let us consider the function $sq : Int \longrightarrow Int$ defined as $sq(x) = x^2$. The type of $List(sq)$ is $List(sq) : List(Int) \longrightarrow List(Int)$. What should be the value of $List(sq)[-2, 1, 3]$? The obvious answer is the list $[(-2)^2, 1^2, 3^2] = [4, 1, 9]$. In the general case, the part that actuates over the morphisms of *List* is the *maplist* function, that distributes a function over the elements of a list. In this case we have defined how *List* actuates over objects and morphisms. Next step is to ask ourselves how does *List* respect the categorical structure, that is, what is the behavior over the composition of morphisms and over the identity morphism? It is expected that

$$List(g \circ f) = List(g) \circ List(f),$$
$$List(id_a) = id_{List(a)}.$$

It is not difficult to check this for *maplist*. Now we are ready for the functor definition. Let $\mathtt{C}$ and $\mathtt{D}$ be categories. A (covariant) *functor* φ from $\mathtt{C}$ to $\mathtt{D}$, denoted $\varphi : \mathtt{C} \longrightarrow \mathtt{D}$, is a mapping that assigns each $\mathtt{C}$-object A to a $\mathtt{D}$-object $\varphi(A)$ and each $\mathtt{C}$-morphism $A \xrightarrow{f} B$ to a $\mathtt{D}$-morphism $\varphi(A) \xrightarrow{\varphi(f)} \varphi(B)$, such that $\varphi(f \circ g) = \varphi(f) \circ \varphi(g)$ and $\varphi(id_A) = id_{\varphi(A)}$. We often write φA and φf instead of $\varphi(A)$ and $\varphi(f)$. For functors $\varphi : \mathtt{C} \longrightarrow \mathtt{D}$ and $\psi : \mathtt{D} \longrightarrow \mathtt{E}$, we can easily see that the composite functor $\psi \circ \varphi : \mathtt{C} \longrightarrow \mathtt{E}$ given by

$$(\psi \circ \varphi)(A \xrightarrow{f} A') = \psi(\varphi A) \xrightarrow{\psi(\varphi f)} \psi(\varphi A')$$

indeed is a functor.

Example 5. Any category $\mathtt{C}$ defines an identity functor $id_{\mathtt{C}} : \mathtt{C} \longrightarrow \mathtt{C}$ given by

$$id_{\mathtt{C}}(A \xrightarrow{f} B) = A \xrightarrow{f} B.$$

Example 6. The (covariant) powerset functor $P : \mathtt{Set} \longrightarrow \mathtt{Set}$ is defined by PA being the powerset of A, i.e. the set of subsets of A, and $Pf(X)$, for $X \subseteq A$, being the image of X under f, i.e. $Pf(X) = \{f(x) \mid x \in X\}$. The contravariant powerset functor $Q : \mathtt{Set}^{op} \longrightarrow \mathtt{Set}$ is defined by QA again being the powerset of A, and further

$$Q(A \xrightarrow{f} B) = QA \xrightarrow{Qf} QB$$

where $Qf(X)$, $X \subseteq A$, is the inverse image of X under the function $f : B \longrightarrow A$.

Example 7. The list functor $List : \mathtt{Set} \longrightarrow \mathtt{Set}$ is defined by $List(A)$ being the set of finite lists with elements in A, i.e. $List(A) = \bigcup_{n \in \mathbf{N}} A^n$, and further for $f : A \longrightarrow B$ we have

$$Listf(L) = [f(a_1), \ldots, f(a_n)]$$

for finite lists $L = [a_1, \ldots, a_n]$ with $a_1, \ldots, a_n \in A$.

A functor $\varphi : \mathtt{C} \longrightarrow \mathtt{D}$ is a (functor) isomorphism if there is a functor $\psi : \mathtt{D} \longrightarrow \mathtt{C}$ such that $\psi \circ \varphi = id_{\mathtt{C}}$ and $\varphi \circ \psi = id_{\mathtt{D}}$.

Example 8. The category $\mathtt{Boo}$ is isomorphic to the category of boolean rings (ring with unit, and each element being idempotent with respect to multiplication, i.e. $a \cdot a = a$) and ring homomorphisms.

Example 9 (T-algebras). Let $T : \mathtt{X} \longrightarrow \mathtt{X}$ be a functor. A T-algebra is a pair (X, h), where X is an $\mathtt{X}$-object and $h : TX \longrightarrow X$ is an $\mathtt{X}$-morphism. A T-homomorphism $f : (X, h) \longrightarrow (X', h')$ between T-algebras is an $\mathtt{X}$-morphism $f : X \longrightarrow X'$ such that the diagram

$$\begin{array}{ccc} TX & \xrightarrow{\;h\;} & X \\ {\scriptstyle Tf}\downarrow & & \downarrow{\scriptstyle f} \\ TX' & \xrightarrow[\;h'\;]{} & X' \end{array}$$

commutes. We denote by $\mathtt{Alg}(T)$ the category consisting of T-algebras as $\mathtt{Alg}(T)$-objects and T-homomorphisms as $\mathtt{Alg}(T)$-morphisms. It can be shown that $\mathtt{Alg}(\Omega)$ (Example 4) is isomorphic to $\mathtt{Alg}(T)$ for some suitable functors T.

Example 10. If $\mathcal{G}$ and $\mathcal{H}$ are groups considered as categories with a single object, then a functor from $\mathcal{G}$ to $\mathcal{H}$ is exactly a group homomorphism.

Example 11. If $\mathcal{P}$ and $\mathcal{Q}$ are posets, a functor from $\mathcal{P}$ to $\mathcal{Q}$ is exactly a nondecreasing map.

Example 12. The list functor $List : \mathtt{Set} \longrightarrow \mathtt{Set}$ ($\mathtt{Set}$ denotes the category of sets) is defined by $List(A)$ being the set of finite lists with elements in A, i.e. $List(A) = \bigcup_{n \in \mathbf{N}} A^n$, and further for $f : A \longrightarrow B$ we have

$$List\, f(L) = [f(a_1), \ldots, f(a_n)]$$

for finite lists $L = [a_1, \ldots, a_n]$ with $a_1, \ldots, a_n \in A$.

3.3 Natural Transformations

In the same way as functors are defined as morphisms between categories, we could think of defining morphisms between functors. The concept of naturality is central in many of the applications of category theory. *Natural transformations* are certain structure-preserving mappings from one functor to another. Maybe, in a first approach to this, it seems abstract to consider morphisms between morphisms of categories. We will show here, how natural transformations appear in a natural way not only in mathematics, but also in programming. Continuing with lists, let us consider that the function that inverts lists, has type $rev : List(S) \longrightarrow List(S)$ where S is a type. Obviously, it is expected that rev inverts any kind of lists, i.e., it is expected that the definition rev is uniform with respect to the type of the elements on the list. One definition of *rev* can be given in the functional program Hope as follows:

- $rev(nil) <= nil$
- $rev(a :: l) <= rev(l) :: a$

Instead of considering *rev* as a list whose type is polymorphic, we can consider it as a collection of functions indexed by the element's type of the list, so we could write $rev_S : List(S) \longrightarrow List(S)$ or even, $rev : List \longrightarrow List$. In the last case, we apply the argument S, that is the type of the elements on the list, to *rev* and to the functor *List*. Note that when we apply *rev* to an argument, we get a function, in this way we can consider *rev* as a morphism from the functor *List* to the functor *List*. In this context we must make sure that, for types S and T, the functions rev_S and rev_T are well related to each other. The relation between these two mappings can be expressed through the commutativity of the following diagram

$$
\begin{array}{ccccc}
S & & List(S) & \xrightarrow{rev_S} & List(S) \\
\Big\downarrow f & & \Big\downarrow List(f) & & \Big\downarrow List(f) \\
T & & List(T) & \xrightarrow[rev_T]{} & List(T)
\end{array}
$$

In this case, the action of *List* over functions is the function *maplist*. It is easy to check that the diagram commutes, and for any $f : S \longrightarrow T$, it expresses a fundamental property of the function *rev*. Now, we can give the definition of natural transformation. Let $\varphi, \psi : \mathtt{C} \longrightarrow \mathtt{D}$ be functors.

Definition 1. *A* natural transformation τ *from* φ *to* ψ*, written* $\tau : \varphi \longrightarrow \psi$ *or* $\varphi \xrightarrow{\tau} \psi$*, assigns to each* C*-object* A *a* D*-morphism* $\tau_A : \varphi A \longrightarrow \psi A$ *such that the diagram*

$$
\begin{array}{ccccc}
A & & \varphi A & \xrightarrow{\tau_A} & \psi A \\
\Big\downarrow f & & \Big\downarrow \varphi f & & \Big\downarrow \psi f \\
A' & & \varphi A' & \xrightarrow[\tau_{A'}]{} & \psi A'
\end{array}
$$

commutes.

Let φ be a functor. The identity natural transformation $\varphi \xrightarrow{id_\varphi} \varphi$ is defined by $(id_\varphi)_A = id_{\varphi A}$. For functors φ and natural transformations τ we often write $\varphi\tau$ and $\tau\varphi$, respectively, to mean $(\varphi\tau)_A = \varphi\tau_A$ and $(\tau\varphi)_A = \tau_{\varphi A}$. It is easy to see that $\eta : id_{\mathtt{Set}} \longrightarrow P$ given by $\eta_X(x) = \{x\}$, and $\mu : P \circ P \longrightarrow P$ given by $\mu_X(\mathcal{B}) = \bigcup \mathcal{B} (= \bigcup_{B \in \mathcal{B}} B)$ are natural transformations. Natural transformations can be composed *vertically* as well as *horizontally*. Let $\varphi, \psi, \vartheta : \mathtt{C} \longrightarrow \mathtt{D}$ be functors and let further $\varphi \xrightarrow{\tau} \psi$ and $\psi \xrightarrow{\sigma} \vartheta$ be natural transformations. The (vertical) *composition* $\varphi \xrightarrow{\sigma \circ \tau} \vartheta$, defined by $(\sigma \circ \tau)_A = \sigma_A \circ \tau_A$, is a natural transformation. In order to define the corresponding horizontal composition, let $\varphi', \psi' : \mathtt{C} \longrightarrow \mathtt{D}$ be functors and let $\varphi' \xrightarrow{\tau'} \psi'$ be a natural transformation. The *star product* (horizontal composition) $\varphi' \circ \varphi \xrightarrow{\tau' \star \tau} \psi' \circ \psi$ is defined by

$$\tau' \star \tau = \tau' \psi \circ \varphi' \tau = \psi' \tau \circ \tau' \varphi. \tag{1}$$

For the identity transformation $id_\varphi \colon \varphi \longrightarrow \varphi$, also written as 1_φ or 1, we have

$$1_\varphi \star 1_\psi = 1_{\varphi \circ \psi}. \tag{2}$$

For a natural transformation $\tau \colon \varphi \longrightarrow \psi$, and a functor ϑ, $\vartheta\tau = 1_\vartheta \star \tau$ and $\tau\vartheta = \tau \star 1_\vartheta$. For natural transformations $\varphi \xrightarrow{\tau} \psi \xrightarrow{\sigma} \vartheta$ and $\varphi' \xrightarrow{\tau'} \psi' \xrightarrow{\sigma'} \vartheta'$ we have the Interchange Law $(\sigma' \circ \tau') \star (\sigma \circ \tau) = (\sigma' \star \sigma) \circ (\tau' \star \tau)$.

3.4 Monads and Kleisli Categories

In the following we include some formal definitions of concepts required.

Definition 2. *Let* $\mathtt{C}$ *be a category. A monad (or triple, or algebraic theory) over* $\mathtt{C}$ *is written as* $\mathbf{\Phi} = (\varphi, \eta, \mu)$, *where* $\varphi : \mathtt{C} \to \mathtt{C}$ *is a (covariant) functor, and* $\eta : id \to \varphi$ *and* $\mu : \varphi \circ \varphi \to \varphi$ *are natural transformations for which* $\mu \circ \varphi\mu = \mu \circ \mu\varphi$ *and* $\mu \circ \varphi\eta = \mu \circ \eta\varphi = id_\varphi$ *hold.*

Definition 3. *A Kleisli category* $\mathtt{C}_{\mathbf{\Phi}}$ *for a monad* $\mathbf{\Phi}$ *over a category* $\mathtt{C}$ *is given with objects in* $\mathtt{C}_{\mathbf{\Phi}}$ *being the same as in* $\mathtt{C}$, *and morphisms being defined as* $hom_{\mathtt{C}_{\mathbf{\Phi}}}(X, Y) = hom_{\mathtt{C}}(X, \varphi Y)$. *Morphisms* $f \colon X \rightharpoonup Y$ *in* $\mathtt{C}_{\mathbf{\Phi}}$ *are thus morphisms* $f \colon X \to \varphi Y$ *in* $\mathtt{C}$, *with* $\eta_X^\varphi : X \to \varphi X$ *being the identity morphism. Composition of morphisms in* $\mathtt{C}_{\mathbf{\Phi}}$ *is defined as*

$$(X \xrightharpoonup{f} Y) \diamond (Y \xrightharpoonup{g} Z) = X \xrightarrow{\mu_Z^\varphi \circ \varphi g \circ f} \varphi Z. \tag{3}$$

Composition in the case of the term monad comes down to substitution, and this brings us immediately to substitution theories in general for monads. Monads can be composed and especially the composition of the powerset monad with the term monad provides groundwork for a substitution theory as a basis for many-valued logic [21]. In the following we will elaborate on powerset monads. The concept of *subfunctors* and *submonads* can be used to provide a technique for constructing new monads from given ones.

Definition 4. *Let* φ *be a set functor. A set functor* φ' *is a subfunctor of* φ, *written* $\varphi' \leq \varphi$, *if there exists a natural transformation* $e \colon \varphi' \longrightarrow \varphi$, *called the inclusion transformation, such that* $e_X \colon \varphi' X \longrightarrow \varphi X$ *are inclusion mappings, i.e.,* $\varphi' X \subseteq \varphi X$. *The conditions on the subfunctor imply that* $\varphi f \mid_{\varphi' X} = \varphi' f$ *for all mappings* $f : X \longrightarrow Y$. *Further,* $\leq$ *is a partial ordering.*

Proposition 1 ([13]). *Let* $\mathbf{\Phi} = (\varphi, \eta, \mu)$ *be a monad over* $\mathtt{Set}$, *and consider a subfunctor* φ' *of* φ, *with the corresponding inclusion transformation* $e : \varphi'$

$\longrightarrow\varphi$, *together with natural transformations* $\eta' : id \longrightarrow \varphi'$ *and* $\mu' : \varphi'\varphi' \longrightarrow \varphi'$ *satisfying the conditions*

$$e \circ \eta' = \eta, \tag{4}$$

$$e \circ \mu' = \mu \circ \varphi e \circ e\varphi'. \tag{5}$$

Then $\mathbf{\Phi}' = (\varphi', \eta', \mu')$ *is a monad, called the submonad of* $\mathbf{\Phi}$, *written* $\mathbf{\Phi}' \preceq \mathbf{\Phi}$.

4 Examples of Monads

Monads have been used in many different areas such as topology or functional programming. The applications and use of monads in computer science is well-known and provides an abstract tool to handle properties of structures. Examples developed in this section have an important role in many applications. Powerset monads and their many-valued extensions are in close connection to fuzzification and are good candidates to represent situations with incomplete or imprecise information. With respect to topological application, the fuzzy filter monad is a key construction when studying convergence structures from a more general point of view. Unless otherwise stated, we assume L to be a completely distributive lattice. For $L = \{0, 1\}$ we write $L = 2$.

Remark 1. Extending functors to monads is not trivial, and unexpected situations may arise. Let the id^2 functor be extended to a monad with

$$\eta_X(x) = (x, x) \text{ and } \mu_X((x_1, x_2), (x_3, x_4)) = (x_1, x_4).$$

Further, the proper powerset functor P_0, where $P_0X = PX \setminus \{\emptyset\}$, as well as $id^2 \circ P_0$ can, respectively, be extended to monads, even uniquely. However, as shown in [15], $P_0 \circ id^2$ cannot be extended to a monad.

4.1 The Term Monad

Notations in this part follow [17], which were adopted also in [15, 11]. Let $\Omega = \bigcup_{n=0}^{\infty} \Omega_n$ be an operator domain, where Ω_n contains n-ary operators. The term functor $T_\Omega \colon \mathtt{SET} \to \mathtt{SET}$ is given as $T_\Omega(X) = \bigcup_{k=0}^{\infty} T_\Omega^k(X)$, where

$$T_\Omega^0(X) = X,$$
$$T_\Omega^{k+1}(X) = \{(n, \omega, (m_i)_{i \leq n}) \mid \omega \in \Omega_n, n \in N, m_i \in T_\Omega^k(X)\}.$$

In our context, due to constructions related to generalised terms [14, 13, 11], it is more convenient to write terms as $(n, \omega, (x_i)_{i \leq n})$ instead of the more common $\omega(x_1, \ldots, x_n)$. It is clear that $(T_\Omega X, (\sigma_\omega)_{\omega \in \Omega})$ is an Ω-algebra, if $\sigma_\omega((m_i)_{i \leq n}) = (n, \omega, (m_i)_{i \leq n})$ for $\omega \in \Omega_n$ and $m_i \in T_\Omega X$. Morphisms $X \xrightarrow{f} Y$ in $\mathtt{Set}$ are extended in the usual way to the corresponding Ω-homomorphisms $(T_\Omega X, (\sigma_\omega)_{\omega \in \Omega}) \xrightarrow{T_\Omega f} (T_\Omega Y, (\tau_\omega)_{\omega \in \Omega})$, where $T_\Omega f$ is given as the Ω-extension of $X \xrightarrow{f} Y \hookrightarrow T_\Omega Y$ associated to $(T_\Omega Y, (\tau_{n\omega})_{(n,\omega) \in \Omega})$. To obtain the term monad, define $\eta_X^{T_\Omega}(x) = x$, and let $\mu_X^{T_\Omega} = id_{T_\Omega X}^{\star}$ be the Ω-extension of $id_{T_\Omega X}$ with respect to $(T_\Omega X, (\sigma_{n\omega})_{(n,\omega) \in \Omega})$.

Proposition 2. *[32]* $\mathbf{T}_\Omega = (T_\Omega, \eta^{T_\Omega}, \mu^{T_\Omega})$ *is a monad.*

4.2 The Powerset Monad

The covariant powerset functor L_{id} is obtained by $L_{id}X = L^X$, i.e. the set of mappings (or L-fuzzy sets) $A : X \to L$, and following [22], for a morphism $f : X \to Y$ in Set, the category of sets and functions, by defining

$$L_{id}f(A)(y) = \bigvee_{f(x)=y} A(x).$$

Further, define $\eta_X : X \to L_{id}X$ by

$$\eta_X(x)(x') = \begin{cases} 1 & if \text{x=x'} \\ 0 & otherwise \end{cases} \tag{6}$$

and $\mu : L_{id} \circ L_{id} \to L_{id}$ by

$$\mu_X(\mathcal{M})(x) = \bigvee_{A \in L_{id}X} A(x) \wedge \mathcal{M}(A).$$

Proposition 3. *[32]* $\mathbf{L}_{id} = (L_{id}, \eta, \mu)$ *is a monad.*

Note that $\mathbf{2}_{id}$ is the usual covariant powerset monad $\mathbf{P} = (P, \eta, \mu)$, where PX is the set of subsets of X, $\eta_X(x) = \{x\}$ and $\mu_X(\mathcal{B}) = \bigcup \mathcal{B}$.

4.3 Powerset Monads with Fuzzy Level Sets

In [12], a number of set functors extending the powerset functor together with their *extension principles* are introduced. By *extension principles* we mean the two possible generalizations of a mapping $f : X \longrightarrow Y$ where X, Y are sets, when working in the fuzzy case according to an optimistic or pessimistic interpretation of the fuzziness degree.

1. Maximal extension principle: $Ff_M : FX \longrightarrow FY$,

$$Ff_M(A)(y) = \begin{cases} \sup\{A(x) \mid f(x) = y \text{ and } A(x) > 0\} & \text{if the set is nonempty} \\ 0 & \text{otherwise} \end{cases}$$

2. Minimal extension principle: $Ff_m : FX \longrightarrow FY$,

$$Ff_m(A)(y) = \begin{cases} \inf\{A(x) \mid f(x) = y \text{ and } A(x) > 0\} & \text{if the set is nonempty} \\ 0 & \text{otherwise} \end{cases}$$

Both extensions Ff_M and Ff_m coincide with the direct image extension in the case of crisp subsets, that is, given $A \in PX$, then $Pf_M(A) = Pf_m(A) = f(A) \in PY$. These maximal and minimal extension principles can be further generalized to the L-fuzzy powersets, just changing the calculations of suprema and infima by the lattice join and meet operators. We will use the set $I = \{x \in X \mid f(x) = y \text{ and } A(x) > 0\}$:

1. Maximal L-fuzzy extension principle: $Lf_M\colon LX \longrightarrow LY$ is

$$Lf_M(A)(y) = \begin{cases} \bigvee_I A(x) & \text{if } I \neq \emptyset \\ 0 & \text{otherwise} \end{cases}$$

2. Minimal L-fuzzy extension principle: $Lf_m\colon LX \longrightarrow LY$,

$$Lf_m(A)(y) = \begin{cases} \bigwedge_I A(x) & \text{if } I \neq \emptyset \\ 0 & \text{otherwise} \end{cases}$$

We can now extend the definition of powersets to powersets with fuzzy level sets. Functors for α-upper L-fuzzy sets and α-lower L-fuzzy sets, denoted L_α and L^α, respectively, are given as follows:

$$L_\alpha X = \{A \in L_{id}X \mid A(x) \geq \alpha \text{ or } A(x) = 0, \text{ for all } x \in X\}$$
$$L^\alpha X = \{A \in L_{id}X \mid A(x) \leq \alpha \text{ or } A(x) = 1, \text{ for all } x \in X\}.$$

For mappings $f\colon X \longrightarrow Y$, we define $L_\alpha f : L_\alpha X \longrightarrow L_\alpha Y$ as the restriction of the mapping given by the minimal L-fuzzy extension principle to the L-fuzzy set $L_\alpha X$. Similarly, $L^\alpha f : L^\alpha X \longrightarrow L^\alpha Y$ is given as the restriction of the mapping given by the maximal L-fuzzy extension principle. L-fuzzy set categories are defined for each of these extended power set functors and the rationality of the extension principle is proved in the categorical sense, i.e. the associated L-fuzzy set categories are shown to be equivalent to the category of sets and mappings. We can easily generalize the fact that (L_{id}, η, μ) is a monad and obtain:

Proposition 4. *[12] $(L^\alpha, \eta^\alpha, \mu^\alpha)$ is a monad.*

For L_α we define:

$$\eta_{\alpha X}(x)(x') = \begin{cases} 1 & \text{if } x = x' \\ 0 & \text{otherwise} \end{cases}$$

$$\mu_{\alpha X}(\mathcal{A})(x) = \begin{cases} \bigwedge_{A \in I} A(x) \wedge \mathcal{A}(A) & \text{if } I = \{A \in L_\alpha X \mid A(x) \wedge \mathcal{A}(A) > 0\} \neq \emptyset \\ 0 & \text{otherwise} \end{cases}$$

Proposition 5. *[12] $(L_\alpha, \eta_\alpha, \mu_\alpha)$ is a monad.*

Remark 2. For mappings $f\colon X \longrightarrow Y$, we could obtain $L^\alpha f$ as $L_{id} f_{|L^\alpha X}$. Thus, L^α become subfunctor of L_{id} and $\mathbf{L}^\alpha = (L^\alpha, \eta^{L^\alpha}, \mu^{L^\alpha})$ is a submonads of $\mathbf{L_{id}}$.

Remark 3. For $L = 2$, $\mathbf{L}_\alpha = \mathbf{L}^\alpha = \mathbf{2_{id}}$.

4.4 The Covariant Double Contravariant Powerset Monad

The contravariant powerset functor L^{id} is the contravariant hom-functor related to L, i.e. $L^{id} = hom(-, L) : \mathtt{Set} \longrightarrow \mathtt{Set}$, which to each set X and mapping $f : X \longrightarrow Y$ assigns the set L^X of all mappings of X into L, and the mappings $hom(f, L)(g) = g \circ f \;\; (g \in L^Y)$, respectively. Note that 2^{id} is the usual contravariant powerset functor, where $2^{id}X = PX$, and morphisms $X \xrightarrow{f} Y$ in $\mathtt{Set}$ are mapped to $2^{id}f$ representing the mapping $M \mapsto f^{-1}[M]$ $(M \in PY)$ from PY to PX. For double powerset functors it is convenient to write $L_{L_{id}} = L_{id} \circ L_{id}$ and $L^{L^{id}} = L^{id} \circ L^{id}$. Note that $L^{L^{id}}$ is a covariant functor. It may be interesting also to note that the *filter*[1] functor is a subfunctor of $2^{2^{id}}$, but not a subfunctor of $2_{2_{id}}$. In the case of $L^{L^{id}}$, for $X \xrightarrow{f} Y$ in $\mathtt{Set}$ and $\mathcal{M} \in L^{L^X}$, we have $L^{L^{id}}f(\mathcal{M}) = \mathcal{M} \circ L^{id}f$, and hence, $L^{L^{id}}f(\mathcal{M})(g) = \mathcal{M}(g \circ f)$.

Proposition 6. *[15] The covariant set functor $LL = L^{id} \circ L^{id}$ can be extended to a monad, considering the following definitions of the natural transformations η^{LL} and μ^{LL}:*

$$\eta_X^{LL}(x)(A) = A(x), \qquad \mu_X^{LL}(\mathcal{U}) = \mathcal{U} \circ \eta_{LX}^{LL}.$$

It is well-known that the proper[2] filter functor F_0 becomes a monad where $\eta^{F_0} : id \longrightarrow F_0$ is the unique natural transformation and $\mu^{F_0} : F_0 \circ F_0 \longrightarrow F_0$ is given by

$$\mu_X^{F_0}(\mathcal{U}) = \bigcup_{R \in \mathcal{U}} \bigcap_{\mathcal{M} \in R} \mathcal{M}$$

i.e. the contraction mapping suggested in [30].

Remark 4. In relation with the functor $2^{2^{id}}$, it can easily be seen that $\mu_X^{2^{2^{id}}}(\mathcal{U}) = \mu_X^{F_0}(\mathcal{U})$.

5 Partially Ordered Monads

Godement in 1958 used monads named as *standard constructions* and Huber in 1961 showed that adjoint pairs give rise to monads. In 1965, Kleisli [28], Eilenberg and Moore [4] proved the converse. Lawvere [31] introduced universal algebra and thereby the term monad. These developments provide all categorical tools for generalized substitutions. In 2000, Gähler develops partially ordered monads [18], where topology and convergence provided underlying theories. Partially ordered monads contain sufficient structure also for modelling rough sets [33] in a generalized setting with set functors. This generalization builds upon a more general powerset functor setting far beyond just strings [27] and relational algebra [37]. Let acSLAT be the category of almost complete

1 A *filter* on a set X is a nonempty set $\mathcal{F}$ of subsets of X such that: (i) $\emptyset \notin \mathcal{F}$, (ii) $A, B \in \mathcal{F} \Rightarrow A \cap B \in \mathcal{F}$, (iii) $A \in \mathcal{F} \;\; A \subseteq B \Rightarrow B \in \mathcal{F}$.

2 $F_0X = FX \setminus \{\emptyset\}$

semilattices, i.e. partially ordered sets $(X, \leq)$ such that the suprema $\sup \mathcal{M}$ of all non-empty subsets $\mathcal{M}$ of X exists. Morphisms $f : (X, \leq) \to (Y, \leq)$ satisfy $f(\sup \mathcal{M}) = \sup f[\mathcal{M}]$ for non-empty subsets $\mathcal{M}$ of X. A *basic triple* ([18]) is a triple $\Phi = (\varphi, \leq, \eta)$, where $(\varphi, \leq) : \mathtt{SET} \to \mathtt{acSLAT}$, $X \mapsto (\varphi X, \leq)$ is a covariant functor, with $\varphi : \mathtt{SET} \to \mathtt{SET}$ as the underlying set functor, and $\eta : \mathrm{id} \to \varphi$ is a natural transformation. If $(\varphi, \leq, \eta^{\varphi})$ and $(\psi, \leq, \eta^{\psi})$ are basic triples, then also $(\varphi \circ \psi, \leq, \eta^{\varphi}\psi \circ \eta^{\psi})$ is a basic triple.

Definition 5. *A partially ordered monad is a quadruple $\Phi = (\varphi, \leq, \eta, \mu)$, such that*

(i) $(\varphi, \leq, \eta)$ is a basic triple.
(ii)$\mu : \varphi\varphi \to \varphi$ is a natural transformation such that (φ, η, μ) is a monad.
(iii)For all mappings $f, g : Y \to \varphi X$, $f \leq g$ implies $\mu_X \circ \varphi f \leq \mu_X \circ \varphi g$, where $\leq$ is defined argumentwise with respect to the partial ordering of φX.
(iv)For each set X, $\mu_X : (\varphi\varphi X), \leq) \to (\varphi X, \leq)$ preserves non-empty suprema.

The usual covariant powerset monad $\mathbf{P} = (P, \eta, \mu)$, can be extended to a partially ordered monad, $(P, \subseteq, \eta, \mu)$, considering as the partial ordering the inclusion, $\subseteq$. Clearly by the properties of the monad, $(P, \subseteq, \eta)$ is a basic triple, μ is a natural transformation and $\mu_X : (PPX), \subseteq) \to (PX, \subseteq)$ preserves non-empty suprema. Given $f, g : Y \longrightarrow PX$ with $f \subseteq g$ e.g. $f(y) \subseteq g(y)$ for all $y \in Y$ implies $\mu_X \circ Pf \subseteq \mu_X \circ Pg$:

$$(\mu_X \circ Pf)(B) = \bigcup_{y \in B \subseteq Y} f(y) \subseteq \bigcup_{y \in B \subseteq Y} g(y) = (\mu_X \circ Pg)(B)$$

The powerset monad, (L_{id}, η, μ) can also be extended to a partially ordered monad, considering the partial order defined as $A \leq A'$, with $A, A' \in L_{id}X$ if $A(x) \leq A'(x)$ for all $x \in X$. Let us see that $\mu_X \circ L_{id}f \leq \mu_X \circ L_{id}g$: provided that $f \leq g$ where $f, g : Y \longrightarrow L_{id}X$.

$$\begin{aligned}
\mu_X^{L_{id}}(L_{id}f(B))(x) &= \bigvee_{A \in L_{id}X} A(x) \wedge L_{id}f(B)(A) \\
&= \bigvee_{A \in L_{id}X} A(x) \wedge \bigvee_{f(y)=A} B(y) \\
&= \bigvee_{A \in L_{id}X} \bigvee_{f(y)=A} A(x) \wedge B(y) \\
&= \bigvee_{y \in Y} f(y)(x) \wedge B(y) \\
&\leq \bigvee_{y \in Y} g(y)(x) \wedge B(y) \\
&= \mu_X^{L_{id}}(L_{id}g(B))(x).
\end{aligned}$$

Finally, also the monad $(L^\alpha, \eta^\alpha, \mu^\alpha)$ can be extended to a partially ordered monad. This result is a generalization of L_{id} being extendable to a partially ordered monad. To provide L_α with the partially ordered monad structure we need to check that if $f, g : Y \longrightarrow L_\alpha X$ are such that $f \leq g$ then $\mu_X \circ L_\alpha f \leq \mu_X \circ L_\alpha g$. In the same way as the case of L_{id}, the partial order is defined as $A \leq A'$, with $A, A' \in L_\alpha X$ meaning $A(x) \leq A'(x)$ for all $x \in X$.

$$\begin{aligned}
\mu_X^{L_\alpha}(L_\alpha f(B))(x) &= \bigwedge_{A \in L_\alpha X, A(x)>0, L_\alpha f(B)(A)>0} A(x) \wedge L_\alpha f(B)(A) \\
&= \bigwedge_{A \in L_\alpha X, A(x)>0, L_\alpha f(B)(A)>0} A(x) \wedge \bigwedge_{y \in Y, f(y)=A, B(y)>0} B(y) \\
&= \bigwedge_{A \in L_\alpha X, A(x)>0, f(y)=A, B(y)>0} A(x) \wedge B(y) \\
&= \bigwedge_{B(y)>0} f(y)(x) \wedge B(y) \\
&\leq \bigwedge_{B(y)>0} g(y)(x) \wedge B(y) \\
&= \mu_X^{L_\alpha}(L_\alpha g(B))(x).
\end{aligned}$$

Note that $f \leq g$ implies $f(y)(x) \wedge B(y) \leq g(y)(x) \wedge B(y)$ for all $x \in X$ and therefore $\mu_X^{L_\alpha}(L_\alpha f(B))(x) \leq \mu_X^{L_\alpha}(L_\alpha g(B))(x)$.

6 Relations, Kleisli Categories and Rough Sets

Rough sets and fuzzy sets are both methods to represent uncertainty. By using partially ordered monads we can find connections between these two concepts. Partially ordered monads are appropriate categorical formalizations and generalizations of rough sets. In this section we introduce relations from a categorical point of view and justify how its composition can be seen within Kleisli categories. Partially ordered monadic reformulation of rough sets based on the powerset partially ordered monad and the fuzzy powerset monad are presented and some properties are studied.

6.1 Crisp Situation

Let us consider a binary relation $R \subseteq X \times Y$. We will use the notation xRy to represent that the element $(x, y) \in R$. Considering P, the crisp powerset functor, we can represent the relation as a mapping $\rho : X \longrightarrow PY$, where

$$\rho(x) = \{y \in Y \text{ such that } xRy\}$$

As regarded as mappings, considering the composition of two relations, $\rho : X \longrightarrow PY$ and $\rho' : Y \longrightarrow PZ$ we clearly see that the conventional composition of mappings can not be done since the domain of ρ' and codomain of ρ are different.

To find the appropriate definition of this composition we have to consider the Kleisli composition as defined previously by (3), i.e. we need to use that P is a monad and has a "flattering" operator, μ:

$$(X \xrightarrow{\rho} Y) \diamond (Y \xrightarrow{\rho'} Z) = X \overset{\mu_Z^P \circ P\rho' \circ \rho}{\longrightarrow} PZ.$$

The reason for this to work is the following proposition:

Proposition 7. *The Kleisli category associated to the crisp powerset monad is equivalent to the category of sets and relations, SetRel.*

Indeed, $\rho : X \longrightarrow PY$ corresponds to a relation $R \subseteq X \times Y$ by the observation $(x, y) \in R$ if and only if $y \in \rho(x)$.

Proposition 8. *Kleisli composition associated to P is given by:*

$$\mu_Z^P \circ P\rho'(\rho(x)) = \bigcup_{y \in \rho(x)} \rho'(y)$$

Clearly Kleisli composition, in this case, corresponds to the usual composition of relations $R \subseteq X \times Y$, $R' \subseteq Y \times Z$, $(x, z) \in R' \circ R$ if and only if $\exists y, y \in \rho(x), z \in \rho'(y)$. Based on indistinguishable relations, *rough sets* are introduced by defining the upper and lower approximation of sets. These approximations represent uncertain or imprecise knowledge. Let us consider a relation R on X, i.e. $R \subseteq X \times X$. We represent the relation as a mapping $\rho_X : X \longrightarrow PX$, where $\rho_X(x) = \{y \in X | xRy\}$. The corresponding inverse relation R^{-1} is represented as $\rho_X^{-1}(x) = \{y \in X | xR^{-1}y\}$. To be more formal, given a subset A of X, the lower approximation of A correspond to the objects that surely (with respect to an indistinguishable relation) are in A. The lower approximation of A is obtained by

$$A^{\downarrow} = \{x \in X | \rho_X(x) \subseteq A\}$$

and the upper approximation by

$$A^{\uparrow} = \{x \in X | \rho_X(x) \cap A \neq \emptyset\}.$$

Let us see now the partially ordered monadic reformulation of rough sets based on the powerset partially ordered monad. In what follows we will assume that the underlying almost complete semilattice has finite infima, i.e. is a join complete lattice. Considering P as the functor in its corresponding partially ordered monad we then immediately have

Proposition 9. *[6] The upper and lower approximations of a subset A of X are given by*

$$A^{\uparrow} = \bigvee_{\rho_X(x) \wedge A > 0} \eta_X(x) = \mu_X \circ P\rho_X^{-1}(A)$$

and

$$A^{\downarrow} = \bigvee_{\rho_X(x) \leq A} \eta_X(x),$$

respectively.

The corresponding R-weakened and R-substantiated sets of a subset A of X are given by

$$A^{\Downarrow} = \{x \in X | \rho_X^{-1}(x) \subseteq A\}$$

and

$$A^{\Uparrow} = \{x \in X | \rho_X^{-1}(x) \cap A \neq \emptyset\}.$$

Proposition 10. *[6] The R-weakened and R-substantiated sets of a subset A of X are given by*

$$A^{\Uparrow} = \mu_X \circ P\rho_X(A)$$

and

$$A^{\Downarrow} = \bigvee_{\rho_X^{-1}(x) \leq A} \eta_X(x),$$

respectively.

Proposition 11. *If $A \subseteq B$ then $A^{\uparrow} \subseteq B^{\uparrow}$, $A^{\downarrow} \subseteq B^{\downarrow}$, $A^{\Uparrow} \subseteq B^{\Uparrow}$, $A^{\Downarrow} \subseteq B^{\Downarrow}$.*

The upper and lower approximations, as well as the R-weakened and R-substantiated sets, can be viewed as $\uparrow_X, \downarrow_X, \Uparrow_X, \Downarrow_X$: $PX \longrightarrow PX$ with $\uparrow_X (A) = A^{\uparrow}$, $\downarrow_X (A) = A^{\downarrow}$, $\Uparrow_X (A) = A^{\Uparrow}$ and $\Downarrow_X (A) = A^{\Downarrow}$. Considering the crisp powerset monad we define equivalence relations (reflexive, symmetric and transitive) by

Definition 6. $\rho_X : X \longrightarrow PX$ *is* reflexive *if* $\eta_X \subseteq \rho_X$, symmetric *if* $\rho_X = \rho_X^{-1}$ *and* transitive *if $y \in \rho(x)$ implies $\rho(y) \subseteq \rho(x)$.*

In what follows, equivalence relations are now connected to upper and lower approximations.

Proposition 12. *The following properties hold:*

(i) If ρ_X is reflexive $A^{\downarrow} \subseteq A$ and $A \subseteq A^{\uparrow}$.
(ii) If ρ_X is symmetric $A^{\downarrow\uparrow} \subseteq A$ and $A \subseteq A^{\uparrow\downarrow}$.
(iii) If ρ_X is transitive $A^{\uparrow\uparrow} \subseteq A^{\uparrow}$ and $A^{\downarrow} \subseteq A^{\downarrow\downarrow}$.

Corollary 1. *If ρ_X is an equivalence relation, $A^{\downarrow\uparrow} = A^{\downarrow}$ and $A^{\uparrow\downarrow} = A^{\uparrow}$.*

Inverse relations in the ordinary case means to mirror pairs around the diagonal. The following propositions relate inverses to the multiplication of the corresponding monads.

Proposition 13. *[6] In the case of P,*

$$\bigvee_{\rho_X(x) \wedge A > 0} \eta_X(x) = \mu_X \circ P\rho_X^{-1}(A)$$

if and only if

$$\rho_X^{-1}(x) = \bigcup_{\eta_X(x) \leq \rho_X(y)} \eta_X(y).$$

6.2 Many-Valued Situation

We will show now how to extend this view of relations to fuzzy relations. In particular it will be interesting the situation where Kleisli composition is defined for composing fuzzy relations. This can be connected to situations where we want to combine different information systems and study rough approximations. Relations can now be extended to fuzzy relations. Let X and Y be nonempty sets. A fuzzy relation R is a fuzzy subset of the cartesian product $X \times Y$. If $X = Y$ we say that R is a binary fuzzy relation on X. $R(x, y)$ is interpreted as the degree of membership of the pair (x, y) in R. If we consider now the generalized powerset monad, $L_{id}X$ is the set of all L-fuzzy sets. An L-fuzzy set A is nothing but a mapping $A : X \longrightarrow L$. As a first step, and in the same way as before we can extend the concept of relation to a fuzzy relation, i.e. a mapping $\rho : X \longrightarrow L_{id}Y$, $\rho(x)$ is nothing but an element in $L_{id}Y$, a mapping $\rho(x) : Y \longrightarrow L$. An element $y \in Y$ will be assigned a membership degree, $\rho(x)(y)$ representing, as a value in L, the degree on which the elements x and y are fuzzy related. Note that this situation extend the classical relations (crisp powerset situation) in the sense that membership values are 1 if the elements are related and 0 otherwise. With respect to the Kleisli category associated to the powerset monad $\mathbf{L_{id}}$, the objects are sets and homomorphisms are given as mappings $X \longrightarrow L_{id}Y$ in `Set`.

Proposition 14. *[8] The Kleisli category associated to* $\mathbf{L_{id}}$ *is equivalent to the category of set and fuzzy relations,* `SetFuzzRel`.

Proposition 15. *[8] Kleisli composition associated to* $\mathbf{L_{id}}$ *is given by:*

$$\mu_Z^{L_{id}}(L_{id}\rho'(\rho(x)))(z) = \bigvee_{y \in Y} \rho'(y)(z) \wedge \rho(x)(y)$$

The previous proposition tells which membership grade we should assign to the composition of two fuzzy relations, i.e. the suprema of the membership grades on the fuzzy relations. This Kleisli composition of fuzzy relations can be connected to situations where we want to combine different information systems and study rough approximations. Similarly to the crisp situation we can now introduce rough set operators for the fuzzy powerset monad. Let $\rho_X : X \longrightarrow L_{id}X$ be a fuzzy relation on X and let $a \in L_{id}X$. The upper and lower approximations are then

$$\uparrow_X (a) = \mu_X \circ L_{id}\rho_X^{-1}(a) \qquad \downarrow_X (a) = \bigvee_{\rho_X(x) \leq a} \eta_X(x)$$

Corresponding generalizations of ρ-weakenedness and ρ-substantiatedness, are given by

$$\Uparrow_X (a) = \mu_X \circ L_{id}\rho_X(a) \qquad \Downarrow_X (a) = \bigvee_{\rho_X^{-1}(x) \leq a} \eta_X(x)$$

Concerning inverse relations, in the case of L_{id} we would accordingly define $\rho_X^{-1}(x)(x') = \rho_X(x')(x)$.

Proposition 16. *[6] In the case of L_{id},*

$$\mu_X \circ L_{id}\rho_X^{-1}(A)(x) = \bigvee_{x' \in X} (\rho_X(x) \wedge A)(x').$$

Consider now the powerset monads with fuzzy level sets, $\mathbf{L}^\alpha$ and $\mathbf{L}_\alpha$. For L^α is similar to L_{id} situation. Let us see how is the situation for L_α

Proposition 17. *[8] In the case of L_α,*

$$\mu_X \circ L_\alpha\rho_X^{-1}(A)(x) = \bigwedge_{x' \in X} (\rho_X(x) \wedge A)(x').$$

Note that in the case of $L = 2$, for the functor 2_α we obtain the classical definition of the upper approximation of a set A. Generalizing from the ordinary power set monad to a wide range of partially ordered monads requires attention to relational inverses and complement. The role of the diagonal clearly changes, and the representation of inverses is an open question. Inverses and complements must be based on negation operators as given by implication operators within basic many-valued logic [23].

7 Rough Monads

In the previous section we have shown how rough sets can be given using partially ordered monads. From a more abstract point of view, we present in this section a generalized view of rough set constructions based on general partially ordered monads. We name these generalizations *rough monads.* Considering the partially ordered powerset monad, we showed in [6] how rough sets operations can be provided in order to complement the many-valued situation. This is accomplished by defining rough monads. Let $\Phi = (\varphi, \leq, \eta, \mu)$ be a partially ordered monad. We say that $\rho_X : X \longrightarrow \varphi X$ is a *Φ-relation* on X, and by $\rho_X^{-1} : X \longrightarrow \varphi X$ we denote its *inverse.* The inverse must be specified for the given set functor φ. For any $f \colon X \longrightarrow \varphi X$, the following condition is required:

$$\varphi f(\bigvee_i a_i) = \bigvee_i \varphi f(a_i)$$

This condition is valid both for P as well as for L_{id}.

Remark 5. Let ρ_X and ρ_Y be relations on X and Y, respectively. Then the mapping $f : X \longrightarrow Y$ is a congruence, i.e. $x' \in \rho_X(x)$ implies $f(x') \in \rho_Y(f(x))$, if and only if $Pf \circ \rho_X \leq \rho_Y \circ f$. Thus, congruence is related to kind of weak naturality.

Let $\rho_X : X \longrightarrow \varphi X$ be a Φ-relation and let $a \in \varphi X$. The upper and lower approximations are then

$$\uparrow_X (a) = \mu_X \circ \varphi\rho_X^{-1}(a) \qquad \downarrow_X (a) = \bigvee_{\rho_X(x) \leq a} \eta_X(x)$$

with the monadic generalizations of ρ-weakenedness and ρ-substantiatedness, for $a \in \varphi X$, being

$$\Uparrow_X (a) = \mu_X \circ \varphi\rho_X(a) \qquad \Downarrow_X (a) = \bigvee_{\rho_X^{-1}(x) \leq a} \eta_X(x)$$

Proposition 18. *[6] If $a \leq b$, then $\Uparrow_X a \leq \Uparrow_X b$, $\downarrow_X a \leq \downarrow_X b$, $\uparrow_X a \leq \uparrow_X b$, $\Downarrow_X a \leq \Downarrow_X b$.*

In the case of $\varphi = P$, i.e. the conventional powerset partially ordered monad, these operators coincide with those for classical rough sets. In this case inverse relations exist accordingly. In the case of fuzzy sets we use the many-valued powerset partially ordered monad based on the many-valued extension of P to L_{id}. Basic properties of relations can now be represented with 'rough monads terminology:

Definition 7. $\rho_X : X \longrightarrow \varphi X$ *is* reflexive *if* $\eta_X \leq \rho_X$, *and* symmetric *if* $\rho = \rho^{-1}$.

Note that in the case of relations for P and L_{id}, if the relations are reflexive, so are their inverses.

Proposition 19. *[6]*

(i) If ρ is reflexive, $a \leq \Uparrow_X (a)$.
(ii) ρ is reflexive iff $\downarrow_X (a) \leq a$.
(iii) ρ_X^{-1} is reflexive iff $a \leq \uparrow_X (a)$.
(iv) If ρ is symmetric, then $\uparrow_X (\downarrow_X (a)) \leq a$.

In the particular case $a = \eta_X(x)$ we have $a \leq \downarrow_X \circ \uparrow_X (a)$. The idea of submonad is similar to the idea of subsets.In this sense, the calculations related to submonads are a way to reduce data in a given information system. Let $\Phi' = (\varphi', \leq, \eta', \mu')$ be a partially ordered submonad of $\Phi = (\varphi, \leq, \eta, \mu)$. Given $a' \in \varphi' X$ we have the following proposition:

Proposition 20. *[8] For $a' \in \varphi' X$,*

$$\uparrow_X (a') = \mu_X \circ \varphi\rho_X^{-1}(a') \qquad \downarrow_X (a') = \bigvee_{\rho_X(x) \leq a'} \eta_X(x)$$

This proposition shows us that rough approximations are well defined wrt submonads, i.e. their definition in the submonad correspond to the one for the monad.

8 Applications

Our theoretical developments are an inspiration for application development. As a first step we have focused on ICT solutions within health care. Information representation based on medical ontologies are usually rather narrow and oriented towards crisp specifications of data information. At the same time, health care ICT solutions call for representation of vagueness and uncertainties both for use within medical records and information databases, as well as for decision support and guideline implementations. We will discuss various fields of health care and possible use of generalized rough sets, and we will in particular develop concrete examples in the area of decision support and, more specifically, decisions related to diagnosis and treatment.

8.1 Drug Interactions

Pharmacological treatment is an excellent area for our experimental purposes where e.g. drug interactions [10] can be favourably described using generalized rough sets. Pharmacological databases provide rich and complete information for therapeutic requirements. In particular, the ATC code with its unique identification of drug compound is the basis e.g. of modelling of generic substitutes and drug interactions. Two drugs are generic substitutes if they have the same ATC code, the same dosages and the same administration route. This is straightforward and precise but the notion of drug-drug interaction is more complicated. In addition, drug-condition interaction adds further complexity as medical conditions themselves are not easy to formalize. Rough sets described by partially ordered monads are able to capture interactions with respect to different granularities in the information hierarchy. The data structure for pharmacologic information is hierarchical in its subdivision according to anatomic, therapeutic and chemical information of the drug compound. National catalogues of drugs aim at being complete with respect to chemical declarations, indications/contraindications, warnings, interactions, side-effects, pharmacodynamics/pharmacokinetics, and pure pharmaceutical information. The Anatomic Therapeutic Chemical (ATC) classification system is a WHO (World Health Organization) standard. The ATC structure can be understood from Table 1 on the classification of verapamil (code C08DA01) for hypertension with stable angina pectoris. Drugs in ATC are, with a very few exceptions, classified according to their main indication of use. The ATC coded is for therapeutic use, while the article code is a unique identifier which is used in the patient's record. For drugs showing therapeutically significant interactions we need to distinguish between types of interactions and to what extent we have evidence for that particular type of interaction. The types of interaction are *recommended combination*, *neutral combination* (no harmful interactions), *risky combination* (should be monitored) and *dangerous combination* (should be avoided). The degrees of evidence are *strong evidence* (internationally), *reasonable belief* (several studies exist), *some indications* (only some studies exist, and results are not conclusive) and *no evidence*. With these qualifications it is clear that a linear quantification cannot be given. Further, the drugs are affected in different ways, according to *no change in effect*, *increases effect*, *reduces effect* and *other* (e.g. a new type of side effect). Interaction type, evidence level, and effect need to be considered in the guideline for respective treatments. In our subsequent discussion we focus on guideline

Table 1. Classification of *verapamil*

C	cardiac and vessel disease medication	1st level main anatomical group
C08	calcium channel blockers	2nd level, therapeutic subgroup
C08D	selective cardiac calcium channel blockers	3rd level, pharmacological subgroup
C08DA	phenylalcylamins	4th level, chemical subgroup
C08DA01	verapamil	5th level

based pharmacologic treatment of hypertension [38]. See also [34] for an implementation of these guidelines for primary care. Typical drugs for hypertension treatment are beta-blockers (C07, C07A) like an atenolol (C07AB03) and diuretics (C03) like thiazides (C03A, C03AA). Atenolol is a selective beta-1-blocker (C07AB). A frequently used thiazide is hydrochlorothiazide (C03AA03). Note that beta-blockers are both therapeutic as well as pharmacological subgroups. Similarly, thiazides are both pharmacological as chemical subgroups. As a basic example concerning interactions consider treatment of hypertension in presence of diabetes. Beta-blockers may mask and prolong beta-blockers insulin-induced hypoglycemia. If the patient shows the medical condition of diabetes without any other medical condition present, then the ACE inhibitor (C09A, C09AA) enalapril (C09AA02) is the first choice for treatment [38].

Drug interactions as relations can be interpreted as mappings $\rho_X^L : X \longrightarrow LX$, based on the many-valued powerset monad (L, η, μ). Let M be a set of medical conditions and let $\rho^L[M]$ be the subrelation of ρ which considers interactions with pharmacological treatments based on these medical conditions in M. We then observe that the clinical usefulness of these interpretations comes down to defining $\rho^L[M]$ so as to correspond to real clinical situations. Operating with these sets then becomes the first step to identify connections to guidelines for pharmacological treatment.

In [26], a software framework for pharmacological information representation is suggested. This framework enables clients to recover information from databases with pharmacological information. In the current implementation, the framework uses ATC codes in the drug metadata. Specifically, the framework provides information about interactions as a set of ATC codes for a particular ATC code (drug). This software framework will be used to recover pharmacological information and related drug interactions, and further using this information in a knowledge-discovery application using the rough set and monad theoretical framework as described in this chapter. The experiment will extract drug information relates also to hypertension treatment [34] from the drug database. Further, to demonstrate that this representation is usable in a realistic situation, the forming of the sets described earlier will indeed take into account a set of medical conditions. These conditions will be described codes from the ICD and corresponding diagnosis encoding system. The hypothesis is that rough monads provide drug interactions with an adequate representation for pharmacological hypertension treatment with respect to an individual and typical patient case.

8.2 Dementia Differential Diagnosis

The differential diagnosis process in the case of dementia involves e.g. to distinguish between dementias of Alzheimer's and vascular type. In the case of Alzheimer's, pharmacological treatment following an early detection can be useful for maintaining acetylcholin in the synapsis between nerve cells. Receptors then remain stimulated thus maintaining activitity and nerve signals. In the scenario of early detection it is important to observe the situations where cognitive problems are encountered and by whom these observations are made. Clearly, the very first observations of cognitive decline are made by relatives (if not self-detected by the patient) or social workers in home care who would forward information about the problems encountered, thus seeking advice firstly from nurses and primary care doctors within their local health care centres. Representatives in social care and nursing will not perform any diagnosis. However, providing some observation and even 'qualified guesses' can speed up the process leading eventually to an accurate diagnosis with possibilities for further pharmacological treatments. It is then important to identify respective information types and rule representations for these professional groups providing everything from 'qualified guesses' to accurate diagnosis. Note that not even autopsy can provide higher diagnostic accuracy than around 80%, so early detection is really hard and challenging. Many-valuedness provides tools for logic transformations between professional groups. Regardless of where decision and/or observations are made, we always need to guarantee consistency when information and knowledge is mapped between ontological domains as understood and used by these professional groups. For further reading on the general logics approach to transformations, see [16]. The intuition of using rough sets and monads is here very natural and even rather obvious. In differential diagnosis we are viewing the set of attributes (symptoms and signs) in a relational setting. Indeed attributes are related not just on powerset level, but also in a 'sets of sets of attributes' fashion. Heteroanamnesis, for instance, is a set of attributes which are grouped according to their interrelations. Thus we are dealing with heteroanamnesis as a set of sets of attributes. Upper and lower approximations are useful as they provide operators transforming a set (or generalized sets), as a relation, to another boundary in some canonic way. Full interpretations are yet to be given, and the pragmatic is still somewhat open, but these developments build upon software developments and real clinical use of these software tools.

9 Conclusion

Rough sets are naturally categorical once we observe how rough set operators can be generalized using other set functors, extendable to partially ordered monads, than just the powerset partially ordered monad representing relations on ordinary sets. The categorical instrumentation reveals many aspects and possibilities for further developments of rough monads, both from theoretical as well as from application points of view. Theoretical developments involve extensions using partially ordered monads and invokes e.g. logical viewpoints that would not appear unless the categorical generalizations are used. Application developments make use of entirely new ways to arrange sets and sets of sets in a wide range of ways. Information pieces and blocks organized as many-valued sets of terms or even filters and ideals opens up an avenue of possibilities for further exploration of intuition combined with formalism.

References

1. Aho, A.V., Hopcroft, J.E., Ullman, J.D.: The Design and Analysis of Computer Algorithms. Addison-Wesley, Reading (1975)
2. Barr, M., Wells, C.: Toposes, Triples and Theories. Springer, Heidelberg (1985)
3. Cattaneo, G., Ciucci, D.: Shadowed sets and related algebraic structures. Fundamenta Iinformaticae 55, 255–284 (2003)
4. Eilenberg, S., Moore, J.C.: Adjoint functors and triples. Illinois J. Math. 9, 381–398 (1965)
5. Eklund, P., Gähler, W.: Partially ordered monads and powerset Kleene algebras. In: Proc. 10th Information Processing and Management of Uncertainty in Knowledge Based Systems Conference (IPMU 2004) (2004)
6. Eklund, P., Galán, M.A.: Monads can be rough. In: Greco, S., Hata, Y., Hirano, S., Inuiguchi, M., Miyamoto, S., Nguyen, H.S., Słowiński, R. (eds.) RSCTC 2006. LNCS, vol. 4259, pp. 77–84. Springer, Heidelberg (2006)
7. Eklund, P., Galán, M.A.: On logic with fuzzy and rough powerset monads. In: Kryszkiewicz, M., Peters, J.F., Rybinski, H., Skowron, A. (eds.) RSEISP 2007. LNCS, vol. 4585, pp. 391–399. Springer, Heidelberg (2007)
8. Eklund, P., Galán, M.A.: Partially ordered monads and rough sets. In: Peters, J.F., Skowron, A. (eds.) Transactions on Rough Sets VIII. LNCS, vol. 5084, pp. 53–74. Springer, Heidelberg (2008)
9. Eklund, P., Galán, M.A., Gähler, W., Medina, J., Ojeda Aciego, M., Valverde, A.: A note on partially ordered generalized terms. In: Proc. of Fourth Conference of the European Society for Fuzzy Logic and Technology and Rencontres Francophones sur la Logique Floue et ses applications (Joint EUSFLAT-LFA 2005), pp. 793–796 (2005)
10. Eklund, P., Galán, M.A., Karlsson, J.: Rough Monadic Interpretations of Pharmacologic Information. In: Proceedings of the 15th International Workshops on Conceptual Structures, ICCS 2007, pp. 108–113. Sheffield, UK (2007)

11. Eklund, P., Galán, M.A., Medina, J., Ojeda Aciego, M., Valverde, A.: A categorical approach to unification of generalised terms. Electronic Notes in Theoretical Computer Science 66(5) (2002), `http://www.elsevier.nl/locate/entcs/volume66.html`
12. Eklund, P., Galán, M.A., Medina, J., Ojeda Aciego, M., Valverde, A.: Set functors, L-fuzzy set categories and generalized terms. Computers and Mathematics with Applications 43, 693–705 (2002)
13. Eklund, P., Galán, M.A., Medina, J., Ojeda-Aciego, M., Valverde, A.: Composing submonads. In: Proc. 31st IEEE Int. Symposium on Multiple-Valued Logic (ISMVL 2001), Warsaw, Poland, May 22-24, pp. 367–372 (2001)
14. Eklund, P., Galán, M.A., Ojeda-Aciego, M., Valverde, A.: Set functors and generalised terms. In: Proc. 8th Information Processing and Management of Uncertainty in Knowledge-Based Systems Conference (IPMU 2000), pp. 1595–1599 (2000)
15. Eklund, P., Gähler, W.: Fuzzy Filter Functors and Convergence. In: Rodabaugh, S.E., Klement, E.P., Höhle, U. (eds.) Applications of category theory to fuzzy subsets. Theory and Decision Library B, pp. 109–136. Kluwer, Dordrecht (1992)
16. Eklund, P., Helgesson, R., Lindgren, H.: Towards refinement of clinical evidence using general logics. In: Rutkowski, L., Tadeusiewicz, R., Zadeh, L.A., Zurada, J.M. (eds.) ICAISC 2008. LNCS, vol. 5097, pp. 1029–1040. Springer, Heidelberg (2008)
17. Gähler, W.: Monads and convergence. In: Proc. Conference Generalized Functions, Convergences Structures, and Their Applications, Dubrovnik (Yugoslavia) 1987, pp. 29–46. Plenum Press (1988)
18. Gähler, W.: General Topology – The monadic case, examples, applications. Acta Math. Hungar. 88, 279–290 (2000)
19. Gähler, W.: Extension structures and completions in topology and algebra, Seminarberichte aus dem Fachbereich Mathematik, Band 70, FernUniversität in Hagen (2001)
20. Gähler, W., Eklund, P.: Extension structures and compactifications, In: Categorical Methods in Algebra and Topology (CatMAT 2000), pp. 181–205 (2000)
21. Galán, M.A.: Categorical Unification, Dissertation, Umeå University (2004)
22. Goguen, J.A.: L-fuzzy sets. J. Math. Anal. Appl. 18, 145–174 (1967)
23. Hájek, P.: Metamathematics of Fuzzy Logic. Kluwer Academic Publishers, Dordrecht (1998)
24. Järvinen, J.: On the structure of rough approximations. Fundamenta Informaticae 53, 135–153 (2002)
25. Järvinen, J.: Lattice theory for rough sets. In: Peters, J.F., Skowron, A., Düntsch, I., Grzymała-Busse, J.W., Orłowska, E., Polkowski, L. (eds.) Transactions on Rough Sets VI. LNCS, vol. 4374, pp. 400–498. Springer, Heidelberg (2007)
26. Karlsson, J.: Interface for accessing pharmacological information. In: Proc. 21st IEEE International Symposium on Computer-Based Medical Systems, Jyväskylä (Finland), June 17-19. IEEE CS Press, Los Alamitos (to appear, 2008)
27. Kleene, S.C.: Representation of events in nerve nets and finite automata. In: Shannon, C.E., McCarthy, J. (eds.) Automata Studies, pp. 3–41. Princeton University Press, Princeton (1956)
28. Kleisli, H.: Every standard construction is induced by a pair of adjoint functors. Proc. Amer. Math. Soc. 16, 544–546 (1965)
29. Kortelainen, J.: A Topological Approach to Fuzzy Sets, Ph.D. Dissertation, Lappeenranta University of Technology, Acta Universitatis Lappeenrantaensis 90 (1999)

30. Kowalsky, H.-J.: Limesräume und Komplettierung. Math. Nachr. 12, 301–340 (1954)
31. Lawvere, F.W.: Functorial semantics of algebraic theories, Dissertation, Columbia University (1963)
32. Manes, E.G.: Algebraic Theories. Springer, Heidelberg (1976)
33. Pawlak, Z.: Rough sets. Int. J. Computer and Information Sciences 5, 341–356 (1982)
34. Persson, M., Bohlin, J., Eklund, P.: Development and maintenance of guideline-based decision support for pharmacological treatment of hypertension. Comp. Meth. Progr. Biomed. 61, 209–219 (2000)
35. Rydeheard, D., Burstall, R.: A categorical unification algorithm. In: Poigné, A., Pitt, D.H., Rydeheard, D.E., Abramsky, S. (eds.) Category Theory and Computer Programming. LNCS, vol. 240, pp. 493–505. Springer, Heidelberg (1986)
36. Salomaa, A.: Two complete axiom systems for the algebra of regular events. J. ACM 13, 158–169 (1966)
37. Tarski, A.: On the calculus of relations. J. Symbolic Logic 6, 65–106 (1941)
38. The sixth report of the joint national committee on prevention detection, evaluation, and treatment of high blood pressure, Technical Report 98-4080, National Institutes of Health (1997)
39. Zadeh, L.A.: Fuzzy sets. Information and Control 8, 338–353 (1965)

Granular Structures and Approximations in Rough Sets and Knowledge Spaces

Yiyu Yao[1], Duoqian Miao[2], and Feifei Xu[1,2]

[1] Department of Computer Science, University of Regina, Regina, Saskatchewan, Canada S4S 0A2
yyao@cs.uregina.ca

[2] Department of Computer Science and Technology, Tongji University, Shanghai, China 201804
miaoduoqian@163.com, xufeifei1983@hotmail.com

Summary. Multilevel granular structures play a fundamental role in granular computing. In this chapter, we present a general framework of granular spaces. Within the framework, we examine the granular structures and approximations in rough set analysis and knowledge spaces. Although the two theories use different types of granules, they can be unified in the proposed framework.

1 Introduction

Granular computing is an emerging field of study focusing on structured thinking, structured problem solving and structured information processing with multiple levels of granularity [1, 2, 7, 8, 13, 17, 19, 20, 21, 22, 24, 25, 29]. Many theories may be interpreted in terms of granular computing. The main objective of this chapter is to examine the granular structures and approximations used in rough set analysis [10, 11] and knowledge spaces [3, 4].

A primitive notion of granular computing is that of granules. Granules may be considered as parts of a whole. A granule may be understood as a unit that we use for describing and representing a problem or a focal point of our attention at a specific point of time. Granules can be organized based on their inherent properties and interrelationships. The results are a multilevel granular structure. Each level is populated by granules of the similar size or the similar nature. Depending on a particular context, levels of granularity may be interpreted as levels of abstraction, levels of details, levels of processing, levels of understanding, levels of interpretation, levels of control, and many more. An ordering of levels based on granularity provides a hierarchical granular structure.

The formation of granular structures is based on a vertical separation of levels and a horizontal separation of granules in each level. It explores the property of loose coupling and nearly-decomposability [14] and searches for a good approximation [12]. Typically, elements in the same granules interact more than elements in different granules. Granules in the same level are relatively independent and granules in two adjacent levels are closely related.

A. Abraham, R. Falcón, and R. Bello (Eds.): Rough Set Theory, SCI 174, pp. 71–84.
springerlink.com

Rough set analysis and knowledge spaces use set-theoretic formulations of granular structures. A granule is interpreted as a subset of a universal set, satisfying a certain condition. In the case of rough set analysis, a granule must be definable with respect to a logic language [23, 28]. In the case of knowledge spaces, a granule must be defined by a surmise relation or a surmise system [3, 4]. Granules are ordered by the set-inclusion relation. The family of granules forms a subsystem of the power set of the universal set. As a consequence, one needs to study approximations of any subset by granules in the subsystem [16, 26].

The above discussion suggests a unified framework for studying granular structures. It serves as a basis of the current study. The rest of the chapter is organized as follows. Section 2 proposes a framework of granular spaces. Sections 3 and 4 examine the granular structures and approximations in rough sets and knowledge spaces.

2 Granular Spaces

A set-theoretic interpretation of granules and granular structures is presented and a framework of granular spaces is introduced.

2.1 A Set-Theoretic Interpretation of Granules

Categorization or classification is one of the fundamental tasks of human intelligence [12]. In the process of categorization, objects are grouped into categories and a name is given to each category. One obtains high-level knowledge about groups of objects. Such knowledge may be applied later to similar objects of the category, since each object is no longer viewed as a unique entity. In order to obtain a useful categorization, one needs to search for both similarity and dissimilarity between objects. While similarity leads to the integration of individuals into categories, dissimilarity leads to the division of larger categories into smaller subcategories.

The idea of categorization immediately leads to a set-theoretic interpretation of granules. A granule may be simply viewed as the set of objects in a category. The process of categorization covers two important issues of granulation, namely, the construction of granules and the naming of granules. The construction of granules explores both the similarity and dissimilarity of objects. Objects in the same categories must be more similar to each other, and objects in different granules are more dissimilar to each other. Consequently, one may view objects in the same category as being indistinguishable or equivalent from a certain point of view. For those objects, it may be more economic to give a name, so that one can talk about the category by its name instead of many individuals. In addition, we can apply the common properties of the objects in the category to similar objects in order to make meaningful inference in the future [12].

With the set-theoretic interpretation of granules, we can apply a set-inclusion relation on granules to form sub-super relations between granules. We can also apply set-theoretic operations on granules to construct new granules. The resulting family of granules forms a multilevel hierarchical structure.

There is another way to arrange granules. Once names are assigned to granules, we may use these names to form more abstract new granules. In this case, the elements of a granule are names to other granules. That is, a granule is treated as a whole by its name in a higher-level granule. A granule therefore plays dual roles. In the current level, it is a set of individuals; in its adjacent higher-level, it is considered as a whole by its name.

2.2 A Formulation of Granules as Concepts

The set-theoretic interpretation of granules can be formally developed based on the notion of concepts, a basic unit of human thought.

The classical view of concepts defines a concept jointly by a set of objects, called the extension of the concept, and a set of intrinsic properties common to the set of objects, called the intension of the concept [15]. The intension reflects the intrinsic properties or attributes shared by all objects (i.e., instances of a concept). Typically, the name of a concept reflects the intension of a concept. The extension of a concept is the set of objects which are concrete examples of a concept. One may introduce a logic language so that the intension of a concept is represented by a formula and the extension is represented by the set of objects satisfying the formula [28].

The language $\mathcal{L}$ is constructed from a finite set of atomic formulas, denoted by $\mathcal{A} = \{p, q, ...\}$. Each atomic formula may be interpreted as basic knowledge. They are the elementary units. In general, an atomic formula corresponds to one particular property of an individual. The construction of atomic formulas is an essential step of knowledge representation. The set of atomic formulas provides a basis on which more complex knowledge can be represented. Compound formulas can be built from atomic formulas by using logic connectives. If ϕ and φ are formulas, then $(\neg\phi)$, $(\phi \wedge \varphi)$, $(\phi \vee \varphi)$, $(\phi \rightarrow \varphi)$, and $(\phi \leftrightarrow \varphi)$ are also formulas.

The semantics of the language $\mathcal{L}$ are defined in the Tarski's style by using the notions of a model and satisfiability [9, 11, 28]. The model is a nonempty domain consisting of a set of individuals, denoted by $U = \{x, y, ...\}$. For an atomic formula p, we assume that an individual $x \in U$ either satisfies p or does not satisfy p, but not both. For an individual $x \in U$, if it satisfies an atomic formula p, we write $x \models p$, otherwise, we write $x \not\models p$. The satisfiability of an atomic formula by individuals of U is viewed as the knowledge describable by the language $\mathcal{L}$. An individual satisfies a formula if the individual has the properties as specified by the formula. To emphasize the roles played by the set of atomic formulas $\mathcal{A}$, the operations $\{\neg, \wedge, \vee\}$ and the set of individuals U, we also rewrite the language $\mathcal{L}$ as $\mathcal{L}(\mathcal{A}, \{\neg, \wedge, \vee\}, U)$.

If ϕ is a formula, the set $m(\phi)$ defined by:

$$m(\phi) = \{x \in U \mid x \models \phi\}, \tag{1}$$

is called the meaning of the formula ϕ. The meaning of a formula ϕ is indeed the set of all objects having the properties expressed by the formula ϕ. In other words, ϕ can be viewed as the description of the set of objects $m(\phi)$. As a result, a concept can be expressed by a pair $(\phi, m(\phi))$, where $\phi \in \mathcal{L}$. ϕ is the intension of a concept while $m(\phi)$ is the extension of a concept. A connection between formulas and subsets of U is established. Similarly, a connection between logic connectives and set-theoretic operations can be stated [11]:

$$\begin{aligned}
&\text{(i)} && m(\neg\phi) = (m(\phi))^c,\\
&\text{(ii)} && m(\phi \wedge \psi) = m(\phi) \cap m(\psi),\\
&\text{(iii)} && m(\phi \vee \psi) = m(\phi) \cup m(\psi),\\
&\text{(iv)} && m(\phi \rightarrow \psi) = (m(\phi))^c \cup (m(\psi))^c,\\
&\text{(v)} && m(\phi \equiv \psi) = (m(\phi) \cap m(\psi)) \cup ((m(\phi))^c \cap (m(\psi))^c),
\end{aligned}$$

where $(m(\phi))^c = U - m(\phi)$ is the complement of $m(\phi)$. Under this formulation, we can discuss granules in terms of intensions in a logic setting and in terms of extension in a set-theoretic setting.

2.3 Granular Spaces and Granular Structures

Each atomic formula in $\mathcal{A}$ is associated with a subset of U. This subset may be viewed as an elementary granule in U. Each formula is obtained by taking logic operations on atomic formulas. The meaning set of the formula can be obtained from the elementary granules through set-theoretic operations. With the language $\mathcal{L}$, for each formula, we can find its meaning set by equation (1). On the other hand, for an arbitrary subset of the universe U, one may not be able to find a formula to precisely represent it. This leads to the introduction of the definability of granules in a logic language. We say a subset or a granule $X \subseteq U$ is definable if and only if there exists a formula ϕ in the language $\mathcal{L}$ such that,

$$X = m(\phi). \tag{2}$$

Otherwise, it is undefinable [23]. By the formulation of the logic language, it follows that each definable granule can be expressed in terms of elementary granules through set-theoretic operations.

A definable granule is represented by a pair $(\phi, m(\phi))$ or simply $m(\phi)$. The family of all definable granules is given by:

$$Def(\mathcal{L}(\mathcal{A}, \{\neg, \wedge, \vee\}, U)) = \{m(\phi) \mid \phi \in \mathcal{L}(\mathcal{A}, \{\neg, \wedge, \vee\}, U)\}, \tag{3}$$

which is a subsystem of the power set 2^U, closed under set complement, intersection and union. Based on these notions, we formally define a granular space by the triplet:

$$(U, \mathcal{S}_0, \mathcal{S}), \tag{4}$$

where

U is the universe,
$\mathcal{S}_0 \subseteq 2^U$ is a family of elementary granules, i.e., $\mathcal{S}_0 = \{m(p) \mid p \in \mathcal{A}\}$,
$\mathcal{S} \subseteq 2^U$ is a family of definable granules, i.e., $\mathcal{S} = Def(\mathcal{L}(\mathcal{A}, \{\neg, \wedge, \vee\}, U))$.

Note that $\mathcal{S}$ can be generated from $\mathcal{S}_0$ through set-theoretic operations.

The family $\mathcal{S}$ of definable granules is called a granular structure. Since the logic language uses logic operations $\neg$, $\wedge$, and $\vee$, the family of definable granules $\mathcal{S}$ is closed under set complement, intersection and union. That is, $\mathcal{S}$ is an σ-algebra. Note that $\mathcal{S}_0$ is not necessarily the basis of the σ-algebra $\mathcal{S}$.

There are additional requirements to make the granular space more practical. For example, the family of elementary granules normally can not be all singleton subsets of U, as a singleton subset is equivalent to its unique object. The set of all granules constructed from the family of elementary granules is normally a superset of the family of elementary granules. Furthermore, it is typically a subset of the power set of U. Otherwise, we do not have the benefits of granulation. It also requires that the union of all the elementary granules covers the universe U. That is, an object satisfies at least one basic formula in $\mathcal{A}$.

From the above discussion, the granular structure can be described by a logic language $\mathcal{L}(\mathcal{A}, \{\neg, \wedge, \vee\}, U)$. To make the granular structure more practical, one may consider a logic language using a subset of logic connectives. As a special case, suppose that the granular structure is closed under set intersection and union but not complement. Since the logic operators $\wedge$ and $\vee$ correspond to set intersection and union, we can use a logic language $\mathcal{L}(\mathcal{A}, \{\wedge, \vee\}, U)$ to describe this type of granular structures. Two more special cases of granular structures are defined by the languages $\mathcal{L}(A, \{\wedge\}, U)$ and $\mathcal{L}(A, \{\vee\}, U)$, respectively [28]. While the former is closed under set intersection, the latter is closed under set union. Note that a granular structure containing U and being closed under set intersection is called a closure system.

3 Rough Set Analysis

The granular space of rough set analysis is a quotient space induced by an equivalence relation.

3.1 Granular Spaces and Granular Structures

Rough set analysis studies relationships between objects and their attribute values in an information table [10, 11]. An information table provides a convenient way to describe a finite set of objects by a finite set of attributes. Formally, an information table can be expressed as:

$$M = (U, At, \{V_a \mid a \in At\}, \{I_a \mid a \in At\}), \tag{5}$$

where U is a finite nonempty set of objects, At is a finite nonempty set of attributes, V_a is a nonempty set of values for an attribute $a \in At$, and $I_a : U \rightarrow V_a$ is an information function.

For a set of attributes $P \subseteq At$, one can define an equivalence relation on the set of objects:

$$xE_P y \Longleftrightarrow \forall a \in P(I_a(x) = I_a(y)). \tag{6}$$

Two objects are equivalent if they have the same values on all attributes in P. Rough set theory is developed based on such equivalence relations.

Let $E \subseteq U \times U$ denote an equivalence relation on U. The pair $apr = (U, E)$ is called an approximation space [10]. The equivalence relation E partitions the set U into disjoint subsets called equivalence classes. This partition of the universe is denoted by U/E. The partition U/E may be considered as a granulated view of the universe. For an object $x \in U$, the equivalence class containing x is given by:

$$[x]_E = \{y \mid yEx\}. \tag{7}$$

Equivalence classes are referred to as elementary granules.

By taking the union of a family of equivalence classes, we can obtain a composite granule. The family of all such granules contains the entire set U and the empty set $\emptyset$, and is closed under set complement, intersection and union. More specifically, the family is an σ-algebra, denoted by $\sigma(U/E)$, with the basis U/E.

The above formulation can be expressed in terms of a logic language. For an attribute-value pair (a, v), where $a \in At$ and $v \in V_a$, we have an atomic formula $a = v$. The meaning of $a = v$ is the following set of objects:

$$m(a = v) = \{x \in U \mid I_a(x) = v\}. \tag{8}$$

It immediately follows that the equivalence class $[x]$ is defined by the formula $\bigwedge_{a \in At} a = I_a(x)$. That is, $[x]$ is a definable granule. One can easily see that the union of a family of equivalence classes is also a definable granule. Thus, the set of all definable granules is $\sigma(U/E)$.

Based on the above discussion, we can conclude that a granular space used in rough set theory is given by:

$$(U, U/E, \sigma(U/E)),$$

where

U is the universe,
$U/E \subseteq 2^U$ is the family of equivalence classes,
$\sigma(U/E) \subseteq 2^U$ is the σ-algebra generated from U/E.

That is, the granular space and granular structure of rough set analysis are a special case of the ones introduced in the last sect.

3.2 Rough Set Approximations

An arbitrary set $A \subseteq U$ may not necessarily be the union of some equivalence classes. This implies that one may not be able to describe A precisely using the logic language $\mathcal{L}$. In order to infer information about such undefinable granules, it is necessary to approximate them by definable granules.

For a subset of objects $A \subseteq U$, it may be approximated by a pair of lower and upper approximations:

$$\begin{aligned} \underline{apr}(A) &= \bigcup\{X \in \sigma(U/E) \mid X \subseteq A\}, \\ \overline{apr}(A) &= \bigcap\{X \in \sigma(U/E) \mid A \subseteq X\}. \end{aligned} \tag{9}$$

The lower approximation $\underline{apr}(A)$ is the union of all the granules in $\sigma(U/E)$ that are subsets of A. The upper approximation $\overline{apr}(A)$ is the intersection of all the granules in $\sigma(U/E)$ that contain A. This is referred to as the subsystem based definition [16, 18, 26]. Since $\sigma(U/E)$ is closed under $\cap$ and $\cup$, the definition is well defined. In addition, $\underline{apr}(A) \in \sigma(U/E)$ is the largest granule in $\sigma(U/E)$ that is contained by A, and $\overline{apr}(A) \in \sigma(U/E)$ is the smallest granule that contains A. That is, the pair $(\underline{apr}(A), \overline{apr}(A))$ is the tightest approximation.

Lower and upper approximations are dual to each other in the sense:

$$\begin{aligned} &\text{(Ia)} \quad \underline{apr}(A) = (\overline{apr}(A^c))^c, \\ &\text{(Ib)} \quad \overline{apr}(A) = (\underline{apr}(A^c))^c. \end{aligned}$$

The set A lies within its lower and upper approximations:

$$\text{(II)} \quad \underline{apr}(A) \subseteq A \subseteq \overline{apr}(A).$$

One can also verify the following properties:

$$\begin{aligned} &\text{(IIIa)} \quad \underline{apr}(A \cap B) = \underline{apr}(A) \cap \underline{apr}(B), \\ &\text{(IIIb)} \quad \overline{apr}(A \cup B) = \overline{apr}(A) \cup \overline{apr}(B). \end{aligned}$$

The lower (upper) approximation of the intersection (union) of a finite number of sets can be obtained from their lower (upper) approximations. However, we only have:

$$\begin{aligned} &\text{(IVa)} \quad \underline{apr}(A \cup B) \supseteq \underline{apr}(A) \cup \underline{apr}(B), \\ &\text{(IVb)} \quad \overline{apr}(A \cap B) \subseteq \overline{apr}(A) \cap \overline{apr}(B). \end{aligned}$$

It is impossible to obtain the lower (upper) approximation of the union (intersection) of some sets from their lower (upper) approximations. Additional properties of rough set approximations can be found in [10] and [27].

3.3 An Example

Suppose $U = \{a, b, c, d, e\}$. Given an equivalence relation:

$$aEa, \quad aEb, \quad bEa, \quad bEb, \quad cEc, \quad cEe, \quad dEd, \quad eEc, \quad eEe,$$

it induces the partition $U/E = \{\{a, b\}, \{c, e\}, \{d\}\}$. We can construct an σ-algebra by taking union of any family of equivalence classes:

$$\sigma(U/E) = \{\emptyset, \{a, b\}, \{c, e\}, \{d\}, \{a, b, c, e\}, \{a, b, d\}, \{c, d, e\}, U\}.$$

The corresponding granular space is:

$$(U, U/E, \sigma(U/E)),$$

where

$$U = \{a, b, c, d, e\},$$

$$U/E = \{\{a, b\}, \{c, e\}, \{d\}\},$$
$$\sigma(U/E) = \{\emptyset, \{a, b\}, \{c, e\}, \{d\}, \{a, b, c, e\}, \{a, b, d\}, \{c, d, e\}, U\}.$$

Consider a subset of objects $A = \{a, b, c, d\}$. It can not be obtained by taking union of some elementary granules. That is, it is an undefinable granule. We approximate it by a pair of subsets from below and above in the σ-algebra $\sigma(U/E)$. From equation (9), we have:

$$\underline{apr}(A) = \bigcup\{X \in \sigma(U/E) \mid X \subseteq A\} = \{a, b\} \cup \{d\} \cup \{a, b, d\} = \{a, b, d\},$$
$$\overline{apr}(A) = \bigcap\{X \in \sigma(U/E) \mid A \subseteq X\} = \{a, b, c, d, e\}.$$

It follows that $\underline{apr}(A) = \{a, b, d\} \subseteq A \subseteq \overline{apr}(A) = \{a, b, c, d, e\}$.

4 Knowledge Space Theory

Knowledge spaces [3, 4, 5, 6] represent a new paradigm in mathematical psychology. It is a systematic approach to the assessment of knowledge by constructing sequences of questions to be asked.

In knowledge spaces, we consider a pair $(Q, \mathcal{K})$, where Q is a finite set of questions and $\mathcal{K} \subseteq 2^Q$ is a collection of subsets of Q. Each element $K \in \mathcal{K}$ is called a knowledge state and $\mathcal{K}$ is the set of all possible knowledge states. Intuitively, the knowledge state of an individual is represented by the set of questions that he is capable of answering. Each knowledge state can be considered as a granule. The collection of all the knowledge states together with the empty set $\emptyset$ and the whole set Q is called a knowledge structure, and may be viewed as a granular knowledge structure in the terminology of granular computing.

There are two types of knowledge structures. One is closed under set union and intersection, and the other is closed only under set union, and the latter is called a knowledge space.

4.1 Granular Spaces Associated to Surmise Relations

One intuitive way to study a knowledge structure is through a surmise relation. A surmise relation on the set Q of questions is a reflexive and transitive relation S. By aSb, we can surmise the mastery of a if a student can correctly answer question b. A surmise relation imposes conditions on the corresponding knowledge structure. For example, aSb means that if a knowledge state contains b, it must also contain a.

Formally, for a surmise relation S on the (finite) set Q of questions, the associated knowledge structure $\mathcal{K}$ is defined by:

$$\mathcal{K} = \{K \mid \forall q, q' \in Q((qSq', q' \in K) \Longrightarrow q \in K)\}. \quad (10)$$

The knowledge structure associated to a surmise relation contains the empty set $\emptyset$, the entire set Q, and is closed under set intersection and union.

For each question q in Q, under a surmise relation, we can find one unique prerequisite question set $R_p(q) = \{q' \mid q'Sq\}$. The family of the prerequisite question sets for all the questions is denoted by $\mathcal{B}$, which is a covering of Q. Each prerequisite set for a question is called an elementary granule. By taking the union of prerequisite sets for a family of questions, we can obtain a knowledge structure $\mathcal{K}$ associated to the surmise relation S. It defines a granular space $(Q, \mathcal{B}, \mathcal{K})$. All knowledge states are called granules in $(Q, \mathcal{B}, \mathcal{K})$.

As a result, we have a granular space based on a surmise relation:

$$(Q, \mathcal{B}, \mathcal{K}),$$

where

Q is a set of question set,
$\mathcal{B} \subseteq 2^Q$ is a family of prerequisite sets, i.e., $\mathcal{B} = \{R_p(q) \mid q \in Q\}$,
$\mathcal{K} \subseteq 2^Q$ is a family of knowledge states.

Note that $\mathcal{B} \subseteq \mathcal{K}$ and each knowledge state can be expressed as a union of some elements of $\mathcal{B}$.

4.2 An Example

Suppose $Q = \{a, b, c, d, e\}$. Given a surmise relation:

$$aSa, \quad aSd, \quad bSb, \quad bSc, \quad bSd, \quad bSe, \quad cSc, \quad cSd, \quad cSe, \quad dSd, \quad eSe,$$

we have a knowledge structure:

$$\mathcal{K} = \{\emptyset, \{a\}, \{b\}, \{a, b\}, \{b, c\}, \{a, b, c\}, \{b, c, e\}, \{a, b, c, e\}, \{a, b, c, d\}, Q\}.$$

It can be easily seen that the knowledge structure is closed under set union and intersection. It is a knowledge structure associated to a surmise relation. As a result, we can find the prerequisite set for each question:

$$\begin{aligned} R_p(a) &= \{a\}, \\ R_p(b) &= \{b\}, \\ R_p(c) &= \{b, c\}, \\ R_p(d) &= \{a, b, c, d\}, \\ R_p(e) &= \{b, c, e\}. \end{aligned}$$

As a result, we have a granular space based on a surmise relation:

$$(Q, \mathcal{B}, \mathcal{K}),$$

where

$Q = \{a, b, c, d, e\}$,
$\mathcal{B} = \{\{a\}, \{b\}, \{b, c\}, \{a, b, c, d\}, \{b, c, e\}\}$,
$\mathcal{K} = \{\emptyset, \{a\}, \{b\}, \{a, b\}, \{b, c\}, \{a, b, c\}, \{b, c, e\}, \{a, b, c, e\}, \{a, b, c, d\}, Q\}$.

A granule in $\mathcal{K}$ can be attained by taking union of some elementary granules in $\mathcal{B}$. For example, $\{a, b\} = \{a\} \cup \{b\}$ and $\{a, b, c, e\} = \{a\} \cup \{b, c, e\}$.

4.3 Granular Spaces Associated to Surmise Systems

Modeling a knowledge structure with a surmise relation is sometimes too restrictive. That is, a question can only have one prerequisite set. In real-life situations, a question may have several prerequisite sets. This leads to the concept of surmise systems.

A surmise system on a (finite) set Q is a mapping σ that associates any element $q \in Q$ to a nonempty collection $\sigma(q)$ of subsets of Q satisfying the following three conditions [4]:

$$
\begin{aligned}
&(1) \quad C \in \sigma(q) \Longrightarrow q \in C,\\
&(2) \quad (C \in \sigma(q), q' \in C) \Longrightarrow \exists C' \in \sigma(q')(C' \subseteq C),\\
&(3) \quad C \in \sigma(q) \Longrightarrow \forall C' \in \sigma(q)(C' \not\subset C),
\end{aligned}
$$

where C is a subset in $\sigma(q)$ called a clause for question q. A surmise system may be interpreted as a neighborhood system. A clause for question q is actually a prerequisite set for q. Each question may have several prerequisite sets, namely, the clauses for question q are not always unique. Condition (1) generalizes the reflexivity condition for a relation, while the second condition extends the notion of transitivity. Condition (3) requires that the clauses for question q are the maximal sets.

Formally, for a surmise system (Q, σ), the associated knowledge structure is given by:

$$\mathcal{K} = \{K \mid \forall q \in Q(q \in K \Longrightarrow \exists C \in \sigma(q)(C \subseteq K))\}, \tag{11}$$

which is closed under set union. Any knowledge structure that is closed under union is called a knowledge space. There is a one-to-one correspondence between surmise systems on Q and knowledge spaces on Q.

There always exists exactly one minimal sub-collection $\mathcal{B}$ of $\mathcal{K}$. For a minimal sub-collection $\mathcal{B}$, any knowledge state in the sub-collection can not be the union of any other knowledge states in $\mathcal{B}$. All the knowledge states in a knowledge space can be obtained by the union of some subsets in the minimal sub-collection. We call a minimal sub-collection a basis of $\mathcal{K}$. It can be easily shown that the basis is a covering of Q. The subsets in the basis are called elementary granules. The corresponding knowledge space defines a granular space $(Q, \mathcal{B}, \mathcal{K})$. The knowledge states in the knowledge space are also called granules.

We have a granular space in a knowledge space:

$$(Q, \mathcal{B}, \mathcal{K}),$$

where

Q *is the question set,*
$\mathcal{B} \subseteq 2^Q$ *is the basis of a knowledge space called elementary granules,*
$\mathcal{K} \subseteq 2^Q$ *is a family of knowledge states in a knowledge space called granules.*

It has been proven that the basis of a knowledge space is the family of all clauses [6]. Each of the clauses is an element of the basis $\mathcal{B}$. Conversely, each element of the basis is a clause for some question. Let $\mathcal{B}_q$ represent the set of all minimal states containing question q in a knowledge space $\mathcal{K}$ with a basis $\mathcal{B}$. Thus, $\mathcal{B}_q \subseteq \mathcal{K}_q$, where $\mathcal{K}_q$ is the set of all states containing q. We also have $\mathcal{B}_q \subseteq \mathcal{B}$ for any question q, and the set $\sigma(q)$ of all the clauses for q is obtained by setting:

$$\sigma(q) = \mathcal{B}_q. \tag{12}$$

Equation (12) specifies how the clauses can be constructed from the basis of a knowledge space.

4.4 An Example

Suppose $Q = \{a, b, c, d, e\}$. Given a surmise system:

$$\begin{aligned}
\sigma(a) &= \{\{a\}\}, \\
\sigma(b) &= \{\{b, d\}, \{a, b, c\}, \{b, c, e\}\}, \\
\sigma(c) &= \{\{a, b, c\}, \{b, c, e\}\}, \\
\sigma(d) &= \{\{b, d\}\}, \\
\sigma(e) &= \{\{b, c, e\}\},
\end{aligned}$$

we can obtain a knowledge structure $\mathcal{K} = \{\emptyset,\ \{a\},\ \{b, d\},\ \{a, b, c\},\ \{b, c, e\}, \{a, b, d\}, \{a, b, c, d\}, \{a, b, c, e\}, \{b, c, d, e\}, Q\}$. It can be easily verified that it is a knowledge space closed under set union.

We have a granular space in the knowledge space:

$$(Q, \mathcal{B}, \mathcal{K}),$$

where

$$\begin{aligned}
Q &= \{a, b, c, d, e\}, \\
\mathcal{B} &= \{\{a\}, \{b, d\}, \{a, b, c\}, \{b, c, e\}\}, \\
\mathcal{K} &= \{\emptyset, \{a\}, \{b, d\}, \{a, b, c\}, \{b, c, e\}, \{a, b, d\}, \{a, b, c, d\}, \{a, b, c, e\}, \{b, c, d, e\}, \\
&\quad Q\}.
\end{aligned}$$

$\mathcal{K}$ can be constructed by taking union of any clauses in the basis $\mathcal{B}$.

4.5 Rough Set Approximations in Knowledge Spaces

The two types of knowledge structures defined by a surmise relation and a surmise system satisfy different properties. They produce different rough-set-like approximations.

Knowledge Structure Defined by a Surmise Relation

Each knowledge state is considered as a granule. For an arbitrary subset of questions $A \subseteq Q$, it may not be a knowledge state. We can use a pair of states from below and above to approximate A. Since the knowledge structure associated to a surmise relation is closed under set intersection and union, we can extend the subsystem based definition [16, 18, 26]. The lower and upper approximations are unique. They are defined by:

$$\underline{apr}(A) = \bigcup\{K \in \mathcal{K} \mid K \subseteq A\},$$
$$\overline{apr}(A) = \bigcap\{K \in \mathcal{K} \mid A \subseteq K\}. \tag{13}$$

If A is a knowledge state, we have $\underline{apr}(A) = A = \overline{apr}(A)$. The lower approximation $\underline{apr}(A)$ is the union of all the knowledge states which are subsets of A. The upper approximation $\overline{apr}(A)$ is the intersection of all the knowledge states which contain A. The knowledge structure associated to a surmise relation is not closed under complement. It follows that the lower and upper approximations are no longer dual to each other.

The physical interpretation of approximations in knowledge spaces may be given as follows. Suppose A is a set of questions that can be answered correctly. Since it is not a knowledge state, we approximate it by two knowledge states. The lower approximation represents the affirmatory mastery of a subset of questions, while the upper approximation describes the possible mastery of a subset of questions.

Knowledge Structure Defined by a Surmise System

A knowledge structure associated to a surmise system is only closed under set union, but not necessarily closed under set intersection. In this case, we can still use a subsystem-based definition [16, 18, 26] to define the lower approximation. However, we must introduce a new definition for upper approximation.

By keeping the interpretation of lower and upper approximations as the greatest knowledge states that are contained in A and the least knowledge states that contain A, we have the following definition:

$$\underline{apr}(A) = \{\cup\{K \in \mathcal{K} \mid K \subseteq A\}\},$$
$$\overline{apr}(A) = \{K \in \mathcal{K} \mid A \subseteq K, \forall K' \in \mathcal{K}(K' \subset K \Longrightarrow A \not\subseteq K')\}. \tag{14}$$

The lower approximation in knowledge spaces is unique while the upper approximation is a family of sets [16]. If A is a knowledge state, we have $\underline{apr}(A) = \{A\} = \overline{apr}(A)$.

5 Conclusion

Granules and granular structures are two fundamental notions of granular computing. A set-theoretic framework of granular spaces is proposed. The framework considers three levels of characterization of a universal set U, namely, the triplet $(U, \mathcal{S}_0, \mathcal{S})$. They are the ground level which is U, the elementary granule level which is a subsystem $\mathcal{S}_0$ of 2^U, and the granular structure level which is a system $\mathcal{S}$ of 2^U. Typically, $\mathcal{S}_0$ is a covering of U and $\mathcal{S}$ is a hierarchical structure generated from $\mathcal{S}_0$. The framework enables us to study rough set analysis and knowledge spaces in a common setting. Moreover, results from the two theories can be applied to each other, which brings more insights into the research of the two theories.

With the proposed framework, we demonstrate that rough set analysis and knowledge spaces both consider a similar type of granular space. Their main differences lie in the construction and interpretation of elementary granules and granular structures. This observation immediately opens up new avenues of research. As an illustration, we introduce the notion of approximations from rough set analysis to knowledge spaces theory. The results are a new type of approximations not considered in rough set analysis.

The framework of granular spaces can also be used to interpret other theories. For example, it can be used to study formal concept analysis [26]. As future research, the potential of the proposed framework will be further explored.

References

1. Bargiela, A., Pedrycz, W.: Granular Computing: An Introduction. Kluwer Academic Publishers, Boston (2002)
2. Bargiela, A., Pedrycz, W.: Toward a theory of granular computing for human-centred information processing. IEEE Transactions On Fuzzy Systems (to appear)
3. Doignon, J.P., Falmagne, J.C.: Spaces for the assessment of knowledge. International Journal of Man-Machine Studies 23, 175–196 (1985)
4. Doignon, J.P., Falmagne, J.C.: Knowledge spaces. Springer, Heidelberg (1999)
5. Duntsch, I., Gediga, G.: A note on the correspondences among entail relations, rough set dependencies, and logical consequence. Mathematical Psychology 43, 393–401 (2001)
6. Falmagne, J.C., Koppen, M., Villano, M., Doignon, J.P., Johanessen, L.: Introduction to knowledge spaces: how to test and search them. Psychological Review 97, 201–224 (1990)
7. Lin, T.Y., Yao, Y.Y., Zadeh, L.A.: Data Mining, Rough Sets and Granular Computing. Physica-Verlag, Heidelberg (2002)
8. Nguyen, S.H., Skowron, A., Stepaniuk, J.: Granular computing: a rough set approach. Computational Intelligence 17, 514–544 (2001)
9. Orlowska, E.: Logical aspects of learning concepts. International Journal of Approximate Reasoning 2, 349–364 (1988)
10. Pawlak, Z.: Rough sets. International Journal of Computer and Information Sciences 11, 341–356 (1982)
11. Pawlak, Z.: Rough Sets. In: Theoretical Aspects of Reasoning about Data. Kluwer Academic Publishers, Dordrecht (1991)
12. Pinker, S.: How the mind works. Norton, New York (1997)
13. Polkowski, L., Semeniuk-Polkowska, M.: On foundations and applications of the paradigm of granular rough computing. International Journal of Cognitive Informatics and Natural Intelligence 2, 80–94 (2008)
14. Simon, H.A.: The Sciences of the Artificial. MIT Press, Cambridge (1996)
15. Solso, R.L., Maclin, M.K., Maclin, O.H.: Cognitive psychology. Allyn and Bacon, Boston (2005)

16. Xu, F.F., Yao, Y.Y., Miao, D.Q.: Rough set approximations in formal concept analysis and knowledge spaces. In: An, A., Matwin, S., Raś, Z.W., Ślęzak, D. (eds.) Foundations of Intelligent Systems. LNCS, vol. 4994, pp. 319–328. Springer, Heidelberg (2008)
17. Yao, J.T.: A ten-year review of granular computing. In: Proceedings of 2007 IEEE International Conference on Granular Computing, pp. 734–739 (2007)
18. Yao, Y.Y.: On generalizing rough set theory. In: Wang, G., Liu, Q., Yao, Y., Skowron, A. (eds.) RSFDGrC 2003. LNCS, vol. 2639, pp. 44–51. Springer, Heidelberg (2003)
19. Yao, Y.Y.: A partition model of granular computing. In: Peters, J.F., Skowron, A., Grzymała-Busse, J.W., Kostek, B.z., Świniarski, R.W., Szczuka, M.S. (eds.) Transactions on Rough Sets I. LNCS, vol. 3100, pp. 232–253. Springer, Heidelberg (2004)
20. Yao, Y.Y.: Granular computing. Computer Science (Ji Suan Ji Ke Xue). 31, 1–5 (2004)
21. Yao, Y.Y.: Perspectives of granular computing. In: Proceedings of 2005 IEEE International Conference on Granular Computing, pp. 85–90 (2005)
22. Yao, Y.Y.: Three perspectives of granular computing. Journal of Nanchang Institute of Technology 25(2), 16–21 (2006)
23. Yao, Y.Y.: A note on definability and approximations. In: Peters, J.F., Skowron, A., Marek, V.W., Orłowska, E., Słowiński, R., Ziarko, W.P. (eds.) Transactions on Rough Sets VII. LNCS, vol. 4400, pp. 274–282. Springer, Heidelberg (2007)
24. Yao, Y.Y.: The art of granular computing. In: Proceeding of the International Conference on Rough Sets and Emerging Intelligent Systems Paradigms, pp. 101–112 (2007)
25. Yao, Y.Y.: The rise of granular computing. Journal of Chongqing University of Posts and Telecommunications (Natural Science Edition) 20(3), 1–10 (2008)
26. Yao, Y.Y., Chen, Y.H.: Rough set approximations in formal concept analysis. In: Peters, J.F., Skowron, A. (eds.) Transactions on Rough Sets V. LNCS, vol. 4100, pp. 285–305. Springer, Heidelberg (2006)
27. Yao, Y.Y., Lin, T.Y.: Generalization of rough sets using modal logic. Intelligent Automation and Soft Computing 2, 103–120 (1996)
28. Yao, Y.Y., Zhou, B.: A logic language of granular computing. In: Proceedings 6th IEEE International Conference on Cognitive Informatics, pp. 178–185 (2006)
29. Zadeh, L.A.: Towards a theory of fuzzy information granulation and its centrality in human reasoning and fuzzy logic. Fuzzy Sets and Systems 90, 111–127 (1997)

On Approximation of Classifications, Rough Equalities and Rough Equivalences

B.K. Tripathy

School of Computing Sciences, V.I.T. University, Vellore, T.N. 632 014, India
tripathybk@rediffmail.com

Summary. In this chapter we mainly focus on the study of some topological aspects of rough sets and approximations of classifications. The topological classification of rough sets deals with their types. We find out types of intersection and union of rough sets, New concepts of rough equivalence (top, bottom and total) are defined, which capture approximate equality of sets at a higher level than rough equality (top, bottom and total) of sets introduced and studied by Novotny and Pawlak [23,24,25] and is also more realistic. Properties are established when top and bottom rough equalities are interchanged. Also, parallel properties for rough equivalences are established. We study approximation of classifications (introduced and studied by Busse [12]) and find the different types of classifications of an universe completely. We find out properties of rules generated from information systems and observations on the structure of such rules. The algebraic properties which hold for crisp sets and deal with equalities loose their meaning when crisp sets are replaced with rough sets. We analyze the validity of such properties with respect to rough equivalences.

1 Introduction

The notion of rough sets was introduced by Pawlak [26,27,28] as an extension of the concept of crisp sets to capture impreciseness. Imprecision in this approach is expressed by the boundary region of a set. In fact, the idea of rough set is based upon approximation of a set by a pair of sets, called the *lower* and *upper approximations* of the set.

In real life situations, fruitful and accurate applications of rough sets require two aspects, called accuracy measure and topological characterization. We shall mainly concentrate on topological characterization of rough sets in this chapter. The other related aspect to be dealt with is approximations of classifications, which are in a sense extensions of the concept of approximation of sets but their characteristics are not exactly same. We shall study the types of union and intersection of rough sets which are also used in dealing with types of classifications. New notions of approximate equalities, called rough equivalences are introduced and their properties are studied. Using this notion, some basic algebraic properties for crisp sets are extended to rough sets. We also touch the topic of rule generation from information systems.

A. Abraham, R. Falcón, and R. Bello (Eds.): Rough Set Theory, SCI 174, pp. 85–133.
springerlink.com

Now, we present the detailed structure of the chapter here. In sect. 2, we establish two theorems on rough approximations which provide necessary and sufficient conditions for equality to hold in two of the properties, where in general inclusions hold true. There are several applications of these results as we shall see in sections 4.3, 5.5 and 9.4.

As mentioned by Pawlak [30], one important difference between the concept of rough set and the classical notion of set is the equality of sets. In classical set theory, two sets are equal if they have exactly the same elements. But a more practically applicable form of equality (approximate equality) called rough equality was introduced in [23,24,25]. Here, two sets may not be equal in the classical sense but they have enough of close features (that is they differ slightly from each other) to be assumed to be approximately equal. These types of equalities of sets refer to the topological structure of compared sets but not to the elements they consist of. In fact two sets can be exactly equal in one knowledge base but approximately equal or not equal in another. The practicality of this notion depends upon the common observation that things are equal or not equal from the point of view of our knowledge about them. Certain properties of these equalities were established by Novotny and Pawlak [23,24,25]. But they have remarked that these properties cease to be true when top and bottom rough equalities are replaced one by the other. In sect. 3, we see that some of these properties are true under replacement and others hold true if some additional conditions are imposed.

A topological characterization of imprecision defined through the lower and upper approximation of sets is the notion of type of rough sets. There are four such types [30]. This method of characterization of imprecision complements the other method of characterization of imprecision through accuracy measures, which expresses degree of completeness of our knowledge about a set. As observed by Pawlak ([30], p. 22), in practical applications of rough sets we combine both kinds of information. As far as the information available, no further study is found in rough set literature on this topic after its introduction. In sect. 4, we study the types of rough sets obtained by union and intersection of rough sets. We shall deal with applications of these results in sects. 5, 7 and 9.

As mentioned above, rough equalities deal with topological structures of the compared sets. In sect. 5, we introduce and study another type of approximate equality, called rough equivalence of sets, which captures topological structures of the compared sets at a higher level than rough equality. By this, we mean that any two sets comparable with the notions of rough equalities (bottom, top and total) are also comparable with the corresponding notion of rough equivalence (bottom, top and total) and the converse is not necessarily true. In fact, there are many practical situations, where we can talk of approximate equality of the compared sets with new notion but can not do so with the old one. More importantly, this new comparison very much matches with our perception of equality depending upon our knowledge about the universe. We illustrate this with some examples. Also, properties rough equivalences, which are in parallel with those for rough equalities along with the corresponding replacement properties are analyzed and established.

To deal with knowledge acquisition under uncertainty, Busse [12] considered the approximations of classifications as a new approach. Some earlier approaches to the acquisition of knowledge and reasoning under uncertainty by expert systems research community are in [1,11,19,44]. Uncertainty may be caused by ambiguous meanings of the terms used, corrupted data or uncertainty in the knowledge itself [12]. One of the popular ways to acquire the knowledge is based upon learning from examples [12]. The information system (a data base-like system) represents what is called an 'instant space' in learning from examples. In the approach of Busse, inconsistencies are not corrected. Instead, produced rules are categorized into certain rules and possible rules. Some other authors who have dealt without correcting inconsistencies in information systems are Mamdani et. al.[19] and Quinlan [35]. Four results were established by Busse on approximation of classifications. In sect. 6, we generalize these results to necessary and sufficient type ones from which, along with the results of Busse many other results can be obtained as corollaries. The types of classifications are thoroughly analyzed and their properties are studied in sect. 7. We find that the eleven numbers of possible types reduce either directly or transitively to the five types considered by Busse. In sect. 8, we present some of the properties of rules generated from information systems and obtain many observations on the structure of such rules.

There are many fundamental algebraic properties of crisp sets with respect to the operations of union, intersection and complementation. All these properties involve equality of two such expressions. When the involved sets are taken to rough sets the equalities bear very little meaning (particularly, after the introduction of the concepts of rough equalities and rough equivalences). To make them more and more meaningful, one has to consider rough equality or rough equivalence in general. In sect. 9, we consider the validity of many of these basic properties with crisp equality being replaced by rough equivalence. Rough equalities being special cases of rough equivalences, we can derive the corresponding validities easily. We shall end the chapter with some concluding remarks and finally provide a bibliography of papers and other related materials, which are referred during the compilation of the materials of the chapter.

2 Rough Sets and Properties of Approximations

In this sect. we shall first introduce the definitions of rough set and related concepts in sect. 2.1. In sect. 2.2 we introduce some properties of lower and upper approximations and establish two theorems related to these properties, which are to be used in later sections.

2.1 Rough Sets

Let U be a universe of discourse and R be an equivalence relation over U. By U/R we denote the family of all equivalence classes of R, referred to as *categories* or *concepts* of R and the equivalence class of an element $x \in U$ is denoted by $[x]_R$. The basic philosophy of rough set is that knowledge is deep-seated in the classificatory abilities of human beings and other species. Knowledge is connected with the variety of classification patterns related to specific parts of real or abstract world, called the universe. Knowledge consists of a family of various classification patterns of a domain of interest, which provide explicit facts about reality-together with the reasoning capacity able to deliver implicit facts derivable from explicit knowledge ([30], p. 2).

There is, however a variety of opinions and approaches in this area, as to how to understand, represent and manipulate knowledge [3,4,6,7,9,15,20,21].

Usually, we do not deal with a single classification, but with families of classifications over U. A family of classifications over U is called a knowledge base over U. This provides us with a variety of classification patterns which constitute the fundamental equipment to define its relation to the environment. More precisely, by a knowledge base we mean a relational system $\mathbf{K}=(U,\boldsymbol{R})$, where U is as above and $\boldsymbol{R}$ is a non-empty family of equivalence relations over U.

For any subset $P(\neq \phi) \subseteq \boldsymbol{R}$, the intersection of all equivalence relations in P is denoted by *IND(P)* and is called the *indiscernilibity relation* over P. By $IND(K)$ we denote the family of all equivalence relations defined in K, that is $IND(K) = \{IND(P) : P \subseteq \boldsymbol{R}, P \neq \phi\}$.

Given any $X \subseteq U$ and $R \in$ *IND(K)*, we associate two subsets, $\underline{R}X = \bigcup\{Y \in U/R : Y \subseteq X\}$ and $\bar{R}X = \bigcup\{Y \in U/R : Y \cap X \neq \phi\}$, called the R-*lower* and R-*upper approximations* of X respectively. The R-*boundary* of X is denoted by $BN_R(X)$ and is given by $BN_R(X) = \bar{R}X - \underline{R}X$. The elements of $\underline{R}X$ are those elements of U which can certainly be classified as elements of X and elements of $\bar{R}X$ are those elements of U which can possibly be classified as elements of X, employing the knowledge of R. We say that X is rough with respect to R if and only if $\underline{R}X \neq \bar{R}X$, equivalently $BN_R(X) \neq \phi$. X is said to be R-*definable* if and only if $\underline{R}X = \bar{R}X$, or $BN_R(X) = \phi$.

2.2 Properties of Approximations

The lower and upper approximations of rough sets have several properties [30]. We shall be using the following four properties in our discussions:

$$\underline{R}X \cup \underline{R}Y \subseteq \underline{R}(X \cup Y) \tag{1}$$

$$\bar{R}(X \cap Y) \subseteq \bar{R}X \cap \bar{R}Y \tag{2}$$

$$\bar{R}(X \cup Y) = \bar{R}(X) \cup \bar{R}(Y) \tag{3}$$

$$\underline{R}(X \cap Y) = \underline{R}(X) \cap \underline{R}(Y) \tag{4}$$

The inclusions in (1) and (2) can be proper [30] and also can be extended to a finite number of sets. These results confirm to the obvious observation that, in general knowledge included in a distributed knowledge base is less than the integrated one. That is, in general, dividing the knowledge base into smaller fragments causes loss of information [30]. This leads to the interesting problem of determining the exact circumstances under which there will be no loss of information even if one distributes the knowledge base or equivalently under what circumstances there will definitely be loss of information. The following two theorems [37] establish necessary and sufficient conditions for the inclusions (1) and (2) to be proper. The corollaries derived from these results provide necessary and sufficient conditions for equalities to hold in (1) and (2). Thus answers to the questions raised above have been obtained. We shall find many applications of these results in this chapter.

Theorem 1. Let $\{E_1, E_2, E_3, ..., E_n\}$ be the partition of any universe U with respect to an equivalence relation R. Then for any finite number of subsets X_1, X_2, X_3, . . ., X_m, of U,

$$\bigcup_{i=1}^{m} \underline{R}(X_i) \subset \underline{R}(\bigcup_{i=1}^{m} X_i) \tag{5}$$

if and only if there exists at least one E_j such that

$$X_i \cap E_j \subset E_j, \text{ for } i = 1, 2, ..., m \textit{ and } \bigcup_{i=1}^{m} X_i \supseteq E_j \tag{6}$$

Proof. The sufficiency follows from the fact that $E_j \not\subset \underline{R}(X_i)$, for $i = 1, 2, ..., m$, but $E_j \subset \underline{R}(\bigcup_{i=1}^{m} X_i)$. Conversely, suppose $\bigcup_{i=1}^{m} \underline{R}(X_i) \subset \underline{R}(\bigcup_{i=1}^{m} X_i)$. As $\underline{R}X$ for any X is the union of E_j's only, there is at least one E_j such that $E_j \subseteq \underline{R}(\bigcup_{i=1}^{m} X_i)$ and $E_j \not\subset \underline{R}(X_i)$ for any $i = 1, 2, ..., m$. So, $E_j \subseteq \bigcup_{i=1}^{m} X_i$, but $E_j \not\subset X_j$, for any $i = 1, 2, ..., m$. Thus $X_i \cap E_j \subset E_j$ and $E_j \subseteq \bigcup_{i=1}^{m} X_i$.

Corollary 1. A necessary and sufficient condition for

$$\bigcup_{i=1}^{m} \underline{R}(X_i) = \underline{R}(\bigcup_{i=1}^{m} X_i) \tag{7}$$

is that there exist no E_j such that

$$X_i \cap E_j \subset E_j, i = 1, 2, ..., m \textit{ and } \bigcup_{i=1}^{m} X_i \supseteq E_j. \tag{8}$$

We shall be using the following example to illustrate the results of this sect.

Example 1. Let us consider an organization having four different sites. For simplicity in computation we assume that there are 20 employees only in the organization who are distributed over four sites.

Further, suppose that these employees are working on different projects $p_i, i = 1, 2, 3, 4$; irrespective of their branch. Some of the employees are involved in more than one project whereas some are not involved in any of the projects. Let the sets E_1, E_2, E_3, E_4 denote employees working at the four sites and X_1, X_2, X_3, X_4 be the set of employees working for the projects p_1, p_2, p_3 and p_4 respectively. Let

$E_1 = \{e_1, e_2, e_3, e_4, e_5\}$
$E_2 = \{e_6, e_7, e_8, e_9, e_{10}\}$
$E_3 = \{e_{11}, e_{12}, e_{13}, e_{14}, e_{15}\}$
$E_4 = \{e_{16}, e_{17}, e_{18}, e_{19}, e_{20}\}$

$X_1 = \{e_1, e_2, e_4, e_7, e_{11}, e_{13}, e_{19}\}$
$X_2 = \{e_4, e_7, e_{11}, e_{12}, e_{15}, e_{19}\}$
$X_3 = \{e_4, e_7, e_{11}, e_{16}, e_{18}, e_{19}\}$ and
$X_4 = \{e_4, e_7, e_{11}, e_{16}, e_{17}, e_{18}, e_{19}, e_{20}\}$

Let us define a relation R over the set of employees U in the organization as $e_i R e_j$ if and only if both e_i and e_j work in the same branch.

The lower approximation of a set $X_i, i = 1, 2, 3, 4$ here provides the fact whether all the employees in a particular site work in a given project or not. Similarly, the upper approximation of these sets provide the fact whether any employee in a particular site works in a project or not. For example, $\underline{R}X_4 = E_4$, says that all the employees in site 4 work in project 4. Similarly, $\bar{R}X_1 = U$ means that some employees of every site work in project 1.

Illustration for Corollary 1. Here, $(\bigcup_{i=1}^{4} X_i) \not\supseteq E_j \; for \; j = 1, 2, 3$ and for E_4,

$$(\bigcup_{i=1}^{4} X_i) \supseteq E_4, \text{ but } X_4 \cap E_4 = E_4.$$

So, the conditions of Corollary 1 are satisfied. Hence we must have the equality true.

In fact, we see that

$$\underline{R}(\bigcup_{i=1}^{4} X_i) = E_4 \text{ and } (\bigcup_{i=1}^{4} \underline{R}X_i) = E_4 \; as \; \underline{R}X_1 = \underline{R}X_2 = \underline{R}X_3 = \phi$$

and $\underline{R}X_4 = E_4$.

Theorem 2. Let $\{E_1, E_2, ..., E_n\}$ be a partition of any universe U with respect to an equivalence relation R. Then for a finite number of subsets $X_1, X_2, ..., X_m$ of U, the necessary and sufficient condition for

$$\bar{R}(\bigcap_{i=1}^{m} X_i) \subset \bigcap_{i=1}^{m} \bar{R}(X_i) \tag{9}$$

is that there exists at least one E_j such that

$$X_i \cap E_j \neq \phi \text{ for } i = 1, 2, ..., m \; and \; (\bigcap_{i=1}^{m} X_i) \cap E_j = \phi \tag{10}$$

Proof. The sufficiency follows from the fact that

$$E_j \not\subset \bar{R}(\bigcap_{i=1}^{m} X_i) \text{ and } \bar{R}(X_i) \supseteq E_j \text{ for } i = 1, 2, ..., m.$$

Conversely, suppose the conclusion is true. Then for some E_j,

$$E_j \subseteq \bigcap_{i=1}^{m} \bar{R}(X_i) \text{ but } E_j \not\subset \bar{R}(\bigcap_{i=1}^{m} X_i).$$

So, $\quad E_j \subseteq \bar{R}(X_i)$ for $i = 1, 2, ..., m$ and $E_j \bigcap (\bigcap_{i=1}^{m} X_i) = \phi$.

That is $\quad E_j \cap X_i \neq \phi, i = 1, 2, ..., m$ and $(\bigcap_{i=1}^{m} X_i) \bigcap E_j = \phi$.

This completes the proof.

Corollary 2. Let $\{E_1, E_2, ..., E_n\}$ be a partition of U with respect to an equivalence relation R. Then for any finite number of subsets $X_1, X_2, ..., X_m$ of U,

$$\bar{R}(\bigcap_{i=1}^{m} X_i) = \bigcap_{i=1}^{m} \bar{R}(X_i) \tag{11}$$

if and only if there is no E_j such that

$$X_i \cap E_j \neq \phi \textit{ for } i = 1, 2, ..., m \textit{ and } (\bigcap_{i=1}^{m} X_i) \cap E_j = \phi. \tag{12}$$

Illustration for Corollary 2

$$\textit{Here } \bigcap_{i=1}^{4} X_i = \{e_4, e_7, e_{11}, e_{19}\}. \text{ So, } E_j \cap \bigcap_{i=1}^{4} X_i \neq \phi \text{ for } j = 1, 2, 3, 4.$$

Also, $X_i \cap E_j \neq \phi$, for $i = 1, 2, 3, 4$ and $j = 1, 2, 3, 4$. Hence conditions of Corollary 2 are satisfied. Also, we see that

$$\bar{R}(\bigcap_{i=1}^{4} X_i) = U = \bigcap_{i=1}^{4} \bar{R}(X_i).$$

3 Rough Equality of Sets

Comparison of sets plays a major role in classical set theory. When we move to the representation of approximate knowledge through rough sets the usual comparisons loose their meaning and in a sense are of no use. To bring about more meaning into such comparisons of rough sets which translate into approximate comparison of knowledge bases, Novotny and Pawlak [23,24,25] introduced three notions of rough equalities (bottom, top and total) and established several of their properties. However, it is mentioned [30] that these properties fail to hold when notions of bottom and top rough equalities are replaced one by the other. We show in this sect. that some of these properties hold under such interchanges and establish suitable conditions under which these interchanges are valid. Some other papers which have dealt with rough equalities are [2,5,8].

Two sets are said to be equal in crisp set theory if and only if they have the same elements. The concept has been extended to define rough equalities of sets by Novotny and Pawlak [23,24,25]. In the next sect. we state these equalities.

3.1 Definitions

Definition 1. Let $K = (U,\boldsymbol{R})$ be a knowledge base, $X, Y \subseteq U$ and $R \in IND(K)$. We say that

(i) Two sets X and Y are *bottom R-equal* ($X{=}_BY$) if $\underline{R}X = \underline{R}Y$;
(ii) Two sets X and Y are *top R-equal* ($X{=}_TY$) if $\bar{R}X = \bar{R}Y$;
(iii) Two sets X and Y are *R-equal* ($X{=}Y$) if ($X{=}_BY$) and ($X{=}_TY$); equivalently, $\underline{R}X = \underline{R}Y$ and $\bar{R}X = \bar{R}Y$.

We have dropped the suffix R in the notations to make them look simpler and easy to use. Also the notations used are different from the original ones. This has been done due to non-availability of the original notations in the symbol set. It can be easily verified that the relations bottom R-equal, top R-equal and R-equal are equivalence relations on *P(U)*, the power set of *U*.

The concept of approximate equality of sets refers to the topological structure of the compared sets but not the elements they consist of. Thus, sets having significantly different elements may be rough equal. In fact, if $X =_B Y$ then $\underline{R}X = \underline{R}Y$ and as $X \supseteq \underline{R}X, Y \supseteq \underline{R}Y$, X and Y can differ only in elements of $X - \underline{R}X$ and $Y - \underline{R}Y$. However, taking the example; $U = \{x_1, x_2, ..., x_8\}$ and $R = \{\{x_1, x_2\}, \{x_3, x_4\}, \{x_5, x_6\}, \{x_7, x_8\}\}$, we see that the two sets $X = \{x_1, x_3, x_5\}$ and $Y = \{x_2, x_4, x_6\}$ are top R-equal, even though $X \cap Y = \phi$.

As noted by Pawlak [30, p.26], this concept of rough equality of sets is of relative character, that is things are equal or not equal from our point of view depending on what we know about them. So, in a sense the definition of rough equality refers to our knowledge about the universe.

3.2 Properties of Rough Equalities

The following properties of rough equalities are well known [30].

$$X{=}_BY \; if \; and \; only \; if \; X \cap Y{=}_BX \; and \; X \cap Y{=}_BY. \tag{13}$$

$$X{=}_TY \; if \; and \; only \; if \; X \cap Y{=}_TX \; and \; X \cap Y{=}_TY. \tag{14}$$

$$If\ X{=_T}X'\ and\ Y{=_T}Y'\ then\ X \cup Y{=_T}X' \cup Y'. \tag{15}$$

$$If\ X{=_B}X'\ and\ Y{=_B}Y'\ then\ X \cap Y{=_B}X' \cap Y'. \tag{16}$$

$$If\ X{=_T}Y\ then\ X \cup -Y{=_T}U. \tag{17}$$

$$If\ X{=_B}Y\ then\ X \cap -Y{=_B}\phi. \tag{18}$$

$$If\ X \subseteq Y\ and\ Y{=_T}\phi\ then\ X{=_T}\phi. \tag{19}$$

$$If\ X \subseteq Y\ and\ X{=_T}U\ then\ Y{=_T}U. \tag{20}$$

$$X{=_T}Y\ if\ and\ only\ if\ -X{=_B}-Y. \tag{21}$$

$$If\ X{=_B}\phi\ or\ Y{=_B}U\ then\ X \cap Y{=_B}\phi. \tag{22}$$

$$If\ X{=_T}U\ or\ Y{=_T}U\ then\ X \cup Y{=_T}U. \tag{23}$$

It has been pointed out that (see for instance [30]) the above properties fail to hold if $=_T$ is replaced by $=_B$ or conversely. However, we have the following observations in connection with this interchange.

I. The properties (19) to (23) hold true under the interchange.
That is we have

$$X \subseteq Y\ and\ Y{=_B}\phi \Rightarrow X{=_B}\phi. \tag{19$'$}$$

$$If\ X \subseteq Y\ and\ X{=_B}U \Rightarrow Y{=_B}U. \tag{20$'$}$$

$$X\ {=_B}Y\ if\ and\ only\ if\ -X{=_T}-Y. \tag{21$'$}$$

$$If\ X{=_T}\phi\ or\ Y{=_T}\phi\ then\ X \cap Y{=_T}\phi\ ,\ and \tag{22$'$}$$

$$If\ X{=_B}U\ or\ Y{=_B}U\ then\ X \cup Y{=_B}U. \tag{23$'$}$$

II. The properties (17) and (18) holds true under the interchange in the following form:

$$If\ X{=_B}Y\ then\ X \cup -Y{=_B}U\ if\ BN_R(Y) = \phi. \tag{17$'$}$$

$$If\ X{=_T}Y\ then\ X \cap -Y{=_T}\phi\ if\ BN_R(Y) = \phi. \tag{18$'$}$$

Proof of (17'). $\underline{R}(X \cup -Y) \supseteq \underline{R}(X) \cup \underline{R}(-Y)$

$$= \underline{R}(Y) \cup (-\bar{R}(Y))$$
$$= \underline{R}(Y) \cup (-\underline{R}(-Y)) \cup BN_R(Y)$$
$$= \underline{R}(Y) \cup (-\underline{R}(Y)) \cap (-BN_R(Y))$$
$$= \underline{R}(Y) \cup (-\underline{R}(Y)) \cap (\underline{R}Y \cup (-BN_R(Y)))$$
$$= U \cap (\underline{R}Y \cup (-BN_R(Y)))$$
$$= \underline{R}Y \cup (-BN_R(Y)))$$
$$= \underline{R}Y \cup (\underline{R}Y \cup (-\bar{R}Y))$$
$$= \underline{R}Y \cup (-\bar{R}Y)$$
$$= (-BN_R(Y))$$
$$= U.$$

So, $X \cup (-Y) =_B U$.

Proof of (18'). $\bar{R}(X \cap -Y) \subseteq \bar{R}(X) \cap \bar{R}(-Y)$

$$= \bar{R}(Y) \cap \bar{R}(-Y))$$
$$= \bar{R}(Y) \cap (-\underline{R}(Y))$$
$$= \bar{R}(Y) \cap ((-\bar{R}(Y)) \cup (-BN_R(Y)))$$
$$= (\bar{R}(Y) \cap ((-\bar{R}(Y)))) \cup (\bar{R}Y \cap BN_R(Y))$$
$$= \phi \cup (\bar{R}Y \cap BN_R(Y))$$
$$= BN_R(Y)$$
$$= \phi.$$

III. (i) The properties (13) and (16) hold under the interchange, if conditions of Corollary 2 hold with m = 2.
(ii) The properties (14) and (15) hold under the interchange, if conditions of Corollary 1 hold with m = 2.
So, we get

$$X =_T Y \text{ if and only if } X \cap Y =_T X \text{ and } X \cap Y =_T Y, \tag{13'}$$

$$X =_B Y \text{ if and only if } X \cup Y =_B X \text{ and } X \cup Y =_B Y, \tag{14'}$$

$$X =_B X' \text{ and } Y =_B Y' \Rightarrow X \cup Y =_B X' \cup Y', \tag{15'}$$

$$X =_T X' \text{ and } Y =_T Y' \Rightarrow X \cap Y =_B X' \cap Y'. \tag{16'}$$

Proof of (13'). $X =_T Y \Rightarrow \bar{R}X = \bar{R}Y \Rightarrow \bar{R}(X \cap Y) = \bar{R}X \cap \bar{R}Y = \bar{R}X = \bar{R}Y$.
So, $X \cap Y =_B X$ and $X \cap Y =_T Y$. The converse is trivial.

Proof of (14'). $X =_B Y \Rightarrow \underline{R}X = \underline{R}Y \Rightarrow \underline{R}(X \cap Y) = \underline{R}X \cup \underline{R}Y = \underline{R}X = \underline{R}Y$.
So, $X \cap Y =_T X$ and $X \cap Y =_T Y$. The converse is trivial.

The proofs of (15') and (16') are similar.

4 Types of Rough Sets

We have mentioned in the introduction there are four important and different topological characterizations of rough sets called their types. In this sect., we shall start with the introduction of these types. The physical interpretation and intuitive meanings of these types can be found in [30].

Type 1: If $\underline{R}X \neq \phi$ and $\bar{R}X \neq U$, then we say that X is *roughly R-definable.*
Type 2: If $\underline{R}X = \phi$ and $\bar{R}X \neq U$, then we say that X is *internally R-undefinable.*
Type 3: If $\underline{R}X \neq \phi$ and $\bar{R}X = U$, then we say that X is *externally R-undefinable.*
Type 4: If $\underline{R}X = \phi$ and $\bar{R}X = U$, then we say that X is *totally R-undefinable.*

The union and intersection of rough sets have importance from the point of distribution of knowledge base and common knowledge respectively. In this context the study of types of union and intersection of different types of rough sets have significance. For example, if two rough sets are roughly R-definable (Type 1), then there are some objects in the universe which can be positively classified, based on the available information to belong to each these sets. Now, one would like to get information about elements in the universe which can be positively classified to be in both. If the intersection is of Type 1/Type 3, then one can obviously conclude this. On the contrary if the intersection is of Type 2/Type 4, then no such element exists. From the table in sect. 4.1 we see that the intersection is Type 1/Type 2. So, it can not be said definitely that the element is in both. In fact this matches with our normal observation. Similarly, for such sets there are some other elements which can be negatively classified without any ambiguity as being outside the sets. Now, what can one say about the union of two such sets ? That is, are there are still some elements which can be negatively classified without any ambiguity being outside the union of their elements ? If the type of the union is Type 1/Type 2, then we are sure of such elements. On the other hand if it is of Type 3/Type 4 no such elements exist. From the table in sect. 4.2 we see that the union is of Type 1/Type 3. So, one can not be sure about some elements being negatively classified as outside the union. This again matches with our observation. In this sect. we shall produce general results on the types of union and intersection of rough sets of different types. We shall also try to reduce the ambiguities in the possible cases under suitable conditions through establishment of theorems.

So far nothing has been said in the literature regarding the type of a rough set which is obtained as union or intersection of different types of rough sets. In the next two sub-sections we obtain the results of union and intersection of any two types of rough sets. This study was initiated in [37].

4.1 Intersection

In the next sub-section we establish and present in a table, the results of intersection of two rough sets of different types. It is interesting to note that out of sixteen cases, as many as nine are unambiguous. The other ambiguous cases are

Table 1. Intersection of different types of rough sets

$\bigcap$	**Type1**	**Type2**	**Type3**	**Type4**
Type1	Type1/Type2	Type2	Type1/Type2	Type2
Type2	Type2	Type2	Type2	Type2
Type3	Type1 / Type2	Type2	Type1 to Type4	Type2/Type4
Type4	Type2	Type2	Type2 / Type4	Type2 / Type4

mainly due to the inclusion (2). Applying Theorem 2 above, some of the ambiguities of the table can be reduced or removed under suitable conditions which are provided by the theorem. These conditions being of necessary and sufficient type, cannot be improved further.

Proofs

We shall denote the entry in i^{th} row and j^{th} column of the table by (i, j). In the proofs, we shall be using (2) and the property that for any two rough sets X and Y

$$\underline{R}(X \cap Y) = \underline{R}X \cap \underline{R}Y \tag{24}$$

We shall provide the proofs for the cases (1,2) and (3,3). The rest of the proofs can be worked out similarly.

Proof of (1,2)
Here, X is of Type 1 and Y is of Type 2. So $\underline{R}X \neq \phi, \bar{R}X \neq U$ and $\underline{R}Y = \phi, \bar{R}Y \neq U$. Hence by (24) $\underline{R}(X \cap Y) = \phi$, and by (2) $\bar{R}(X \cap Y) \subseteq \bar{R}X \cap \bar{R}Y \neq U$. So, $X \cap Y$ is of Type 2.

Proof of (3,3)
Let both X and Y be of Type 3.

Then $\underline{R}X \neq \phi$, $\bar{R}X = U$ and $\underline{R}Y \neq \phi$, $\bar{R}Y = U$. Now, by (24) $\underline{R}(X \cap Y)$ may or may not be ϕ and by (2) $\bar{R}(X \cap Y)$ may or may not be U. $X \cap Y$ can be of any of the four Types.

Examples

In this sect. we provide examples to show that the ambiguous cases in the table can actually arise for (3). The other cases can be justified similarly. We continue with the same example of sect. 3.

Examples for (3,3)
Let $X = \{e_1, e_2, ..., e_{10}, e_{14}, e_{19}\}$ and $Y = \{e_4, e_9, e_{11}, e_{12}, e_{13}, e_{14}, e_{15}, e_{17}\}$. Then X and Y are of Type 3 as $\underline{R}X = E_1 \cup E_2$, $\bar{R}X = E_1 \cup E_2 \cup E_3 \cup E_4$,

Table 2. Union of different types of rough sets

$\cup$	**Type1**	**Type2**	**Type3**	**Type4**
Type1	Type1/Type3	Type1/Type3	Type3	Type3
Type2	Type1/Type3	Type1 to Type4	Type3	Type3/Type4
Type3	Type3	Type3	Type3	Type3
Type4	Type3	Type3/Type4	Type3	Type3/Type4

$\underline{R}Y = E_3$ and $\bar{R}Y = E_1 \cup E_2 \cup E_3 \cup E_4$. But $X \cap Y = \{e_4, e_9, e_{14}\}$. So that $\underline{R}(X \cap Y) = \phi$, $\bar{R}(X \cap Y) = E_1 \cup E_2 \cup E_3$ and hence, $X \cap Y$ is of Type 2.

Again, considering $X = \{e_1, e_2, ..., e_{10}, e_{14}, e_{19}\}$ and $Y = \{e_1, e_2, e_7, e_{14}, e_{20}\}$, both X and Y are of Type 3 as $\underline{R}X = E_1 \cup E_2$, $\bar{R}X = E_1 \cup E_2 \cup E_3 \cup E_4$, $\underline{R}Y = E_1$ and $\bar{R}Y = E_1 \cup E_2 \cup E_3 \cup E_4$. But $X \cap Y = \{e_1, e_2, , e_7, e_{14}\}$. So that $\underline{R}(X \cap Y) = E_1$ and $\bar{R}(X \cap Y) = E_1 \cup E_2 \cup E_3$. Hence, $X \cap Y$ is of Type 1.

Also, taking $X = \{e_1, e_2, ..., e_{10}, e_{14}, e_{19}\}$ and $Y = \{e_4, e_9, e_{14}, e_{16}, e_{17}, e_{18}, e_{19}, e_{20}\}$, both X and Y are of Type 3 as $\underline{R}X = E_1 \cup E_2$, $\bar{R}X = E_1 \cup E_2 \cup E_3 \cup E_4$, $\underline{R}Y = E_4$ and $\bar{R}Y = E_1 \cup E_2 \cup E_3 \cup E_4$. But $X \cap Y = \{e_4, e_9, e_{14}, e_{19}\}$. So that $\underline{R}(X \cap Y) = \phi$ and $\bar{R}(X \cap Y) = E_1 \cup E_2 \cup E_3 \cup E_4$. Hence, $X \cap Y$ is of Type 4.

Finally, taking $X = \{e_1, e_2, ..., e_{10}, e_{14}, e_{19}\}$ and $Y = \{e_1, e_6, ..., e_{10}, e_{11}, e_{16}, ..., e_{20}\}$, both X and Y are of Type 3 as $\underline{R}X = E_1 \cup E_2$, $\bar{R}X = E_1 \cup E_2 \cup E_3 \cup E_4$, $\underline{R}Y = E_2 \cup E_4$ and $\bar{R}X = E_1 \cup E_2 \cup E_3 \cup E_4$. But $X \cap Y = \{e_1, e_6, e_7, e_8, e_9, e_{10}, e_{19}\}$. So that $\underline{R}(X \cap Y) = E_2$ and $\bar{R}(X \cap Y) = E_1 \cup E_2 \cup E_3 \cup E_4$. Hence, $X \cap Y$ is of Type 3.

4.2 Union

In this sub-sect. we establish and present in a table, the results of union of two rough sets of different types. Like the cases of intersection, here also nine cases are unambiguous. The other ambiguous cases are mostly due to the inclusion (1). Applying Theorem 1 above, some of the ambiguities in the table can be reduced or removed under suitable conditions which are provided by the theorem. These conditions being of necessary and sufficient type, cannot be improved further.

Proofs

We shall denote the entry in i^{th} row and j^{th} column of the table by (i, j) to represent the different possible cases. In the proof, we shall be using (1) and the property that for any two rough sets X and Y

$$\bar{R}(X \cup Y) = \bar{R}X \cup \bar{R}Y \tag{25}$$

We shall provide the proof for the cases (1,2) and (2,2). The rest of the proofs can be worked out similarly.

Proof of (1,2)
Let X be of Type 1 and Y be of Type 2. Then $\underline{R}X \neq \phi$,$\bar{R}X \neq U$ and $\underline{R}Y = \phi$, $\bar{R}Y \neq U$. So, by (1) $\underline{R}(X \cup Y)$ is not ϕ. But, by (25), $\bar{R}(X \cup Y)$ may or may not be U. So, $X \cup Y$ can be of Type 1 or of Type 3.

Proof of (2,2)
Let X and Y be of Type 2. Then $\underline{R}X = \phi, \bar{R}X \neq U$ and $\underline{R}Y = \phi, \bar{R}Y \neq U$. By (1) $\underline{R}(X \cup Y)$ may or may not be ϕ and by (25) $\bar{R}(X \cup Y)$ may or may not be U. So, $X \cup Y$ can be of any of the four Types.

Examples

Below, we provide examples to show that all the possibilities in the ambiguous cases can actually arise for (2,2). The other cases can be justified similarly. We continue with the same Example of sect. 3.

Examples for (2,2)
Let $X = \{e_4, e_9, e_{14}\}$ and $Y = \{e_9, e_{14}, e_{19}\}$. Then both X and Y are of Type 2 as $\underline{R}X = \phi$, $\bar{R}X = E_1 \cup E_2 \cup E_3$, $\underline{R}Y = \phi$ and $\bar{R}Y = E_2 \cup E_3 \cup E_4$. But $X \cup Y = \{e_4, e_9, e_{14}, e_{19}\}$. So that $\underline{R}(X \cup Y) = \phi$, $\bar{R}(X \cup Y) = E_1 \cup E_2 \cup E_3 \cup E_4 = U$ and hence, $X \cup Y$ is of Type 4.

Again considering $X = \{e_4, e_9, e_{14}\}$ and $Y = \{e_4, e_9\}$, both X and Y are of Type 2 as $\underline{R}X = \phi$, $\bar{R}X = E_1 \cup E_2 \cup E_3$, $\underline{R}Y = \phi$ and $\bar{R}Y = E_1 \cup E_2$. But $X \cup Y = \{e_4, e_9, e_{14}\}$. So that $\underline{R}(X \cup Y) = \phi$ and $\bar{R}(X \cup Y) = E_1 \cup E_2 \cup E_3$. Hence, $X \cup Y$ is of Type 2.

Also, taking $X = \{e_4, e_9, e_{14}\}$ and $Y = \{e_6, e_7, e_8, e_{10}, e_{14}\}$, both X and Y are of Type 2 as $\underline{R}X = \phi$, $\bar{R}X = E_1 \cup E_2 \cup E_3$, $\underline{R}Y = \phi$ and $\bar{R}Y = E_2 \cup E_3$. But $X \cup Y = \{e_4, e_6, e_7, e_8, e_9, e_{10}, e_{14}\}$. So that $\underline{R}(X \cup Y) = E_2$ and $\bar{R}(X \cup Y) = E_1 \cup E_2 \cup E_3$. Hence, $X \cup Y$ is of Type 1.

Finally, taking $X = \{e_4, e_9, e_{14}\}$ and $Y = \{e_4, e_6, e_7, e_8, e_{10}, e_{19}\}$, both X and Y are of Type 2 as $\underline{R}X = \phi$, $\bar{R}X = E_1 \cup E_2 \cup E_3$, $\underline{R}Y = \phi$ and $\bar{R}X = E_1 \cup E_2 \cup E_4$. But $X \cup Y = \{e_4, e_6, e_7, e_8, e_9, e_{10}, e_{14}, e_{19}\}$. So that $\underline{R}(X \cup Y) = E_2$ and $\bar{R}(X \cup Y) = E_1 \cup E_2 \cup E_3 \cup E_4$. Hence, $X \cup Y$ is of Type 3.

4.3 Application of Theorems 1 and 2

As we have seen in sect. 3, there are a number of ambiguous entries in the union and intersection tables. However, if the conditions of corollaries 1 and 2 are satisfied, equalities hold in (1) and (2) and as a result the number of ambiguities decreases. This provides a much more convenient and improved situation. The conditions being of necessary and sufficient types cannot be improved further, under the circumstances.

Table for Intersection

As observed above, there were seven ambiguous cases in the table for intersection. Now, if hypotheses of Corollary 2 are satisfied with m = 2, then the number

Table 3. Intersection of different types of rough sets after applying Corollary 2

⋂	**Type 1**	**Type 2**	**Type 3**	**Type 4**
Type 1	Type 1/Type 2	Type 2	Type 1/Type 2	Type 2
Type 2	Type 2	Type 2	Type 2	Type 2
Type 3	Type 1/Type 2	Type 2	Type 3/Type 4	Type 4
Type 4	Type 2	Type 2	Type 4	Type 4

reduces to four. In the new table presented below, we find that there is no ambiguous entry having all four Types.

Table for Union

As in case of intersection, there were seven ambiguous cases in the union table also. Now, if the hypotheses of Corollary 1 are satisfied with m = 2, then the number reduces to four. As in case of intersection, there are no ambiguous entries in the improved table, which we present below.

Table 4. Union of different types of rough sets after applying Corollary 1 with $m = 2$

⋃	**Type 1**	**Type 2**	**Type 3**	**Type 4**
Type 1	Type 1/Type 3	Type 1/Type 3	Type 3	Type 3
Type 2	Type 1/Type 3	Type 2/Type 4	Type 3	Type 4
Type 3	Type 3	Type 3	Type 3	Type 3
Type 4	Type 3	Type 4	Type 3	Type 4

5 Rough Equivalence of Sets

A new concept of rough equivalence is to be introduced in this sect. As mentioned in the introduction, this concept captures approximate equality of sets at a higher level than rough equality. In parallel to rough equalities (bottom, top and total) we shall deal with three corresponding types of rough equivalences. Obviously, these concepts deal with topological structures of the lower and upper approximations of the sets. The rough equalities depend upon the elements of the approximation sets but on the contrary rough equivalences depend upon only the structure of the approximation sets. As shall be evident from the definitions, rough equalities (bottom, top and total) imply the corresponding rough equivalences (bottom, top and total) but the converse is not true. However, we shall see through a real life example that the new concepts are very much used by us to infer imprecise information.

5.1 Definitions

I. We say that two sets X and Y are *bottom R-equivalent* if and only if both $\underline{R}X$ and $\underline{R}Y$ are ϕ or not ϕ together (we write, X is b_eqv. to Y). We put the restriction here that for bottom R-equivalence of X and Y either both $\underline{R}X$ and $\underline{R}Y$ are equal to U or none of them is equal to U.

II. We say that two sets X and Y are *top R-equivalent* if and only if both $\bar{R}X$ and $\bar{R}Y$ are U or not U together (we write, X is t_eqv. to Y). We put the restriction here that for top R-equivalence of X and Y either both $\bar{R}X$ and $\bar{R}Y$ are equal to ϕ or none of them is equal to ϕ.

III. We say that two sets X and Y are *R-equivalent* if and only if X and Y are bottom R-equivalent and top R-equivalent (we write, X is eqv. to Y). We would like to note here that when two sets X and Y are R-equivalent, the restrictions in **I** and **II** become redundant.

For example, in case **I**, if one of the $\underline{R}X$ and $\underline{R}Y$ are equal to U then the corresponding upper approximation must be U and for rough equivalence it is necessary that the other upper approximation must also be U. Similarly, the other case.

5.2 Elementary Properties

I. It is clear from the definition above that in all cases (bottom,top,total) R-equality implies R-equivalence.

II. Obviously, the converses are not true.

III. Bottom R-equivalence, top R-equivalence and R-equivalence are equivalence relations on P(U).

IV. The concept of approximate equality of sets refers to the topological structure of compared sets but not to the elements they consist of.

If two sets are roughly equivalent then by using our present knowledge, we may not be able to say whether two sets are approximately equal as described above, but, we can say that they are approximately equivalent. That is both the sets have or not have positive elements with respect to R and both the sets have or not have negative elements with respect to R.

5.3 Example 2

Let us consider all the cattle in a locality as our universal set U. We define a relation R over U by xRy if and only if x and y are cattle of the same kind. Suppose for example, this equivalence relation decomposes the universe into disjoint equivalence classes as given below.

$C = \{\text{Cow, Buffalo, Goat, Sheep, Bullock}\}$.

Let P_1 and P_2 be two persons in the locality having their set of cattle represented by X and Y.

We cannot talk about the equality of X and Y in the usual sense as the cattle can not be owned by two different people.

Similarly we can not talk about the rough equality of X and Y except the trivial case when both the persons do not own any cattle.

We find that rough equivalence is a better concept which can be used to decide the equality of the sets X and Y in a very approximate and real sense.

There are four different cases in which we can talk about equivalence of P_1 and P_2.

Case I. $\bar{R}X, \bar{R}Y$ are not U and $\underline{R}X, \underline{R}Y$ are ϕ. That is P_1 and P_2 both have some kind of cattle but do not have all cattle of any kind in the locality. So, they are equivalent.

Case II. $\bar{R}X, \bar{R}Y$ are not U and $\underline{R}X, \underline{R}Y$ are not ϕ. That is P_1 and P_2 both have some kind of cattle and also have all cattle of some kind in the locality. So, they are equivalent.

Case III. $\bar{R}X, \bar{R}Y$ are U and $\underline{R}X, \underline{R}Y$ are ϕ. That is P_1 and P_2 both have all kinds of cattle but do not have all cattle of any kind in the locality. So, they are equivalent.

Case IV. $\bar{R}X, \bar{R}Y$ are U and $\underline{R}X, \underline{R}Y$ are not ϕ. That is P_1 and P_2 both have all kinds of cattle and also have all cattle of some kind in the locality. So, they are equivalent.

There are two different cases under which we can talk about the non - equivalence of P_1 and P_2.

Case V. One of $\bar{R}X$ and $\bar{R}Y$ is U and the other one is not. Then, out of P_1 and P_2 one has cattle of all kinds and other one dose not have so. So, they are not equivalent. Here the structures of $\underline{R}X$ and $\underline{R}Y$ are unimportant.

Case VI. Out of $\underline{R}X$ and $\underline{R}Y$ one is ϕ and other one is not. Then, one of P_1 and P_2 does not have all cattle of any kind, whereas the other one has all cattle of some kind. So, they are not equivalent. Here the structures of $\bar{R}X$ and $\bar{R}Y$ are unimportant.

It may be noted that we have put the restriction for top rough equivalence that in the case when $\bar{R}X$ and $\bar{R}Y$ are not equal to U, it should be the case that both are ϕ or not ϕ together. It will remove the cases when one set is ϕ and the other has elements from all but one of the equivalence classes but does not have all the elements of any class completely being rough equivalent. Taking the example into consideration it removes cases like when a person has no cattle being rough equivalent to a person, who has some cattle of every kind except one.

Similarly, for bottom rough equivalence we have put the restriction that when $\underline{R}X$ and $\underline{R}Y$ are not equal to ϕ, it should be the case that both are U or not U together.

5.4 General Properties

In this sect. we establish some properties of rough equivalences of sets, which are parallel to those stated in sect. 4.2. Some of these properties hold, some are partially true and some do not hold at all. For those properties, which do hold partially or do not hold at all, we shall provide some sufficient conditions for the conclusion to be true. Also, we shall verify the necessity of such conditions. The sufficient conditions depend upon the concepts of different rough inclusions (Pawlak [30], p.27) and rough comparabilities which we introduce below.

Definition 2. Let $K = (U, \boldsymbol{R})$ be a knowledge base, $X, Y \subseteq U$ and $R \in IND(K)$. Then

(i) We say that X is *bottom R-included* in $Y (X \sqsubseteq_{BR} Y)$ if and only if $\underline{R}X \subseteq \underline{R}Y$.

(ii) We say that X is *top R-included* in $Y (X \sqsubseteq_{TR} Y)$ if and only if $\bar{R}X \subseteq \bar{R}Y$.

(iii) We say that X is *R-included* in $Y (X \sqsubseteq_R Y)$ if and only if $X \sqsubseteq_{BR} Y$ and $X \sqsubseteq_{TR} Y$.

We shall drop the suffixes R from the notations above in their use of make them simpler.

Definition 3

(i) We say X, $Y \subseteq U$ are *bottom rough comparable* if and only if $X \sqsubseteq_B Y$ or $Y \sqsubseteq_B X$ holds.

(ii) We say X, $Y \subseteq U$ are *top rough comparable* if and only if $X \sqsubseteq_T Y$ or $Y \sqsubseteq_T X$ holds.

(iii) We say X, $Y \subseteq U$ are *rough comparable* if and only if X and Y are both top rough comparable and bottom rough comparable.

Property 1

(i) If $X \cap Y$ is *b_eqv* to X and $X \cap Y$ is *b_eqv* to Y then X is *b_eqv* to Y.

(ii) The converse of (i) is not necessarily true.

(iii) The converse is true if in addition X and Y are bottom rough comparable.

(iv) The condition in (iii) is not necessary.

Proof

(i) Since $\underline{R}(X \cap Y)$ and $\underline{R}X$ are ϕ or not ϕ together and $\underline{R}(X \cap Y)$ and $\underline{R}Y$ are ϕ or not ϕ together, $\underline{R}(X \cap Y)$ being common we get that $\underline{R}X$ and $\underline{R}Y$ are ϕ or not ϕ together. Hence X is bottom equivalent to Y.

(ii) The cases when $\underline{R}X$ and $\underline{R}Y$ are both not ϕ but $\underline{R}(X \cap Y) = \phi$ the converse is not true.

(iii) We have $\underline{R}(X \cap Y) = \underline{R}X \cap \underline{R}Y = \underline{R}X$ or $\underline{R}Y$, as the case may be. Since X and Y are bottom rough comparable.

So, $X \cap Y$ is *b_eqv* to X and $X \cap Y$ is *b_eq* to Y.

(iv) We provide an example to show that this condition is not necessary. Let us take $U = \{x_1, x_2, .., x_8\}$ and the partition induced by an equivalence relation R be $\{\{x_1, x_2\}, \{x_3, x_4\}, \{x_5, x_6\}, \{x_7, x_8\}\}$.

Now, for $X = \{x_1, x_2, x_3, x_4\}$ and $Y = \{x_3, x_4, x_5, x_6\}$, we have $\underline{R}X = X \neq \phi, \underline{R}Y = Y \neq \phi, X \cap Y = \{x_3, x_4\}$ and $\underline{R}(X \cap Y) = \{x_3, x_4\} \neq \phi$. So, $X \cap Y$ is *b_eqv* to both X and Y. But X and Y are not bottom rough comparable.

Property 2

(i) If $X \cup Y$ is *t_eqv* to X and $X \cup Y$ is *t_eqv* to Y then X is *t_eqv* to Y.

(ii) The converse of (i) may not be true.

(iii)A sufficient condition for the converse of (i) to be true is that X and Y are top rough comparable.
(iv) The condition in (iii) is not necessary.

Proof

(i) Similar to part(i) of property 1.

(ii) The cases when $\bar{R}(X) \neq U$ and $\bar{R}(Y) \neq U$ but $\bar{R}(X \cup Y) = U$, the converse is not true.

(iii) Similar to part(iii) of property 1.

(iv) We take the same example as above to show that this condition is not necessary. Here, we have $\bar{R}X = X \neq U, \bar{R}Y = Y \neq U, \bar{R}(X \cup Y) = \{x_1, x_2, x_3, x_4, x_5, x_6\} \neq U$. So, X is *t_eqv* to Y.Also, $X \cup Y$ is *t_eqv* to both X and Y. But X and Y are not top rough comparable.

Property 3

(i) If X is *t_eqv* to X' and Y is *t_eqv* to Y' then it may or may not be true that $X \cup Y$ is *t_eqv* to $X' \cup Y'$.

(ii) A sufficient condition for the result in (i) to be true is that X and Y are top rough comparable and X' and Y' are top rough comparable.

(iii) The condition in (ii) is not necessary for result in (i) to be true.

Proof

(i) The result fails to be true when all of $\bar{R}(X)$, $\bar{R}(X')$, $\bar{R}(Y)$ and $\bar{R}(Y')$ are not U and exactly one of $X \cup Y$ and $X' \cup Y'$ is U.

(ii) We have $\bar{R}(X) \neq U$, $\bar{R}(X') \neq U$, $\bar{R}(Y) \neq U$ and $\bar{R}(Y') \neq U$. So, under the hypothesis,$\bar{R}(X \cup Y) = \bar{R}X \cup \bar{R}Y = \bar{R}(X)$ or $\bar{R}(Y)$, which is not equal to U. Similarly, $\bar{R}(X' \cup Y') \neq U$. Hence, $X \cup Y$ is *t_eqv* to $X' \cup Y'$.

(iii) Continuing with the same example, taking $X = \{x_1, x_2, x_3\}$, $X' = \{x_1, x_2, x_4\}$, $Y = \{x_4, x_5, x_6\}$ and $Y' = \{x_3, x_5, x_6\}$, we find that $\bar{R}X = \{x_1, x_2, x_3, x_4\} = \bar{R}X' \neq U$ and $\bar{R}Y = \{x_3, x_4, x_5, x_6\} = \bar{R}Y' \neq U$. So, X and Y are not top rough comparable. X' and Y' are not top rough comparable. But, $\bar{R}(X \cup Y) = \{x_1, x_2, x_3, x_4, x_5, x_6\} = \bar{R}(X' \cup Y')$. So, $X \cup Y$ is top equivalent to $X' \cup Y'$.

Property 4

(i) X is *b_eqv* to X' and Y is *b_eqv* to Y' may or may not imply that $X \cap Y$ is *b_eqv* to $X' \cap Y'$.

(ii) A sufficient condition for the result in (i) to be true is that X and Y are bottom rough comparable and X' and Y' are bottom rough comparable.

(iii) The condition in (ii) is not necessary for result in (i) to be true.

Proof

(i) When all of $\underline{R}(X)$, $\underline{R}(X')$, $\underline{R}(Y)$ and $\underline{R}(Y')$ are not ϕ and exactly one of the $X \cap Y$ and $X' \cap Y'$ is ϕ, the result fails.

(ii) Now, under the hypothesis, we have $\underline{R}(X \cap Y) = \underline{R}(X) \cap \underline{R}(Y) = \underline{R}(X)$ or $\underline{R}(Y) \neq \phi$, Similarly, $\underline{R}(X' \cap Y') \neq \phi$. So, $X \cap Y$ is *b_eq* to $X' \cap Y'$.

(iii) Continuing with the same example and taking $X = \{x_1, x_2, x_3\}$, $X' = \{x_3, x_4, x_5\}$, $Y = \{x_3, x_7, x_8\}$ and $Y' = \{x_5, x_7, x_8\}$, we find that $\underline{R}X \neq \phi$, $\underline{R}X' \neq \phi$, $\underline{R}Y \neq \phi$ and $\underline{R}Y' \neq \phi$. So, X is *b_eqv* to X' and Y is *b_eqv* to Y'. Also, $\underline{R}(X \cap Y) = \phi$ and $\underline{R}(X' \cap Y') = \phi$. So, $X \cap Y$ is *b_eqv* to $X' \cap Y'$. However, X and Y are not bottom rough comparable and so are X' and Y'.

Property 5
(i) X is *t_eqv* to Y may or may not imply that $X \cup (-Y)$ is *t_eqv* to U.
(ii) A sufficient condition for result in (i) to hold is that $X {=}_B Y$.
(iii) The condition in (ii) is not necessary for the result in (i) to hold.

Proof
(i) The result fails to hold true when $\bar{R}(X) \neq U$, $\bar{R}(Y) \neq U$ and still $\bar{R}(X \cup (-Y)) = U$.
(ii) As $X {=}_B Y$, we have $\underline{R}X = \underline{R}Y$. So, $-\underline{R}X = -\underline{R}Y$. Equivalently, $\bar{R}(-X) = \bar{R}(-Y)$. Now, $\bar{R}(X \cup -Y) = \bar{R}(X) \cup \bar{R}(-Y) = \bar{R}(X) \cup \bar{R}(-X) = \bar{R}(X \cup -X) = \bar{R}(U) = U$. So, $X \cup -Y$ is *t_eqv* to U.
(iii) Continuing with the same example and taking $X = \{x_1, x_2, x_3\}$, $Y = \{x_2, x_3, x_4\}$ we get $-Y = \{x_1, x_5, x_6, x_7, x_8\}$. So that $\underline{R}X = \{x_1, x_2\}$ and $\underline{R}Y = \{x_3, x_4\}$. Hence, it is not true that $X {=}_B Y$. But, $X \cup -Y = \{x_1, x_2, x_3, x_5, x_6,$
$x_7, x_8\}$. So, $\bar{R}(X \cup -Y) = U$. That is, $X \cup -Y$*t_eqv* to U.

Property 6
(i) X is *b_eqv* to Y may or may not imply that $X \cap (-Y)$ is *b_eqv* to ϕ.
(ii) A sufficient condition for the result in (i) to hold true is that $X {=}_T Y$.
(iii) The condition in (ii) is not necessary for the result in (i) to hold true.

Proof
(i) The result fails to hold true when $\underline{R}(X) \neq \phi$, $\underline{R}(Y) \neq \phi$ and $\underline{R}(X) \cap \underline{R}(-Y) = \phi$.
(ii) As $X {=}_T Y$, we have $\bar{R}X = \bar{R}Y$. So, $-\bar{R}X = -\bar{R}Y$. Equivalently, $\underline{R}(-X) = \underline{R}(-Y)$. Now, $\underline{R}(X \cap -Y) = \underline{R}(X) \cap \underline{R}(-Y) = \underline{R}(X) \cap \underline{R}(-X) = \underline{R}(X \cap -X) = \underline{R}(\phi) = \phi$. Hence, $X \cap -Y$ is *b_eqv* to ϕ.
(iii) Continuing with the same example by taking $X = \{x_1, x_2, x_3\}$, $Y = \{x_1, x_2, x_5\}$ we have $-Y = \{x_3, x_4, x_6, x_7, x_8\}$. So, X is *b_eqv* to Y. But X is not top equal to Y. However, $X \cap -Y = \{x_3\}$ and so, $\underline{R}(X \cap -Y) = \phi$. Hence, $X \cap -Y$ is *b_eqv* to ϕ.

Property 7. If $X \subseteq Y$ and Y is *b_eqv* to ϕ then X is *b_eqv* to ϕ.

Proof. As Y is *b_eqv* to ϕ, we have $\underline{R}(Y) = \phi$. So, if $X \subseteq Y$, $\underline{R}(X) \subseteq \underline{R}(Y) = \phi$.

Property 8. If $X \subseteq Y$ and X is *t_eqv* to U then Y is *t_eqv* to U.

Proof. The proof is similar to that of Property 7.

Property 9. X is *t_eqv* to Y if and only if $-X$ is *b_eqv* to $-Y$.

Proof. The proof follows from the property, $\underline{R}(-X) = -\bar{R}(X)$.

Property 10. X is *b_eqv* to ϕ, Y is *b_eqv* to $\phi \Rightarrow X \cap Y$ is *b_eqv* to ϕ.

Proof. The proof follows directly from the fact that under the hypothesis the only possibility is $\underline{R}(X) = \underline{R}(Y) = \phi$.

Property 11. If X is *t_eqv* to U or Y is *t_eqv* to U then $X \cup Y$ is *t_eqv* to U.

Proof. The proof follows directly from the fact that under the hypothesis the only possibility is $\bar{R}(X) = \bar{R}(Y) = U$.

5.5 Replacement Properties

In this sect. we shall consider properties obtained from the properties of sect. 5.4 by interchanging top and bottom rough equivalences. We shall provide proofs whenever these properties hold true. Otherwise, sufficient conditions are to be established under which these properties are valid. In addition, we shall test if such conditions are also necessary for the validity of the properties. Invariably, it has been found that such conditions are not necessary. We shall show it by providing suitable examples.

Property 12

(i) If $X \cap Y$ is *t_eqv* to X and $X \cap Y$ is *t_eqv* to Y then X is *t_eqv* Y.

(ii) The converse of (i) is not necessarily true.

(iii) A sufficient condition for the converse of (i) to hold true is that conditions of Corollary 2 hold with m = 2.

(iv) The condition in (iii) is not necessary.

Proof

(i) Here $\bar{R}X$ and $\bar{R}(X \cap Y)$ are U or not U together and $\bar{R}Y$ and $\bar{R}(X \cap Y)$ are U or not U together being common, we get $\bar{R}X$ and $\bar{R}(Y)$ are U or not U together. So, X is *t_eqv*Y.

(ii) The result fails when $\bar{R}X$ and $\bar{R}(X) = U\bar{R}(Y)$ and $\bar{R}(X \cap Y) \neq U$.

(iii) Under the hypothesis, we have $\bar{R}(X \cap Y) = \bar{R}(X) \cap \bar{R}(Y)$. If X is *t_eqv* to Y then both $\bar{R}X$ and $\bar{R}Y$ are equal to U or not equal to U together. So, accordingly we get $\bar{R}(X \cap Y)$ equal to U or not equal to U. Hence the conclusion follows.

(iv) We see that the sufficient condition for the equality to hold when m = 2 in Corollary 2 is that there is no E_j such that $X \cap E_j \neq \phi$, $Y \cap E_j \neq \phi$ and $X \cap Y \cap E_j = \phi$.

Let us take U and the relation as above. Now, taking $X = \{x_1, x_3, x_6\}$, $Y = \{x_3, x_5, x_6\}$. The above sufficiency conditions are not satisfied as $\{x_5, x_6\} \cap$

$X \neq \phi$, $\{x_5, x_6\} \cap Y \neq \phi$ and $\{x_5, x_6\} \cap X \cap Y = \phi$. However, $\bar{R}X = \{x_1, x_2, x_3, x_4, x_5, x_6\} \neq U$.

Property 13

(i) $X \cup Y$ is *b_eqv* to X and $X \cup Y$ is *b_eqv* to Y then X is *b_eqv* to Y.

(ii) The converse of (i) is not necessarily true.

(iii) A sufficient condition for the converse of (i) to hold true is that the condition of Corollary 1 holds for m = 2.

(iv) The condition in (iii) is not necessary.

Proof

(i) $\underline{R}X$ and $\underline{R}(X \cup Y)$ are ϕ or not ϕ together and $\underline{R}Y$ and $\underline{R}(X \cup Y)$ are ϕ or not ϕ together. Since $\underline{R}(X \cup Y)$ is common, $\underline{R}X$ and $\underline{R}Y$ are ϕ or not ϕ together. So, X is *b_eqv* to Y.

(ii) Suppose X and Y are such that $\underline{R}X$ and $\underline{R}Y$ are both ϕ but $\underline{R}(X \cup Y) \neq \phi$. Then X is *b_eqv* to Y but $X \cup Y$ is not *b_eqv* to any one of X and Y.

(iii) Suppose X is *b_eqv* to Y. Then $\underline{R}X$ and $\underline{R}Y$ are ϕ or not ϕ together. If the conditions are satisfied then $\underline{R}(X \cup Y) = \underline{R}X \cup \underline{R}Y$. So, if both $\underline{R}X$ and $\underline{R}Y$ are ϕ or not ϕ together then $\underline{R}(X \cup Y)$ is ϕ or not ϕ accordingly and the conclusion holds.

(iv) Let us take U as above. The classification corresponding to the equivalence relation be given by $\{\{x_1, x_2\}, \{x_3, x_4, x_5\}, \{x_6\}, \{x_7, x_8\}\}$.

Let $X = \{x_1, x_3, x_6\}$, $Y = \{x_2, x_5, x_6\}$. Then $\underline{R}(X) \neq \phi$, $\underline{R}(Y) \neq \phi$ and $\underline{R}(X \cup Y) \neq \phi$. The condition in (iii) is not satisfied as taking $E = \{x_1, x_2\}$ we see that $X \cap E \subset E$, $Y \cap E \subset E$ and $X \cup Y \supseteq E$.

Property 14

(i) X is *b_eqv* to X' and Y is *b_eqv* to Y' may not imply $X \cup Y$ is *b_eqv* to $X' \cup Y'$.

(ii) A sufficient condition for the conclusion of (i) to hold is that the conditions of corollary 2 are satisfied for both X, Y and X', Y' separately with m = 2.

(iii) The condition in (ii) is not necessary for the conclusion in (i) to be true

Proof

(i) When $\underline{R}X$, $\underline{R}Y$,$\underline{R}X'$, $\underline{R}Y'$ are all ϕ and out of $X \cup Y$ and $X' \cup Y'$ one is ϕ but the other one is not ϕ, the result fails to be true.

(ii) Under the additional hypothesis, we have $\underline{R}(X \cup Y) = \underline{R}X \cup \underline{R}Y$ and $\underline{R}(X' \cup Y') = \underline{R}X' \cup \underline{R}Y'$. Here both $\underline{R}X$ and $\underline{R}X'$ are ϕ or not ϕ together and both $\underline{R}Y$ and $\underline{R}Y'$ are ϕ or not ϕ together. If all are ϕ then both $\underline{R}(X \cup Y)$ and $\underline{R}(X' \cup Y')$ are ϕ. So, they are *b_eqv*. On the other hand, if at least one pair is not ϕ then we get both $\underline{R}(X \cup Y)$ and $\underline{R}(X' \cup Y')$ are not ϕ and so they are *b_eqv*.

(iii) The condition is not satisfied means there is E_i with $X \cap E_i \subset E_i$, $Y \cap E_i \subset E_i$ and $X \cup Y \supseteq E_i$; there exists E_j (not necessarily different from E_i) such that $X' \cap E_j \subset E_j$, $Y' \cap E_j \subset E_j$ and $X' \cup Y' \supseteq E_j$.

Let us consider the example, $U = x_1, x_2, ..., x_8$ and the partition induced by an equivalence relation R be $\{\{x_1, x_2\}, \{x_3, x_4\}, \{x_5, x_6\}\{x_7, x_8\}\}$. $X = \{x_1, x_5\}$,

$Y = \{x_3, x_6\}$, $X' = \{x_1, x_4\}$ and $Y' = \{x_3, x_7\}$. Then $\underline{R}X = \underline{R}X' = \underline{R}Y = \underline{R}Y' = \phi$. Also, $\underline{R}(X \cup Y) \neq \phi$, $\underline{R}(X' \cup Y') \neq \phi$. So, X is *b_eqv* to X', Y is *b_eqv* to Y' and $X \cup Y$ is *b_eqv* to $X' \cup Y'$. However, $X' \cap \{x_3, x_4\} \subset \{x_3, x_4\}$, $Y' \cap \{x_3, x_4\} \subset \{x_3, x_4\}$ and $X' \cup Y' \supseteq \{x_3, x_4\}$. So, the condition are not satisfied.

Property 15

(i) X is *t_eqv* to X' and Y is *t_eqv* to Y' may not necessarily imply that $X \cap Y$ is *t_eqv* to $X' \cap Y'$.

(ii) A sufficient condition for the conclusion in (i) to hold is the conditions of corollary 1 are satisfied for both X, Y and X', Y' separately with m = 2.

(iii) The condition in (ii) is not necessary for the conclusion in (i) to hold.

Proof

(i) When $\bar{R}X = \bar{R}X' = \bar{R}Y = \bar{R}Y' = U$ and out of $\bar{R}(X \cap Y)$, $\bar{R}(X' \cap Y')$ one is U whereas the other one is not U the result fails to be true.

(ii) If the conditions of corollary 1 are satisfied for X, Y and X', Y' separately then the case when $\bar{R}X = \bar{R}X' = \bar{R}Y = \bar{R}Y' = U$, we have $\bar{R}(X' \cap Y') = \bar{R}X' \cap \bar{R}Y' = U$ and $\bar{R}(X \cap Y) = \bar{R}X \cap \bar{R}Y = U$. In other cases, if $\bar{R}X$ and $\bar{R}X'$ not U or $\bar{R}Y$ and $\bar{R}Y'$ not U then as $\bar{R}(X' \cap Y') \neq U$ and $\bar{R}(X \cap Y) \neq U$. So, in any case $X \cap Y$ and $X' \cap Y'$ are *t_eqv* to each other.

(iii) We continue with the same example. The conditions are not satisfied means there is no E_j such that $X \cap E_j \neq \phi$, $Y \cap E_j \neq \phi$ and $X \cap Y \cap E_j = \phi$ or $X' \cap E_j \neq \phi$, $Y' \cap E_j \neq \phi$ and $X' \cap Y' \cap E_j = \phi$. Taking $X = \{x_1, x_5\}$, $Y = \{x_3, x_5\}, X' = \{x_1, x_4\}$ and $Y' = \{x_2, x_4\}$ we have $X \cap \{x_5, x_6\} \neq \phi$, $Y' \cap \{x_5, x_6\} \neq \phi$ and $X \cap Y \cap \{x_5, x_6\} = \phi$. $X' \cap \{x_3, x_4\}, Y' \cap \{x_3, x_4\} \neq \phi$ and $X' \cap Y' \cap \{x_3, x_4\}$. So, the conditions are violated. But $\bar{R}X \neq U$, $\bar{R}X' \neq U$, $\bar{R}Y \neq U$, $\bar{R}Y' \neq U$. So, X is *t_eqv* and Y is *t_eqv* Y'. Also, $\bar{R}(X \cap Y) \neq U$ and $\bar{R}(X' \cap Y') \neq U$. Hence, $X \cap Y$ is *t_eqv* to $X' \cap Y'$.

Remark

We would like to make the following comments in connection with the properties 16 to 19, 21 and 22:

(i) We know that $\underline{R}U = U$. So, bottom R-equivalent to U can be considered under the case that $\underline{R}U \neq \phi$.

(ii) We know that $\bar{R}\phi = \phi$. So, top R-equivalent to ϕ can be considered under the case that $\bar{R}\phi \neq U$.

The proofs of the properties 16, 17, 18 and 19 are trivial and we omit them.

Property 16. X is *b_eqv* to Y may or may not imply that $X \cup -Y$ is *b_eqv* to U.

Property 17. X is *t_eqv* to Y may or may not imply that $X \cap -Y$ is *t_eqv* to ϕ.

Property 18. If $X \subseteq Y$ and Y is *t_eqv* to ϕ then X is *t_eqv* to ϕ.

Property 19. If $X \subseteq Y$ and X is *b_eqv* to U then Y is *b_eqv* to U.

Property 20. X is *b_eqv* to Y if and only if $-X$ is *t_eqv* to $-Y$.

Proof. Follows from the identity $\bar{R}(-X) = -\bar{R}(X)$.

The proofs of the following two properties are also trivial.

Property 21. X is *t_eqv* to ϕ and Y is *t_eqv* to $\phi \Rightarrow X \cap Y$ is *t_eqv* to ϕ.

Property 22. X is *b_eqv* to U and Y is *b_eqv* to $U \Rightarrow X \cup Y$ is *b_eqv* to U.

6 Approximation of Classifications

Approximation of classifications is a simple extension of the definition of approximation of sets. Let $F = \{X_1, X_2, ..., X_n\}$ be a family of non empty sets, which is a classification of U in the sense that $X_i \cap X_j = \phi$ for $i \neq j$ and

$$\bigcup_{i=1}^{n} X_i = U.$$

Then $\underline{R}F = \{\underline{R}X_1, \underline{R}X_2, ..., \underline{R}X_n\}$ and $\bar{R}F = \{\bar{R}X_1, \bar{R}X_2, ..., \bar{R}X_n\}$ are called the R-*lower* and R-*upper approximations* of the family F, respectively.

Grzymala-Busse [12] has established some properties of approximation of classifications. But, these results are irreversible in nature. Pawlak [30, p.24] has remarked that these results of Busse establish that the two concepts, approximation of sets and approximation of families of sets (or classifications) are two different issues and the equivalence classes of approximate classifications cannot be arbitrary sets. He has further stated that if we have positive example of each category in the *approximate classification* then we must have also negative examples of each category. In this sect., we further analyze these aspects of theorems of Busse and provide physical interpretation of each one of them by taking a standard example.

One primary objective is to extend the results of Busse by obtaining necessary and sufficient type theorems and show how the results of Busse can be derived from them. The results of Busse we discuss here are in their slightly modified form as presented by Pawlak [30]. Some more work in dealing with incomplete data are due to Busse [13,14].

6.1 Theorems on Approximation of Classifications

In this sect., we shall establish two theorems which have many corollaries generalizing the four theorems established by Busse [12] in their modified forms [30]. We shall also provide interpretations for most of these results including those of Busse and illustrate them through a simple example of toys [30].

Example 3. Suppose we have a set of toys of different colours red, blue, yellow and different shapes square, circular, triangular. We define the first description as a classification of the set of toys and represent the second description as an equivalence relation R. We say for two toys x and y, xRy if x and y are of the same shape.

We shall use the following notations for representational convenience :

$N_n = \{1, 2, ..., n\}$ and for any $I \subset N_n$, by I^c we mean the complement of I in N_n.

Theorem 3. Let $F = \{X_1, X_2, ..., X_n\}$ be a classification of U and let R be an equivalence relation on U. Then for any $I \subset N_n$,

$$\bar{R}(\bigcup_{i \in I}(X_i)) = U \textit{ if and only if } \underline{R}(\bigcup_{j \in I^c}(X_j)) = \phi.$$

Proof. We have

$$\underline{R}(\bigcup_{j \in I^c}(X_j)) = \phi \Leftrightarrow \underline{R}(U - \bigcup_{i \in I^c}(X_i)) = \phi \Leftrightarrow -\bar{R}\bigcup_{i \in I}(X_i) = \phi \Leftrightarrow \bar{R}(\bigcup_{i \in I}(X_i)) = U.$$

This completes the proof.

Corollary 3. Let $F = \{X_1, X_2, ..., X_n\}$ be a classification of U and let R be an equivalence relation on U. Then for $I \subset N_n$,

$$\textit{if } \bar{R}(\bigcup_{i \in I}(X_i)) = U \textit{ then } \underline{R}X_j = \phi \textit{ for each } j \in I^c.$$

Proof. By the above theorem, using the hypothesis we get

$$\underline{R}(\bigcup_{j \in I^c}(X_j)) = \phi.$$

As

$$\underline{R}X_j \subseteq \underline{R}(\bigcup_{i \in I^c}(X_j))$$

for each $j \in I^c$, the conclusion follows.

Interpretation

Suppose, in a classification of a universe, there is no negative element for the union of some elements of the classification taken together with respect to an equivalence relation. Then for all other elements of the classification there is no positive element with respect to the equivalence relation. Referring to the example, if we have circular or triangular toys of all different colours then all the toys of no particular colour are rectangular in shape.

Corollary 4. Let $F = \{X_1, X_2, ..., X_n\}$ be a classification of U and R be an equivalence relation on it. Then for each $i \in N_n$, $\bar{R}X_i = U$ if and only if

$$\underline{R}(\bigcup_{j \neq i}(X_j) = \phi.$$

Proof. Taking $I = \{i\}$, in Theorem 3 we get this.

Corollary 5. Let $F = \{X_1, X_2, ..., X_n\}$ be a classification of U and let R be an equivalence relation on U. Then for each $i \in N_n$, $\underline{R}X_i = \phi$ if and only if

$$\bar{R}(\bigcup_{j \neq i} X_j) = U.$$

Proof. Taking $I = \{i\}^c$, in Theorem 3 we get this.

Corollary 6. [30, Proposition 2.6] Let $F = \{X_1, X_2, ..., X_n\}$ be a classification of U and let R be an equivalence relation on U. If there exists $i \in N_n$ such that $\bar{R}X_i = U$ then for each j other than i in N_n, then $\underline{R}X_j = \phi$.

Proof. From Corollary 4., $\bar{R}X_i = U$

$$\Rightarrow \underline{R}(\bigcup_{j \neq i} X_j) = \phi.$$

$$\Rightarrow \underline{R}X_j = \phi \text{ for each } j \neq i.$$

Interpretation
Suppose in a classification of a universe, there are positive elements of one member of the classification with respect to a equivalence relation. Then there are negative elements of all other members of the classification with respect to the equivalence relation.

Taking the above example into consideration if all red toys are of triangular shape (say) then for toys of circular and rectangular shape at least one colour is absent.

Corollary 7. [30, proposition 2.8] Let $F = \{X_1, X_2, ..., X_n\}$ be a classification of U and let R be an equivalence relation on it. If for all $i \in N_n$, $\bar{R}X_i = U$ holds then $\underline{R}X_i = \phi$ for all $i \in N_n$.

Proof. If for some i, $1 \leq i \leq n$, $\underline{R}X_i \neq \phi$, then by Corollary 6 $\bar{R}X_j \neq U$ for some $j(\neq i)$ in N_n ; which is a contradiction.

This completes the proof.

Interpretation
Suppose in a classification of a universe, there is no negative element of one member of the classification with respect to an equivalence relation. Then for all other members of the classifications there is no positive element with respect to the equivalence relation.

Referring to the example, if there are triangular toys of all different colours then for any other shape (circular or rectangular) all the toys of no particular colour are of that shape.

Theorem 4. Let $F = \{X_1, X_2, ..., X_n\}$ be a partition of U and R be an equivalence relation on U. Then for any $I \subset N_n$,

$$\underline{R}(\bigcup_{i \in I} X_i) \neq \phi \quad if\ and\ only\ if \bigcup_{j \in I^C} \bar{R}(X_j) \neq U.$$

Proof. *(Sufficiency)* By property of lower approximation,

$$\bar{R}(\bigcup_{j \in I^c} X_j) = (\bigcup_{j \in I^c} \bar{R}X_j) \neq U.$$

So, there exists $[x]_R$ for some $x \in U$ such that

$$[x]_R \cap (\bigcup_{j \in I^c} X_j) = \phi.$$

Hence,

$$\underline{R}(\bigcup_{i \in I} X_i) \neq \phi.$$

(Necessity) Suppose,

$$\underline{R}(\bigcup_{i \in I} X_i) \neq \phi.$$

Then there exists $x \in U$ such that

$$[x]_R \subseteq (\bigcup_{i \in I} X_i).$$

Thus, $[x]_R \cap X_j = \phi$ for $j \notin I$. So, $x \notin \bar{R}X_j$, for $j \notin I$. Hence

$$(\bigcup_{j \in I^c} \bar{R}X_j) \neq U.$$

Corollary 8. Let $F = \{X_1, X_2, ..., X_n$ be a classification of U and let R be an equivalence relation on U. Then for $I \subset N_n$,

$$if\ \underline{R}(\bigcup_{i \in I} X_i) \neq \phi \ \ then\ \bar{R}X_j \neq U\ for\ each\ j \in I^c.$$

Proof. By Theorem 4,

$$\underline{R}(\bigcup_{i \in I} X_i) \neq \phi$$

$$\Rightarrow (\bigcup_{j \in I^c} \bar{R}X_j) \neq U$$

$\Rightarrow \bar{R}X_j \neq U$ for each $j \in I^c$.

This completes the proof.

Interpretation
Suppose in a classification of a universe, there are positive elements for the union of some elements of the classification taken together with respect to an equivalence relation. Then for all other elements of the classification there are negative elements with respect to the equivalence relation. Referring to the same example, if all toys of red colour are rectangular or triangular in shape then circular toys of at least one colour is absent.

Corollary 9. Let $F = \{X_1, X_2, ..., X_n\}$ be a partition of U and R be an equivalence relation on U. Then for each $i \in N_n$,

$$\underline{R}X_i \neq \phi \text{ if and only if } (\bigcup_{j \neq i} \bar{R}X_j) \neq U.$$

Proof. Taking $I = \{i\}$ in Theorem 4 we get this.

Corollary 10. Let $F = \{X_1, X_2, ..., X_n\}$ be a classification of U and R be an equivalence relation on U. Then for all i, $1 \leq i \leq n$, $\bar{R}X_i \neq U$ if and only if

$$\underline{R}(\bigcup_{j \neq i} X_j) \neq \phi.$$

Proof. Taking $I = \{i\}^C$ in Theorem 4. we get this. Also, this result can be obtained as a contrapositive of Corollary 9.

Corollary 11. [30, proposition 2.5] Let $F = \{X_1, X_2, ..., X_n\}$ be a classification of U and let R be an equivalence relation on U. If there exist $i \in N_n$ such that $\underline{R}X_i \neq \phi$ then for each $j(\neq i) \in N_n$, $\underline{R}X_j \neq U$.

Proof. By Corollary 9,

$$\underline{R}X_i \neq \phi \Rightarrow (\bigcup_{j \neq i} \bar{R}X_j) \neq U \Rightarrow \bar{R}X_j \neq U,$$

for each $j \neq i$, $1 \leq i \leq n$.

Interpretation
Suppose in a classification of a universe, there are positive elements of one member of classification with respect to an equivalence relation. Then there are negative elements of all other numbers of the classification with respect to equivalence relation. Taking the example into consideration if all red toys are of triangular shape (say) then for toys for circular or rectangular shape at least one colour is absent.

Corollary 12. [30, proposition 2.7] Let $F = \{X_1, X_2, ..., X_n\}, n > 1$ be a classification of U and let R be an equivalence relation on U. If for all $i \in N_n$, $\underline{R}X_i \neq \phi$ holds then $\bar{R}X_i \neq U$ for all $i \in N_n$.

Proof. As $\underline{R}X_i \neq \phi$ for all $i \in N_n$, we have

$$\underline{R}(\bigcup_{j \neq i} X_j) \neq \phi \; for \; all \; i \in N_n. \; So, \; by \; Corollary \; 10 \; \bar{R}X_i \neq U \; for \; all \; i \in N_n.$$

Interpretation
Suppose in a classification, there is a positive element in each member of the classification with respect to an equivalence relation. Then there is a negative element in each member of the classification with respect to the equivalence relation.

Referring to the example, if all toys of red colour are triangular, all the toys of green colour are circular and all the toys of blue colour are rectangular in shape then there is no green colour toy of triangular shape and so on.

7 Some Properties of Classifications

In this sect. we shall establish some properties of measures of uncertainty [12] and discuss in detail on properties of classifications with two elements and three elements.

7.1 Measures of Uncertainty

The following definitions are taken from Grzymala-Busse [12].

Definition 4. Let $F = \{X_1, X_2, ..., X_n\}$ be a classification of U and R be an equivalence relation on U. Then the *accuracy of approximation* of F by R, denoted by $\beta_R(F)$ and is defined as

$$\beta_R(F) = (\sum_{i=1}^{n} \mid \underline{R}X_i \mid)/(\sum_{i=1}^{n} \mid \bar{R}X_i \mid). \tag{26}$$

Definition 5. Let F and R be as above. Then the *quality of approximation of* F by R is denoted by $\gamma_R(F)$ and is defined as

$$\gamma_R(F) = (\sum_{i=1}^{n} \mid \underline{R}X_i \mid)/ \mid U \mid . \tag{27}$$

The accuracy of classification expresses the percentage of possible correct decision when classifying objects employing the knowledge R. The quality of classification expresses the percentage of objects which can be correctly classified to classes of F employing knowledge R.

Let R_1 and R_2 be any two equivalence relations on U. F_1 and F_2 be the classification of U generated by R_1 and R_2 respectively.

Definition 6

(i) We say that R_2 *depends in degree* k on R_1 in U and denote it by

$$R_1 \longrightarrow^k R_2 \;\; if \; and \; only \; if \;\; \gamma_{R_1}(F_2) = k. \tag{28}$$

(ii) We say that R_2 *totally depends* on R_1 in U if and only if $k = 1$.
(iii) We say that R_2 *roughly depends* on R_1 in U if and only if $0 < k < 1$.
(iv) We say that R_2 *totally independent* on R_1 in U if and only if $k = 0$.
(v) We say F_2 *depends in degree* k on F_1 in U, written as

$$F_1 \longrightarrow^k F_2 \;\; if \; and \; only \; if \;\; R_1 \longrightarrow^k R_2.$$

Property 23. For any R-definable classification F in U, $\beta_R(F) = \gamma_R(F) = 1$ So, if a classification F is R-definable then it is totally independent on R.

Proof. For all R-definable classifications F,$\underline{R}F = \bar{R}F$. So, by definition $\beta_R(F) = 1$. Again, by property of upper approximation and as F is a classification of U, we have

$$\sum_{i=1}^{n} | \bar{R}X_i | \geq \sum_{i=1}^{n} | X_i | = | \bigcup_{i=1}^{n} X_i | = | U | .$$

Also,

$$\sum_{i=1}^{n} | \underline{R}X_i | \leq \sum_{i=1}^{n} | X_i | = | \bigcup_{i=1}^{n} X_i | = | U | .$$

But, for R-definable classifications

$$\sum_{i=1}^{n} | \underline{R}X_i | = \sum_{i=1}^{n} | \bar{R}X_i | .$$

Hence,

$$\sum_{i=1}^{n} | \underline{R}X_i | = | U | .$$

So, we get $\gamma_R(F) = 1$.

Property 24. For any classification F in U and an equivalence relation R on U, $\beta_R(F) \leq \gamma_R(F) \leq 1$.

Proof. Since $| \underline{R}X_i | \leq | X_i |$, we have

$$\sum_{i=1}^{n} | \underline{R}X_i | \leq \sum_{i=1}^{n} | X_i | = | U | .$$

So, $\gamma_R(F) \leq 1$. Again, as shown above,

$$\sum_{i=1}^{n} | \bar{R}X_i | \geq | U | .$$

Hence,

$$\beta_R(F) = (\sum_{i=1}^{n} | \underline{R}X_i |)/(\sum_{i=1}^{n} | \bar{R}X_i |) \leq (\sum_{i=1}^{n} | \underline{R}X_i |)/ | U | = \gamma_R(F).$$

7.2 Classification Types

In this sect. we present Types of classifications and their rough definability as stated by Busse [12]. As mentioned, classifications are of great interest in the process of learning from examples, rules are derived from classifications generated by single decisions.

Definition 7. Let R be an equivalence relation on U. A classification $F = \{X_1, X_2, ..., X_n\}$ of U will be called *roughly R-definable, weak* in U if and only if there exists a number $i \in N_n$ such that $\underline{R}X_i \neq \phi$.

It can be noted from Corollary 9 that for a R-definable weak classification F of U, there exists $j(\neq i) \in N_n$ such that $\bar{R}X_j \neq U$.

Definition 8. Let R and F be as above. Then F will be called *roughly R-definable strong* in U (Type 1) if $\underline{R}X_i \neq \phi$ and only if $i \in N_n$ for each. By Corollary, in roughly R-definable strong classification in U, $\bar{R}X_i \neq U$ for each $i \in N_n$.

Definition 9. Let R and F be as above. Then F will be called *internally R-undefinable weak* in U if and only if $\underline{R}X_i = \phi$ for each $i \in N_n$ and there exists $j \in N_n$ such that $\bar{R}X_j \neq U$.

Definition 10. Let R and F be as above. Then F will be called *internally R-undefinable strong* in U (Type 2) if and only if $\underline{R}X_i = \phi$ and $\bar{R}X_i \neq U$ for each $i \in N_n$.

It has been observed by Busse [12] that due to Corollary 10 no *externally P-undefinable* set X exists in U. So, extension of Types of rough sets, classified on Type 3 is not possible to the case of classifications.

Definition 11. Let R and F be as above. Then F will be called *totally R-undefinable* in U (Type 4) if and only if $\underline{R}X_i = \phi$ and $\bar{R}X_i = U$ for each $i \in N_n$.

7.3 Classifications with Two or Three Elements

As remarked by Pawlak [30], approximation of sets and approximation of families of sets are two different issues and equivalence classes of approximate classification cannot be arbitrary sets, although they are strongly related. The concepts of compliments in case of sets and in case of classifications are different, which may lead to new concepts of negation in the case of binary and multivalued logics.

In this sect. we shall analyze the structure and properties of classifications having 2 elements and classifications having 3 elements. This may shed some light on the above statement of Pawlak. We shall use $T-i$ to represent $Type-i$, i=1,2,3,4 from this sect. onwards.

Classifications with Two Elements

Let $\chi = \{X_1, X_2\}$. Then $X_2 = X_1^C$. Since complements of T-1/T-4 rough sets are T-1/T-4 respectively and T-2/T-3 rough sets have complements of T-3/T-2 respectively, ([30], proposition 2.4), out of 16 possibilities for χ with respect to Typing only four alternates are possible. Namely, {T-1,T-1}, {T-2, T-3}, {T-3, T-2} and {T-4, T-4}. Again, the second and third possibilities are similar. So, there are only three distinct alternates. Hence, we have.

Property 25. A classification with two elements is roughly R-definable weak or of T-1 or of T-4 only.

Classifications with Three Elements

In a classification with 3 elements, say $\{X_1, X_2, X_3\}$ there are supposed to be 64 possibilities. But we shall show that only 8 of these possibilities can actually occur and other possibilities are not there.

Property 26. In a classification $F = \{X_1, X_2, X_3\}$ of U there are 8 possibilities for F with respect to Types of X_1, X_2, X_3. These are {T-1,T-1, T-1}, {T-1,T-1,T-2}, {T-1,T-2,T-2}, {T-2,T-2,T-2}, {T-2,T-2,T-4}, {T-2,T-4,T-4}, {T-3,T-2,T-2} and { T-4,T-4,T-4}.

Proof. We shall consider four cases depending upon the Type of X_1.
Case 1. X_1 is of T-1. Then $X_2 \cup X_3$ being compliment of X_1, must be of T-1. So, from the table of sect. 5.2, three cases arise for X_2 and X_3, that is {T-1,T-1}, {T-1,T-2}, and {T-2,T-2}.

Case 2. X_1 is of T-2. Complement of T-2 being Type -3, $X_2 \cup X_3$ is of T-3. Now, from the table of sect. 5.2, there are nine cases for X_2 and X_3. Out of these {T-1,T-1} and {T-1,T-2} have been covered in Case 1. {T-1,T-3} cannot occur as {T-2 ∪ T-3} = T-3 and $(\text{T-1})^C$ = T-1. Similarly, {T-1,T-4} cannot occur as {T-2 ∪ T-4} = T-3 and $(\text{T-4})^C$ = T-4. So, four cases remains {T-2,T-2}, {T-2,T-4} and {T-4,T-4}.

Case 3. X_1 is of T-3, $X_2 \cup X_3$ must be of T-2. Referring to the table of sect. 5.2. There is only one possibility for X_2, X_3 that is {T-2,T-2} which has been covered in Case 2.

Case 4. X_1 is of T-4. Then $X_2 \cup X_3$ must be of T-4. Referring to the table of sect. 5.2, there are three possibilities for X_2 and X_3. Out of these cases {T-2,T-2} and {T-2,T-4} have been considered in Case 2. So, only one case remains {T-4,T-4}.

This completes the proof of the property.

7.4 Further Types of Classifications

First we present two theorems which shows that the hypothesis in the theorems of Busse, as presented in Corollary 7 and Corollary 12 can be further relaxed to get the conclusions. However, even these hypothesis are to be shown as not necessary for the conclusions to be true.

Theorem 5. Let $F = \{X_1, X_2, ..., X_n\}$, where $n > 1$ be a classification of U and let R be an equivalence relation on U. If there exists p and q, $1 \leq p$, $q \leq n$ and $p \neq q$ such that $\underline{R}X_p \neq \phi$, $\underline{R}X_q \neq \phi$ then for each $i \in N_n$, $\bar{R}X_i \neq U$.

Proof. Since $\underline{R}X_p \neq \phi$, by Corollary 11, $\bar{R}X_i \neq U$ for $i \neq p$ and since $\underline{R}X_q \neq \phi$, by the same Corollary, $\bar{R}X_i \neq U$ for $i \neq q$. So, from these two we get $\bar{R}X_i \neq U$ for all i as $p \neq q$.

Note 1. The above condition is not necessary. Let us consider $U = \{x_1, x_2, ..., x_8\}$ and R be an equivalence relation on U with equivalence classes $X_1 = \{x_1, x_3, x_5\}$, $X_2 = \{x_2, x_4\}$ and $X_3 = \{x_6, x_7, x_8\}$. Then taking the classification $\{Z_1, Z_2, Z_3\}$ defined by $Z_1 = \{x_2, x_4\}$, $Z_2 = \{x_1, x_3, x_6\}$ and $Z_3 = \{x_5, x_7, x_8\}$, we find that $\bar{R}Z_1 \neq U$, $\bar{R}Z_2 \neq U$, $\bar{R}Z_3 \neq U$. But $\bar{R}Z_1 \neq \phi$, $\bar{R}Z_2 \neq \phi$, $\bar{R}Z_3 \neq \phi$.

Theorem 6. $F = \{X_1, X_2, ..., X_n\}$, where $n > 1$ be a classification of U and let R be an equivalence relation on U. If there exists p and q $1 \leq p, q \leq n$ and $p \neq q$ such that $\bar{R}X_p = \bar{R}X_q = U$ then for each $i \in N_n$, $\underline{R}X_i = \phi$.

Proof. Since $\bar{R}X_p = U$, by Corollary 6, $\underline{R}X_i = \phi$ for $i \neq p$ and since $\underline{R}X_q = U$, by the same Corollary, $\underline{R}X_i = \phi$ for $i \neq q$. So, from these two we get $\underline{R}X_i = \phi$ for all i as $p \neq q$.

Note 2. The above condition is not necessary. Let us consider U, R and X_1, X_2 and X_3 as in the above note. We take the classification defined by Z_1, Z_2, Z_3 defined by $Z_1 = \{x_2, x_6\}$, $Z_2 = \{x_1, x_3, x_4\}$ and $Z_3 = \{x_5, x_7, x_8\}$. We find that $\underline{R}Z_1 = \phi$, $\underline{R}Z_2 = \phi$ and $\underline{R}Z_3 = \phi$ whereas $\bar{R}Z_1 \neq U$, $\bar{R}Z_2 \neq U$ and $\bar{R}Z_3 \neq U$.

Observation 1. In Corollary 11, we have $\bar{R}X_j \neq U$ for all $j \neq i$ if $\exists X_i$ such that $\underline{R}X_i \neq \phi$. It is easy to observe that $\bar{R}X_i$ may or may not be U under the circumstances.

Observation 2. In Corollary 6, we have $\underline{R}X_j = \phi$ for all $j \neq i$ if $\exists X_i$ such that $\bar{R}X_i = U$. It is easy to observe that $\underline{R}X_i$ may or may not be ϕ under the circumstances.

For any classification $F = \{X_1, X_2, ..., X_n\}$ of U we have the following possibilities with respect to lower and upper approximations.

Basing upon the above table, possible combinations for classifications are (i, j); $i = 1, 2, 3, 4$ and $j = 5, 6, 7, 8$.

Out of these, several cases have been considered by Busse [12]. We shall examine all the possibilities closely. In fact we have the following table of combinations.

We shall be using the following abbreviations in Table 6:

Roughly R-definable weak 2 = RRdW2
Internally R-undefinable weak 2 = IRudW2
Internally R-undefinable weak 3 = IRudW3
Roughly R-definable weak 1 = RRdW1
Internally R-definable weak 1 = IRdW1
Totally R-undefinable weak 3 = TRudW3
Externally R-undefinable = ERud
Totally R-undefinable weak 1 = TRudW1

Table 5. Possibilities w.r.t. lower and upper approximations

F	**Lower** $\neq \phi$	**Lower** $= \phi$	**Upper** $\neq U$	**Upper** $= U$
$\forall$	1	2	5	6
$\exists$	3	4	7	8

Table 6. Possible combinations

	5	**6**	**7**	**8**
1	T-1	Not Possible	T-1	Not Possible
2	T-2	T-4	IRudW3	TRudW3
3	RRdW2	Not Possible	RRdW1	ERud
4	IRudW2	T-4	IRdW1	TRudW1

The cases (1,6) and (1,8) are not possible by Corollary 12. The case (3,6) is not possible by Corollary 7. The case (1,7) reduces to (1,5) by Corollary 12. The case (1,5) has been termed as roughly R-definable strong classification by Busse and we call it T-1 as all the elements of the classifications are of T-1. So far as row-1 of the table is concerned, the only possible classification is roughly R-definable strong or T-1.

The case (2,5) has been termed by Busse as internally R-undefinable strong. We call it T-2 as all the elements of the classifications are of T-2. The case (2,6) has been termed as totally R-undefinable by Busse and we call it T-4 as all the elements of the classification are of T-4. The case (2,7) has been termed as internally R-undefinable weak by Busse.

The Characterisation. We have the following conventions in connection with types of classifications:

(I) *Internal definability:* Lower approximation $\neq \phi$
(II) *Internal undefinibility:* Lower approximation $= \phi$

(III) *External definibility:* Upper approximation $\neq U$
(IV) *External undefinability:* Upper approximation $= U$

Also, from the set of elements in a classification, if we have $\exists$ some element satisfying a typing property, it leads to a weak type. On the other hand, if a typing property is true $\forall$ element, it leads to a strong type.
So, we have the following general types of classifications.

(I) *Roughly R-definable* $\Leftrightarrow$ Internally R-definable and Externally R-definable
(II) *Internally R-undefinable* $\Leftrightarrow$ Internally R-undefinable and Externally R-definable
(III) *Externally R-undefinable* $\Leftrightarrow$ Internally R-definable and Externally R-undefinable
(IV) *Totally R-undefinable* $\Leftrightarrow$ Internally R-undefinable and Externally R-undefinable

In case (I) we have one strong type, we call it T-1. This is the case when $\forall i$, $\underline{R}X_i \neq \phi$ and $\forall j$, $\underline{R}X_j \neq U$.
Also there are two weak types. We set them as:

(i) *Roughly R-definable* (weak -1) if and only if $\exists i$, $\underline{R}X_i \neq \phi$ and $\exists j$, $\underline{R}X_j \neq U$ and
(ii) *Roughly R-definable* (weak -2) if and only if and $\exists i$, $\underline{R}X_i \neq \phi$ and $\forall j$, $\underline{R}X_j \neq U$.

In case (II) we have one strong type, we call it T-2. This is the case when $\forall i$, $\underline{R}X_i \neq \phi$ and $\forall j$, $\bar{R}X_j \neq U$.
Also there are three weak types. We set them as:

(i) *Internally R-definable* (weak -1) if and only if $\exists i$, $\underline{R}X_i = \phi$ and $\exists j$, $\bar{R}X_j \neq U$.
(ii) *Internally R-definable* (weak -2) if and only if $\exists i$, $\underline{R}X_i = \phi$ and $\forall j$, $\bar{R}X_j \neq U$.
(iii) *Internally R-definable* (weak -3) if and only if $\forall i$, $\underline{R}X_i = \phi$ and $\exists j$, $\bar{R}X_j \neq U$.

In case (III) we have one strong type, we call it Externally R-undefinable only as there is no weak type possible in this case. This is the case when $\exists i$, $\underline{R}X_i \neq \phi$ and $\exists j$, $\bar{R}X_j = U$.
In case (IV) we have one strong type, we call it T-4. This is the case when $\forall i$, $\underline{R}X_i = \phi$ and $\exists j$, $\bar{R}X_j = U$.
Also there are two weak types. We set them as

(i) *Totally r-undefinable* (weak -1) if and only if $\exists i$, $\underline{R}X_i = \phi$ and $\exists j$, $\bar{R}X_j = U$.
(ii) *Totally R-undefinable* (weak -2) if and only if $\forall i$, $\underline{R}X_i = \phi$ and $\exists j$, $\bar{R}X_j = U$.

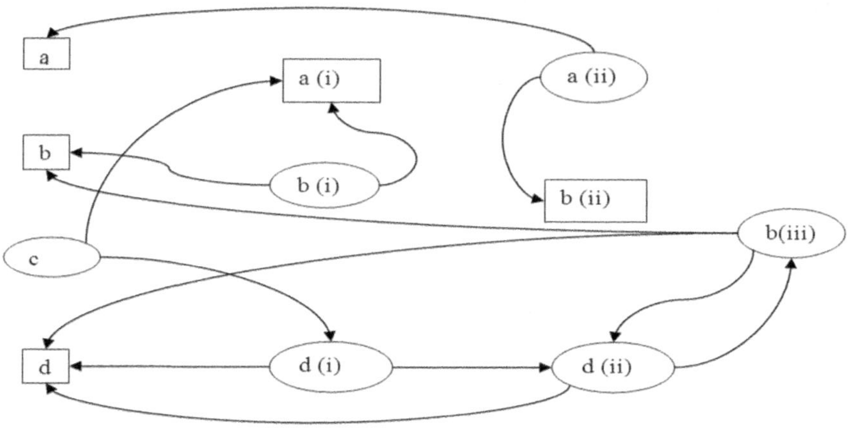

Fig. 1. A schematic representation of Busse's cases

Out of these eleven possibilities only five have been considered by Busse [12]. In Fig. 1, we represent the cases considered by Busse inside rectangles and those not considered by him inside ellipses. The arrows show the reduction of the six cases not considered by Busse to those considered by him either directly or transitively.

7.5 Application

A new approach to knowledge acquisition under uncertainty based on rough set theory was presented by Busse [12]. The real world phenomena are represented

Table 7. An example of inconsistent information system

Q	c_1	c_2	d_1	d_2	d_3	d_4	d_5	d_6
x_1	v_1	w_1	0	0	0	0	0	0
x_2	v_1	w_2	1	0	0	0	0	0
x_3	v_1	w_1	0	0	0	1	2	1
x_4	v_1	w_2	1	0	0	1	2	1
x_5	v_2	w_2	0	0	0	0	0	0
x_6	v_2	w_2	0	1	1	1	1	1
x_7	v_3	w_1	1	1	0	0	1	0
x_8	v_3	w_1	1	1	1	1	2	1
x_9	v_3	w_1	1	1	1	1	2	2

by information system, where inconsistencies are included. For this he considered the example of opinion of six doctors d_1, d_2, d_3, d_4, d_5 and d_6 on nine patients x_1, x_2,..., x_9 based upon the result of two tests c_1 and c_2. On the basis of values of tests, experts classify patients as being on some level of disease. The information system is represented in a tabular form, which is clearly inconsistent.

The classification, generated by the set C of conditions is equal to $\{\{x_1, x_3\}, \{x_2, x_4\}, \{x_5, x_6\}, \{x_7, x_8, x_9\}\}$.

If we denote the classification X_i generated by the opinion of doctor d_i, $i = 1, 2, 3, 4, 5, 6$ then

$X_1 = \{\{x_1, x_3, x_5, x_6\}, \{x_2, x_4, x_7, x_8, x_9\}\}$,
$X_2 = \{\{x_1, x_2, x_3, x_4, x_5\}, \{x_6, x_7, x_8, x_9\}\}$,
$X_3 = \{\{x_1, x_2, x_3, x_4, x_5, x_7\}, \{x_6, x_8, x_9\}\}$,
$X_4 = \{\{x_1, x_2, x_5, x_7\}, \{x_3, x_4, x_6, x_8, x_9\}\}$,
$X_5 = \{\{x_1, x_2, x_5\}, \{x_3, x_4, x_8, x_9\}, \{x_6, x_7\}\}$
and
$X_6 = \{\{x_1, x_2, x_5, x_7\}, \{x_3, x_4, x_6, x_8\ \{x_9\}\}$

It is easy to see that the above classifications are of type C-definable, roughly C-definable strong, roughly C-definable weak, totally C-undefinable, internally C-undefinable and internally C-undefinable weak respectively.

8 Rule Generation

By rules on information systems we mean conditional statements that specify actions under conditions. The following notations are used :

Constants: 0,1
Atomic expression: $a := v \equiv \{\rho(x, a) = v : x \in U\}$
Boolean Operations: $\neg, \vee, \wedge$
$0 \equiv$ Empty set.
$1 \equiv U$.

8.1 Definitions

(i) Rules computed from lower approximations are *certain rules.*
(ii) Rules computed from upper approximations are *possible rules.*

The following properties hold for rule generation.

I. Necessary and sufficient condition for a classification χ to induce certain rules is $\underline{C}\chi \neq \phi$.
II. The number of certain rules is equal to the number of non-empty lower approximations in the classification.
III. The number of possible rules is equal to the number of non-empty boundaries in the classification.

8.2 Observations

I. For C-definable classifications, all the rules are certain rules.
II. For roughly C-definable strong and roughly C-definable weak classifications both certain and possible rules exist.
III. For totally C-undefinable, internally C-undefinable strong and internally C-undefinable weak classifications there are no certain rules.
IV. For roughly C-definable strong sets the number of certain rules is equal to the number of elements in the classification.
V. All types of classifications other than C-definable classifications have the property that there is at least one possible rule.
VI. For roughly C-definable weak classifications there is at least one certain rule.
VII. For totally C-undefinable classifications, there is no certain rule. The number of possible rules is equal to the number of elements in the classification.
VIII. For intrenally C-undefinable strong classifications, there is no certain rule. The number of possible rules is at most equal to the number of elements in the classification.
IX. For internally C-undefinable weak classifications, there is no certain rule. There is no guarantee about the existence of possible rules.

8.3 Examples

Let us see how some certain and possible rules can be generated from the example 7.5.

(I)X_1 is C-definable and hence all the rules corresponding to it are certain rules. In fact, the rules are
(i)$((c_1 = v_1) \wedge (c_2 = w_1)) \vee ((c_1 = v_2) \wedge (c_2 = w_2)) \Rightarrow (d_1 = 0)$ and
(ii) $((c_1 = v_1) \wedge (c_2 = w_2)) \vee ((c_1 = v_3) \wedge (c_2 = w_1)) \Rightarrow (d_1 = 1)$

(II)X_2 is roughly C-definable strong. So, it has both type of rules,
(i)$((c_1 = v_1) \Rightarrow (d_2 = 0)$ (Certain rule) and
(ii) $((c_1 = v_2) \wedge (c_2 = w_2)) \Rightarrow (d_2 = 0)$ (Possible rule).

(III)X_5 is internally C-undefinable strong. So, it has no certain rules. As it has three elements, by Observation VIII it can have at most three possible rules. In fact the rules are
(i)$(c_1 = v_1) \vee (c_1 = v_2) \vee (c_2 = w_2) \Rightarrow (d_5 = 0)$
(ii) $((c_1 = v_2) \vee (c_1 = v_3) \Rightarrow (d_5 = 1)$ and
(iii) $(c_1 = v_1) \vee (c_1 = v_3) \vee (c_2 = w_1) \Rightarrow (d_5 = 2)$.

9 Rough Equivalence of Algebraic Rules

We have several algebraic properties with respect to the set theoretic operations of union, intersection and complementation. Ordinary equality when the sets involved are taken to be rough sets bears little meaning and does not comply with common sense reasoning. So, rough equality or rough equivalence seems to be a possible solution. In this sect. we continue with rough equivalence and verify the validity of rough equivalence of left and right hand sides of these properties. This study was initiated in [39].

9.1 Associative Rule

The two Associative laws for crisp sets are:

For any three sets A, B and C,

$$A \cup (B \cup C) = (A \cup B) \cup C \tag{29}$$

and

$$A \cap (B \cap C) = (A \cap B) \cap C \tag{30}$$

Now, it is interesting to verify whether the left and right hand side of (29) and (30) match with their Types. For this, we consider four different cases depending upon Types of A with B and C being of any of the four Types. We take it as case i, when A is of T-i, $i = 1, 2, 3, 4$. It is observed that in all these cases the left hand side and right hand side of the above equalities match with each other as is evident from the corresponding tables. First we consider the four cases for union and than for intersection. Tables 1 and 2 are used to derive the tables below.

Union

Table 8. Union: case 1

$\cup$	**T-1**	**T-2**	**T-3**	**T-4**
T-1	T-1/T-3	T-1/T-3	T-3	T-3
T-2	T-1/T-3	T-1/T-3	T-3	T-3
T-3	T-3	T-3	T-3	T-3
T-4	T-3	T-3	T-3	T-3

Table 9. Union: case 2

$\cup$	**T-1**	**T-2**	**T-3**	**T-4**
T-1	T-1/T-3	T-1/T-3	T-3	T-3
T-2	T-1/T-3	T-1/T-2/T-3/T-4	T-3	T-3/T-4
T-3	T-3	T-3	T-3	T-3
T-4	T-3	T-4	T-3	T-3/T-4

Table 10. Union: case 3

∪	**T-1**	**T-2**	**T-3**	**T-4**
T-1	T-3	T-3	T-3	T-3
T-2	T-3	T-3	T-3	T-3
T-3	T-3	T-3	T-3	T-3
T-4	T-3	T-3	T-3	T-3

Table 11. Union: case 4

∪	**T-1**	**T-2**	**T-3**	**T-4**
T-1	T-3	T-3	T-3	T-3
T-2	T-3	T-3/T-4	T-3	T-3/T-4
T-3	T-3	T-3	T-3	T-3
T-4	T-3	T-3/T-4	T-3	T-3/T-4

Table 12. Intersection: case 1

∩	**T-1**	**T-2**	**T-3**	**T-4**
T-1	T-1/T-2	T-2	T-1/T-2	T-2
T-2	T-2	T-2	T-2	T-2
T-3	T-1/T-2	T-2	T-1/T-2	T-2
T-4	T-2	T-2	T-2	T-2

Table 13. Intersection: case 2

∩	**T-1**	**T-2**	**T-3**	**T-4**
T-1	T-2	T-2	T-2	T-2
T-2	T-2	T-2	T-2	T-2
T-3	T-2	T-2	T-2	T-2
T-4	T-2	T-2	T-2	T-2

Table 14. Intersection: case 3

$\cap$	**T-1**	**T-2**	**T-3**	**T-4**
T-1	T-1/T-2	T-2	T-1/T-2	T-2
T-2	T-2	T-2	T-2	T-2
T-3	T-1/T-2	T-2	T-1/T-2/T-3/T-4	T-2/T-4
Type 4	T-2	T-2	T-2/T-4	T-2/T-4

Table 15. Intersection: case 4

$\cap$	**T-1**	**T-2**	**T-3**	**T-4**
T-1	T-2	T-2	T-2	T-2
T-2	T-2	T-2	T-2	T-2
T-3	T-2	T-2	T-2/T-4	T-2/T-4
T-4	T-2	T-2	T-2/T-4	T-2/T-4

Table 16. Double negations for different types of rough sets

A	**T-1**	**T-2**	**T-3**	**T-4**
$(A)^C$	T-1	T-3	T-2	T-4
$((A)^C)^C$	T-1	T-2	T-3	T-4

Intersection

9.2 Complement and Double Negation

The Types of complement of rough sets of different Types have been obtained by (Pawlak [30], Theorem 2.4). Using this, it is easy to compute the double negations of different Types of rough sets as provided in the following table and see that the complementation law holds for rough equivalence.

9.3 De Morgan's Theorems

De Morgan's Theorems for crisp sets state that for any two sets X and Y,

$$(X \cup Y)^C = (X)^C \cap (Y)^C \tag{31}$$

and

$$(X \cap Y)^C = (X)^C \cup (Y)^C \tag{32}$$

Also, when both X and Y are rough sets of different types we observe that both sides of (31) and (32) are rough equivalent as is evident from the following tables.

Table for both the sides of (31) is:

Table 17

Table 17. De Morgan's union for different types of rough sets

	T-1	**T-2**	**T-3**	**T-4**
T-1	T-1/T-2	T-1/T-2	T-2	T-2
T-2	T-1/T-2	T-1/T-2/T-3/T-4	T-2	T-2/T-4
T-3	T-2	T-2	T-2	T-2
T-4	T-2	T-2/T-4	T-2	T-2/T-4

Table for both the sides of (32) is:

Table 18. De Morgan's intersection for different types of rough sets

	T-1	**T-2**	**T-3**	**T-4**
T-1	T-1/T-3	T-3	T-1/T-3	T-3
T-2	T-3	T-3	T-3	T-3
T-3	T-2	T-2	T-2	T-2
T-4	T-3	T-3	T-3/T-4	T-3/T-4

9.4 Distributive Property

The two distributive properties for crisp sets state that, for any three sets A, B and C,

$$A \cup (B \cap C) = (A \cup B) \cap (A \cup C) \tag{33}$$

and

$$A \cap (B \cup C) = (A \cap B) \cup (A \cap C) \tag{34}$$

We have the following observations with respect to the rough equivalence of Left hand side and Right hand side of (33) and (34):

When A is T-2, the left hand side and right hand side of (33) are rough equivalent and the case is similar for (34) when A is of T-3.

In other cases, we have following observations:

(i) When A is of T-1, left hand side of (33) is of T-1 or T-3, whereas right hand side can be any of the four types. The result remains same even by using our Corollaries 1 and 2.

So left hand side and right hand side are not rough equivalent of any kind.

When A is of T-1, left hand side of (34) is of T-1 or T-2, whereas right hand side can be any of the four types.

The result remains unchanged even by using our Corollaries. So left hand side and right hand side are not rough equivalent of any kind.

(ii) When A is of T-2, both left hand sides and right hand sides of (33) can be any of the four types.

So, they are not rough equivalent.

When A is of T-2, left hand side of (34) is of T-2, whereas right hand side can be any of the four types.

So, left hand sides and right hand sides of (34) are not rough equivalent.

However, left hand side is Bottom Rough equivalent to its right hand side when condition of Corollaries 1 and 2 are satisfied.

(iii) When A is of T-3, left hand side of (33) is of T-3, whereas right hand side can be any of the four types.

So, left hand sides and right hand sides are not rough equivalent.

However, left hand side is Top Rough equivalent to its right hand side when condition of Corollaries 1 and 2 are satisfied.

When A is of T-3, left hand side and right hand side of (34) can be of any of the four types.

So, left hand sides and right hand sides are not rough equivalent.

(iv) When A is of T-4, left hand side of (33) is of T-3 or T-4, whereas right hand side can be any of the four types.

So, again left hand sides and right hand sides of (34) are not rough equivalent.

However, left hand side is Top Rough equivalent to its right hand side when Corollaries 1 and 2 are used.

When A is of T-4, left hand side of (34) is of T-2 or T-4, whereas right hand side can be any of the four types.

However, left hand side is Top Rough equivalent to its right hand side when Corollaries 1 and 2 are used. So, in general the left hand side and right hand side of distributive properties are not rough equivalent.

9.5 Idempotent and Absorption Property

Idempotent Property

The two idempotent properties for crisp sets state that for any set A,

$$A \cap A = A \tag{35}$$

and

$$A \cup A = A \tag{36}$$

When A is a Rough set, it is clear from the diagonal entries of the union and intersection table in Sect. 3 that (35) holds good with Type matching only when A is of T-2. For rest of types, the left hand side is not Rough equivalent to its right hand side. However we observe that in (35), for A of T-1, the left hand side is Top Rough equivalent to its right hand side. For A of T-3, the left hand side is Top Rough equivalent to its right hand side when conditions of Corollary 2 are satisfied. For A of T-4, the left hand side is Bottom Rough equivalent to its right hand side.

When A is a Rough set, the left hand side and right hand side of (36) are rough equivalent only when A is of T-3 and for rest of the types, the left hand side is not Rough equivalent to its right hand side. However we observe that in (36), for A of T-1, the left hand side is Bottom Rough equivalent to its Right Hand Side. For A of T-2, the left hand side is Bottom Rough equivalent to its right hand side, when conditions of Corollary 1 are satisfied. For A of T-4, the left hand side is Top Rough equivalent to its right hand side.

Absorption Property

The two absorption properties for crisp sets state that for any two sets A and B

$$A \cup (A \cap B) = A \tag{37}$$

and

$$A \cap (A \cup B) = A \tag{38}$$

Taking A and B as Rough sets, we find that when A is of T-3, both the sides of (37) are of T-3 and when A is of T-2, both the sides of (38) are of T-2. Hence, the left hand side and rough hand side are rough equivalent. In the rest of the cases left hand side and right hand sides of (37) and (38) are not rough equivalent. In fact the following properties hold good:

(i) When A is of T-1, left hand side of (37) is of T-1 or T-3.
So, left hand side is Bottom Rough equivalent to its right hand side.
(ii) When A is of T-2, left hand side of (37) is any of the four types. However, left hand side is Bottom Rough equivalent to its right hand side when condition of Corollaries 1 and 2 are satisfied.
(iii) When A is of T-4, left hand side of (37) is of T-3 or T-4.
So, left hand side is Top Rough equivalent to its right hand side.
(iv) When A is of T-1, left hand side of (38) is of T-1 or T-2.
So, left hand side is Top Rough equivalent to its right hand side.
(v) When A is of T-3, left hand side of (38) is any of the four types.
So, left hand side is Top Rough equivalent to its right hand side when conditions of Corollaries 1 and 2 are satisfied.
(vi) When A is of T-4, left hand side of (38) is of T-2 or T-4.

However, left hand side is Bottom Rough equivalent to its right hand side.

So, Left hand side and Right hand sides of absorption rules are not rough equivalent in general.

9.6 Kleene's Property

The Kleene's property states that for any two sets A and B ,

$$(A \cup A^C) \cup (B \cap B^C) = A \cup A^C \tag{39}$$

and

$$(A \cup A^C) \cap (B \cap B^C) = B \cap B^C \tag{40}$$

We show below that for Rough sets A and B, both sides of (39) and (40) match with each other with respect to types. Due to symmetry of the operations of union and intersection, it is enough to consider the ten cases; case (i, j) being A of Type i and B of Type j; i, j =1,2,3,4 and $j \geq i$.

Proof of (39)

In cases (1,1), (1,2), (1,3) and (1,4) both the left hand side and right hand side of (39) are of Type 1 or Type 3.

In cases (2,2), (2,3), (2,4), (3,3) and (3,4) both the left hand side and right hand side of (39) are of Type 3.

Finally, in case of (4, 4) both the left hand side and right hand side of (39) are of Type 3 or Type 4.

Proof of (40)

In case of (1,1) both the left hand side and right hand side of (40) are of Type 1 or Type 2.

In cases (1,2), (1,3), (2,2), (2,3) and (3,3) both the left hand side and right hand side of (40) are of Type 2.

In cases (1,4), (2,4), (3,4) and (4,4) both the left hand side and right hand side of (40) are of Type 2 or Type 4.

Hence, from the above observations, it is clear that the left hand side and right hand sides of Kleene's property are rough equivalent.

9.7 Maximum and Minimum Elements' Properties

It is obvious that both ϕ and U are crisp sets with respect to any equivalence relation defined over the universe.

Maximum Element

The Maximum element property for crisp sets state that for any set A

$$A \cup U = U \tag{41}$$

and

$$A \cap U = A \tag{42}$$

For any Rough set A, as A is a subset of U both (41) and (42) hold true. So, the rough equivalence of both side are obvious.

Minimum Element

The Minimum element property for crisp sets state that for any set A,

$$A \cup \phi = A \tag{43}$$

and

$$A \cap \phi = \phi \tag{44}$$

For any Rough set A, as ϕ is a subset of A both (43) and (44) hold true. So, the rough equivalence of both side is automatically satisfied.

9.8 Complementary Laws

The complementary laws for crisp sets state that for any set A,

$$A \cup A^C = U \tag{45}$$

and

$$A \cap A^C = \phi \tag{46}$$

For any Rough set A, A is a subset of U and also A^C is a subset of U. So, both (45) and (46) hold true. Hence, the rough equivalence of both sides is automatically satisfied.

10 Conclusions

Study of topological classification of sets under consideration provides some insight into how the boundary regions are structured. This has a major role in practical application of rough sets. In this chapter we studied some properties of topological classification of sets starting with types of rough sets, then we moved to find properties of types of union and intersection of rough sets. The concept of rough equivalences of sets introduced by Tripathy and Mitra [38], which captures approximate equality of rough sets at a higher level than rough equalities of Novotny and Pawlak [23,24,25] was discussed in detail. Some real life examples were considered in support of the above claim. Properties of rough equalities which were noted to be not true when bottom and top rough equalities are interchanged, were dealt with and established along with parallel properties for rough equivalences. Approximation of classifications of universes was introduced and studied by Busse [12]. The types of classifications were studied completely by us in this chapter. Also, theorems of Busse establishing properties of approximation of classifications wee completely generalized to their necessary and sufficient type form. From these results new results could be obtained as corollaries. All these results were interpreted with the help of simple examples. Complete characterizations of classifications having 2 or 3 elements are done. A characterization of a general classification having n elements is still awaited. Such a solution would shed light on negation in case of multivalued logic. Continuing with the study of rough equivalences, the algebraic properties involving rough sets were analyzed and established.

References

1. Aikins, J.S.: Prototyping Knowledge for Expert Systems. Artificial Intelligence 20, 163–210 (1983)
2. Banerjee, M., Chakraborty, M.K.: Rough Consequence and Rough Algebra in Rough Sets. In: Ziarko, W.P. (ed.) Fuzzy Sets and Knowledge Discovery, RSKD 1993, Banf, Canada, pp. 196–207. Springer, Heidelberg (1994)
3. Bobrow, D.G.: A Panel on Knowledge Representation. In: Proceeding of Fifth International Joint Conference on Artificial Intelligence, Carnegie - Melon University, Pittsburgh, PA (1977)
4. Bobrow, D.G., Wingograd, T.: An Overview of KRL: A Knowledge Representation Language. Journal of Cognitive Sciences 1, 3–46 (1977)
5. Bonikowski, Z.: A Certain Conception of the Calculus of Rough sets. Notre Dame Journal of Formal Logic 33, 412–421 (1992)
6. Brachman, R.J., Smith, B.C.: Special Issue of Knowledge Representation. SIGART News Letter 70, 1–138 (1980)
7. Brachman, R.J., Levesque, H.J. (eds.): Reading in Knowledge Representation. Morgan Kaufmann Publishers Inc., San Francisco (1986)
8. Chakraborty, M.K., Banerjee, M.: Rough Dialogue and Implication Lattices. Fundamenta Informetica 75(1-4), 123–139 (2007)
9. Davis, R., Leant, D.: Knowledge-Based Systems in Artificial Intelligence. McGraw-Hill, New York (1982)
10. Dubois, D., Prade, H.: Rough fuzzy sets and fuzzy rough sets. International Journal on General Systems 17(2-3), 191–208 (1990)
11. Grzymala-Busse, J.W.: On the Reduction of Knowledge representation Systems. In: Proceeding of 6th International Workshop on Expert Systems and their Applications, Avignon, France, April 28-30, vol. 1, pp. 463–478 (1986)
12. Grzymala-Busse, J.: Knowledge Acquisition under Uncertainty - a rough set approach. Journal of Intelligence and Robotics Systems 1, 3–16 (1988)
13. Grzymala-Busse, J.W.: Characteristic relations for incomplete data: A generalization of the indiscernibility relation. In: Peters, J.F., Skowron, A. (eds.) Transactions on Rough Sets IV. LNCS, vol. 3700, pp. 58–68. Springer, Heidelberg (2005)
14. Grzymala-Busse, J.W.: Incomplete data and generalization of indiscernibility relation, definability, and approximations. In: Ślęzak, D., Wang, G., Szczuka, M.S., Düntsch, I., Yao, Y. (eds.) RSFDGrC 2005. LNCS (LNAI), vol. 3641, pp. 244–253. Springer, Heidelberg (2005)
15. Halpern, J. (ed.): Theoretical Aspect of Reasoning about Knowledge, Proceeding of the 1986 Conference. Morgan Kaufman, Los Altos (1986)
16. Kryszkiewicz, M.: Rough set approach to Incomplete Information System. In: Proceedings of the second annual joint Conference on Information Sciences, Wrightsville Beach, NC, September 28 - October 01, pp. 194–197 (1995)
17. Kryszkiewicz, M.: Properties of Incomplete Information System in the frame work of Rough set. In: Polkowski, L., Skowron, A. (eds.) Rough set in Knowledge Discovery - I: methods and applications. studies in Fuzziness and Soft Computing, vol. 18, pp. 422–450. Physica-Verlag, Heidelberg (1998)
18. Kryszkiewicz, M.: Rules in Incomplete Information System. Information Sciences 113, 271–292 (1999)
19. Mamdani, A., Efstathion, Pang, D.: Inference under Uncertainty, Expert System. In: Proceeding of Fifth Tech. Conference of British Computer Society, Specialist Group on Expert Systems, pp. 181–194 (1985)

20. Minski, M.: A Framework for Representation Knowledge. In: Winston, P. (ed.) The Psychology of Computer Vision, pp. 211–277. McGraw-Hill, New York (1975)
21. Newell, A.: The Knowledge Level. Artificial Intelligence 18, 87–127
22. Nguyen, H.S., Nguyen, S.H.: Rough Set and Association Rule Generation. Fundamenta Informetica 40(4), 383–405 (1999)
23. Novotny, M., Pawlak, Z.: Characterization of Rough Top Equalities and Rough Bottom Equalities. Bulletin Polish Acad. Sci. Math. 33, 91–97 (1985)
24. Novotny, M., Pawlak, Z.: On Rough Equalities. Bulletin Polish Acad. Sci. Math. 33, 99–104 (1985)
25. Novotny, M., Pawlak, Z.: Black Box Analysis and Rough Top Equality. Bulletin Polish Academy of Sci. Math. 33, 105–113 (1985)
26. Pawlak, Z.: Rough Sets - Basic Notations, Institute Comp. Sci., Polish Academy of Sciences, Rep. No. 431, Warshaw (1981)
27. Pawlak, Z.: Rough sets. International Journal of Information and Computer Science 11, 341–356 (1982)
28. Pawlak, Z.: Rough Classification. International Journal of Man Machine Studies 20, 469–483 (1983)
29. Pawlak, Z.: Rough sets and Fuzzy sets. Fuzzy sets and system 17, 90–102 (1985)
30. Pawlak, Z.: Rough sets - Theoretical aspects of reasoning about data. Kluwer academic publishers, Dordrecht (1991)
31. Pawlak, Z., Skowron, A.: Rudiments of rough sets. Information Sciences - An International Journal 177(1), 3–27 (2007)
32. Pawlak, Z., Skowron, A.: Rough sets: Some extensions. Information Sciences - An International Journal 177(1), 28–40 (2007)
33. Pawlak, Z., Skowron, A.: Rough sets and Boolean reasoning. Information Sciences - An International Journal 177(1), 41–73 (2007)
34. Pomykala, J., Pomykala, J.A.: The stone algebra of rough set. Bulletin of Polish Academy of Sciences Mathematics 36, 495–508 (1988)
35. Quinlan, J.R.: Consistency and Plausible Reasoning. In: Proceeding of 8th IJCAI, Karlsruhe, West Germany, pp. 137–144 (1983)
36. Steforowski, J., Tsoukias, A.: Incomplete Information Tables and Rough Classifications. Computational Intelligence 17, 545–566 (2001)
37. Tripathy, B. K., Mitra, A.: Topological Properties of Rough Sets and Applications (Communicated to International Journal of Granular Computing, Rough Sets and Intellegent Systems.)
38. Tripathy, B.K., Mitra, A., Ojha, J.: On Rough Eqalites and Rough Equivalence of Sets. In: RSCTC 2008, Akron, USA, October 23-25 (accepted for presentation in 2008)
39. Tripathy, B.K., Mitra, A.: Comparitive Study of Knowledge Representation using Rough sets and their Algebric properties. In: Proceeding of the National Conference on Research Prospects in Knowledge Mining, Anamalai University, India, March 2-3, pp. 1–6 (2008)
40. Tripathy. B. K., Ojha. J., Mohanty. D.:On Approximation of classification and rough comparision of Rough Sets (Communicated to Foundations of Computing and Decision Sciences) (2008)
41. Waiilewska. A.: Topological Rough Algebra in Rough set and Data Mining. Analysis of Imperfect Knowledge, 411–425
42. Waiilewska, A., Vigneson, L.: Rough Equality Algebra. In: Proceedings of the second annual joint Conference on Information Sciences, Wrightsville Beach, NC, October 28 - October 01, pp. 26–36 (1995)

43. Waiilewska, A., Vigneson, L.: Rough Algebra and Automated Deduction. In: Polkowski, L., Skowron, A. (eds.) Rough set in Knowledge Discovery - I: methods and applications. studies in Fuzziness and Soft Computing, vol. 18, pp. 261–275. Physica-Verlag, Heidelberg (1998)
44. Yager, R.R.: Approximate Reasoning as a basis for Rule Based Expert Systems. IEEE transaction on Systems, Man and Cybernetic 14, 636–643 (1984)
45. Zadeh, L.A.: Fuzzy sets. Information and control 8, 338–353 (1965)

Part II

Rough Set Data Mining Activities

Rough Clustering with Partial Supervision

Rafael Falcón[1], Gwanggil Jeon[2], Rafael Bello[1], and Jechang Jeong[2]

[1] Computer Science Department, Central University of Las Villas (UCLV)
Carretera Camajuaní km 5.5, Santa Clara, Cuba
{rfalcon,rbellop}@uclv.edu.cu

[2] Department of Electronics and Computer Engineering, Hanyang University
17 Haengdang-dong, Seongdong-gu, Seoul, Korea
{windcap315,jjeong}@ece.hanyang.ac.kr

Summary. This study focuses on bringing two rough-set-based clustering algorithms into the framework of partially supervised clustering. A mechanism of partial supervision relying on either fuzzy membership grades or rough memberships and non-memberships of patterns to clusters is envisioned. Allowing such knowledge-based hints to play an active role in the discovery of the overall structure of the dataset has proved to be highly beneficial, this being corroborated by the empirical results. Other existing rough clustering techniques can successfully incorporate this type of auxiliary information with little computational effort.

Keywords: rough clustering, partial supervision, knowledge-based hints, membership grades, rough c-means.

1 Introduction

For many years, clustering [1] has been regarded as an essential component of data mining activities as well as an active research topic. This mechanism of unsupervised learning leads to the discovery of knowledge structures (clusters) which are one of the primary sources of information about the underlying dynamics of a dataset.

The wealth of descriptive capabilities contributed by fuzzy set theory has given rise to the fuzzy clustering domain [2], one of the most rapidly evolving areas within Intelligent Data Analysis. By means of fuzzy sets, it is now possible to precisely represent the behavior of borderline patterns, which are quite common in real-life problems, thus arriving at a more elaborate picture of the suite of clustering activities. Fuzzy clustering provides an additional conceptual enhancement by allowing the pattern to be located to several clusters and with various membership degrees [3].

This very idea has been pursued from the standpoint of rough set theory [4, 5]. As a well-settled methodology for coping with uncertainty (mostly emerging in the form of data inconsistencies), the introduction of the lower and upper approximations as distinctive features of any cluster enables us to provide the degree of flexibility about membership of patterns to clusters already achieved by

A. Abraham, R. Falcón, and R. Bello (Eds.): Rough Set Theory, SCI 174, pp. 137–161.
springerlink.com

fuzzy set theory. The pioneering works in the field of Lingras and West [6, 7, 8, 9] among others [10], have opened a new direction in rough set research.

Recently, granular computing has brought traditional clustering techniques to a new level [11] by laying information granules (arising in the form of fuzzy sets, rough sets, shadowed sets and the like) in a privileged position as the main vehicle through which the optimization scheme is carried out rather than continue leaning upon traditional data patterns. In this context, fuzzy clustering has largely profited from considering several tips coming from the outside world and embedding them into the standard optimization workflow of the most popular clustering engines. These knowledge-based hints have turned clustering into a semi-supervised fashion of discovering knowledge. While one can witness many successful studies outfitting fuzzy clustering with partial supervision [12, 13, 14, 15, 16, 17, 18], to our knowledge this innovative idea has not been taken yet to the realm of rough clustering techniques.

In this chapter we are going to show how rough clustering algorithms can take advantage of partial supervision mechanisms. The assurance that an object certainly belongs to a cluster or doesn't belong to it at all can be effectively used as a hint for driving the entire discovery of the knowledge structures. Membership grades of patterns to clusters can serve the same purpose. The proposed partially supervised algorithms behave well in presence of imbalanced datasets, i.e. unevenly distributed classes. The presented ideas can be extended to other approaches of the same kind with little computational effort.

The chapter has been structured as follows. Section 2 elaborates on the augmentation of fuzzy clustering techniques with some sort of supervised information concerning a subset of labeled data patterns. Some existing rough-set-based clustering approaches are outlined in Sect. 3 whereas the proposed modifications to embrace the corresponding knowledge-based hints are presented next. The benefits brought about by partial supervision are empirically analyzed in Sect. 5 in presence of synthetic and real-world datasets. Finally, some concluding remarks are stated.

2 Clustering with Partial Supervision

In clustering scenarios, determining the underlying structure in data is a process usually accomplished by means of an objective function whose minimization drives the whole clustering scheme. The strength of this unsupervised method lies in that no pattern needs to be flagged with its corresponding class. However, the structural search in the data set could be substantially enhanced by the careful use of some domain knowledge about the classification problem. One can find many real-life scenarios for which a more active engagement of human experts in the clustering process is highly desirable and the term "clustering with partial supervision" has been coined for designating this novel manner of seeking knowledge structures.

Partial supervision in fuzzy clustering manifests as a subset of labeled patterns, i.e. some columns of the partition matrix (holding the set of membership grades for some predefined patterns) are provided a priori. A mixture of labeled and unlabeled objects may be found in many practical situations. From that point on, the ensuing optimization activities are strongly influenced by the external knowledge-based tips and the overall structure to be uncovered is expected to conform to the suite of membership grades already provided for the selected objects [11].

Techniques of this type in the context of c-means clustering were first discussed by Pedrycz [12], who came up with an augmented objective function including a term for minimizing the differences between the provided membership grades and the partition matrix under construction.

Kersten [13] realized that the clustering of signal vectors could benefit from utilizing some signal quality measures often available, such as the signal-to-noise ratio (SNR). This fact led to the modification of the Fuzzy C-Means (FCM) [19] and Fuzzy C-Medoids (FCMED) [13] approaches in order to consider this type of auxiliary information. As a consequence, faster convergence rates and more accurate cluster prototypes in moderate SNR environments were obtained.

A different approach was introduced in [14]. It was pointed out by Bensaid et al that two typical problems endured by clustering techniques, i.e. (1) the a priori determination of the optimal number of clusters and (2) the assignment of meaningful labels to the automatically revealed knowledge structures fade away when some data from each class is labeled beforehand. A semi-supervised fuzzy c-means algorithm (SSFCM) was crafted and found particularly suited to image segmentation 'cause a human expert often examines an image and selects few clearly defined pixels from different classes, which can subsequently be employed to guide the modification towards a realistic labeling of the remaining pixels. As an additional advantage of the proposed approach, it overcomes the intrinsic tendency of FCM to recommend solutions that equalize cluster populations. This is especially important when one deals with imbalanced datasets, where one or more classes are poorly described in terms of the number of representative patterns.

The SSFCM method, though largely motivated by the standard fuzzy c-means, is not to be regarded as a true generalization of it because the core optimization activities of the latter have been redesigned, therefore leading to a different objective function. The partition matrix U will have the form $U = [U^d|U^u]$ where U^d holds the set of labeled patterns and U^u stands for the (most likely larger) set of unlabeled objects of the dataset. The first set of cluster prototypes is computed by relying upon the labeled data alone. This results in an initial number of cluster centroids that have no influence coming from the unlabeled objects. Since the knowledge-based hints are assumed to be reliable and must be preserved throughout the clustering process, only the columns of the U^u matrix are updated in an specific manner. Subsequent updates for the cluster prototypes are realized in the traditional way, i.e. all columns of U are utilized to recompute the cluster prototypes after the first pass of the algorithm. The authors were fully cognizant about the fact that, in practice, only a small subset of labeled data can be provided by the experts with an acceptable degree of confidence. In order to counter the influence exercised by the (larger) unlabeled portion of the data set, it became a must to stress the importance of the external knowledge than that of the still unknown objects. This was attained by introducing a set of weights $\mathbf{w} = (w_1, \ldots, w_{n_d})$, being n_d the number of labeled patterns. Being this so, it is possible to weigh each pattern differently or to apply a common weighing scheme to all labeled entities. The $\mathbf{w}$ vector is application-dependent and must be set up in a careful way. Bearing all this in mind, the SSFCM was conveniently devised and its performance proved to be higher than other fully supervised approaches.

Assuming that the labeled portion of the partition matrix U must be kept fixed impels us to seek for the most suitable membership grades for the unlabeled examples. While traditional alternate optimization techniques provide a fast response to the want for optimal membership grades and cluster prototypes, this sort of methods are very sensitive to initialization and usually get caught in local optima. That's why evolutionary programming is considered in [15] as an appealing choice for coming up with a good clustering result, this being thought of in terms of the highest possible classification rate of the labeled examples and, at the same time, achieving compact clusters on the basis of the unlabeled patterns. These two criteria are mixed in the objective function ruling the ensuing optimization activities. Each chromosome(individual) of the population encodes a prospective matrix U^u, that is, the unlabeled portion of the partition matrix. Though more time-consuming than SSFCM, the evolutionary semi-supervised fuzzy clustering (ESSFC) approach proved to outperform SSFCM and a fully supervised classifier over two benchmark data sets.

The above algorithms perform on the assumption that the knowledge-based hints (membership grades) fueled by the expert are in direct correspondence to the array of classes describing the problem. A wider setting envisioning classes as meta-concepts encompassing the actual clusters is presented in [16]. No longer domain knowledge tips can be regarded as membership grades but as class (in the broader sense) information for some or all patterns in the data set. A penalty term added to the objective function guarantees that no object is assigned to a cluster belonging to a class different than the pattern's. This idea is closer to the one encountered in [17] but enunciated from a different perspective.

3 Some Rough Clustering Schemes

This Sect. is devoted to outline the main features of some rough clustering schemes found in literature. A thorough analysis on their weaknesses and strengths is beyond the aim of this chapter. It is also presupposed that the reader is acquainted with the fundamentals of rough set theory [4, 5]. Though exhibiting some remarkable differences concerning design issues and performance, several aspects blend together to make up the common denominator of these rough clustering approaches, among them:

- The crisp notion of cluster is gracefully extended to embrace a wider concept in which some objects are located at the lower approximation of a cluster (thus implying full membership to it) while others are laid at its upper approximation, this way making room for managing uncertainty about the membership of patterns to clusters.

- The rough sets are "artificially" constructed, meaning that a formal departure from the traditional rough set approach governed by equivalence relations is consciously developed. Some generalizations of rough sets relaxing the assumptions of the underlying equivalence relation [20, 21] become the sound foundation upon which the lower and upper approximations of any set X are shaped. This is done with due respect to some fundamental properties of rough sets, which ought to be preserved if we are to end up with interpretable results.

The so-built rough set model is guaranteed to satisfy the following properties:

1. A member of any lower approximation is also a member of the corresponding upper approximation.
2. An object must belong to at most one lower approximation.
3. If an object is not member of any lower approximation, then it must belong to two or more upper approximations.

3.1 Rough Set Genome

Genetic Algorithms (GAs) [22] were seamlessly combined with rough sets to bring forth a rough set genome [7] for classification tasks in an unsupervised context. Each individual (chromosome) of the population is encoded in such a way that it represents the entire classification scheme.

Let $U = \{u_1, \ldots, u_n\}$ be the set of objects to be partitioned into m classes given by $U/P = \{X_1, \ldots, X_m\}$. For each object in the dataset, a gene is added to the chromosome. This gene, in turn, is comprised of a $2m$-length binary string. The first m bits are associated with the lower approximations of the different clusters to be created while the remaining ones correspond to the upper approximations. This means, for example, that only a single bit in the first half of the binary string could be on at every moment of the chromosome's life cycle, in perfect compliance with the aforementioned rough set basic properties to be preserved.

Figure 1 displays some examples of invalid genes within an individual that can not be generated during the optimization process. To ensure this, the authors leaned upon an already implemented GA library [23] which makes it possible to enumerate a set of valid gene values. All the standard genetic operators (mutation, crossover, inversion and selection) will then only create genomes that have these valid values.

One of the key aspects of the configuration of any evolutionary algorithm is the fitness function , mainly intended for quantitatively assessing the quality of a given individual. Lingras chose the within-group-error measure proposed by Sharma and Werner [24] as shown in expression (1) below.

	Lower			Upper		
	$\underline{A}(X_1)$	$\underline{A}(X_2)$	$\underline{A}(X_3)$	$\overline{A}(X_1)$	$\overline{A}(X_2)$	$\overline{A}(X_3)$
$invalidGene_1$	1	0	0	0	0	0
$invalidGene_2$	0	1	0	0	1	1
$invalidGene_3$	1	0	1	0	0	0
$invalidGene_4$	0	0	0	1	0	0
$invalidGene_5$	0	0	0	0	0	0

Fig. 1. Some examples of invalid genes for an individual in the RSG approach

$$\Delta = \sum_{i=1}^{m} \sum_{u_h, u_k \in X_i} d(u_h, u_k) \tag{1}$$

where $d(u_h, u_k)$ is any distance measure between patterns u_h and u_k and can be chosen according to the problem at hand. In his study, Lingras makes a distinction between three different cases one would encounter during the optimization:

1. Both u_h and u_k lie at the same lower approximation $\underline{B}(X_i)$.
2. u_h is in a lower approximation $\underline{B}(X_i)$ and u_k is in the corresponding upper approximation $\overline{B}(X_i)$ and case 1 is not applicable.
3. Both u_h and u_k belong to the same upper approximation $\overline{B}(X_i)$ and the past two cases are not applicable.

The above options give rise to three corresponding within-group-error measures $\Delta_1, \Delta_2, \Delta_3$ which, additionally, do not carry the same importance for the global clustering process. That is, it is more important the quest for consistency among patterns known to be full members of the lower approximation than among patterns placed outside this confidence region. So, the final within-group-error measure is nothing but a weighted sum of the aforesaid factors:

$$\Delta_{total} = w_1 \times \Delta_1 + w_2 \times \Delta_2 + w_3 \times \Delta_3 \tag{2}$$

with $w_1 > w_2 > w_3$ for obvious reasons. However, the final expression for computing the fitness of an individual reads as in expression (3):

$$\mathrm{f} = p \times precision + \frac{e}{\Delta_{total}} \tag{3}$$

where p, e are additional parameters and the precision measure (percentage of objects classified in lower approximations) is incorporated to the fitness function in an attempt to offset the natural tendency of the GAs, which will try to assign most of the objects to upper approximations since $w_3 < w_1$. Maximizing f translates into a less total error and a greater precision.

Overall, five parameters are now in place which need to be properly tuned for the algorithm to behave well. This adaptation is heavily dependent on the application's characteristics.

Another choice for devising fitness functions relative to the overall classification task would be to adapt some well-known cluster validity indices to the rough set framework, as done in [10] with the Davies-Bouldin and Dunn indices [25].

The rough set genome (RSG) clustering scheme was applied to the classification of 350 highway sections in the province of Alberta, Canada into commuter/business, long distance and recreational. It was expected to specify lower and upper approximations for these classes. The feasibility of the approach was empirically demonstrated in this case, but shortly after it was obvious that RSG did not scale well to large datasets, such as the web mining tasks undertaken with data catering to first and second year computer science students, who were meant to be tagged as studious, crammers or workers. RSG managed to group 1,264 visits occurring during a two-week period around midterm for the course but it proved unfit for coping with the 8,442 visits in a sixteen-week-period [8]. This is mostly caused by the number of times the distance function is invoked along the algorithm's execution. In this latter case, it was in the order of the 21 trillion times for a hundred individuals evolving across 1000 generations. In a short time, a more powerful and less demanding rough clustering algorithm would be designed.

3.2 Rough C-Means

The partitive C-Means [6] algorithm is perhaps one of the most popular clustering techniques. Despite the fact that it provides a crisp classification of patterns, which might not be appropriate in many real-life scenarios, conventional applications resort to its low computational cost and demonstrated efficiency in order to come up with a global view of the intrinsic relationships lying among data. Moreover, Lingras and Huang illustrated the computational advantages of the C-Means approach for large datasets [26].

In C-Means, randomly selected objects are used as the cluster prototypes during the algorithm's first iteration. The objects are then assigned to clusters on the basis of their distance to the closest prototype. The newly formed clusters are then used to determine new centroids. The process goes on until the clusters stabilize, i.e. no more new assignments from patterns to clusters are made.

Extending classical C-Means to the rough set framework is straightforward once we have found how to have the basic rough set properties described at the beginning of this Sect. met. Algorithm 1 depicts the entire workflow of Rough C-Means (RCM).

Algorithm 1. Rough C-Means

procedure RCM(clusters c, dist measure d, parameters $w_{low}, w_{up}, threshold$)
  repeat
    Assign initial centroids v_i for the c clusters
    for each object (pattern) x_k in the dataset **do**
      Compute distance to all cluster centroids $D = \{d_{jk}\}, \quad j \in \{1, \ldots, c\}$
      $d_{ik} \leftarrow$ the minimal distance in D
      for each $j \in \{1, \ldots, c\}$, $j \neq i$ **do**
        if $d_{jk} - d_{ik} < threshold$ **then**
          Assign x_k to both upper approximations $x_k \in \overline{B}X_i, x_k \in \overline{B}X_j$
          and x_k can not be a member of any lower approximation
        else
          Assign x_k to the lower approximation $x_k \in \underline{B}X_i$
        end if
      end for
    end for
    Compute new cluster centroids v_i according to (4)
  until there are no more new assignments of objects
  Output $\underline{B}X_i, \overline{B}X_i$ for each cluster $i \in \{1, \ldots, c\}$
end procedure

$$v_i = \begin{cases} w_{low} \dfrac{\sum_{x_k \in \underline{B}X_i} x_k}{|\underline{B}X_i|} + w_{up} \dfrac{\sum_{x_k \in BNX_i} x_k}{|BNX_i|}, & \text{if } \underline{B}X_i \neq \emptyset \wedge BNX_i \neq \emptyset \\ \dfrac{\sum_{x_k \in BNX_i} x_k}{|BNX_i|}, & \text{if } \underline{B}X_i = \emptyset \wedge BNX_i \neq \emptyset \\ \dfrac{\sum_{x_k \in \underline{B}X_i} x_k}{|\underline{B}X_i|}, & \text{otherwise} \end{cases} \tag{4}$$

where BNX_i represents the boundary region of the cluster X_i, w_{low} and w_{up} are the relative importance of the lower and upper approximations, respectively. These values are often chosen so that $w_{low} + w_{up} = 1$. Besides, the "threshold" parameter is critical in the performance of the algorithm because it is involved in determining whether a pattern will be assigned to a lower approximation or to two or more upper approximations. Usually, $threshold \in [0, 0.5]$. A small threshold value will provoke that many objects will be assigned to any of the existing lower approximations. This might lead to a more accurate overall classification but at the cost of disregarding the topological aspects of the data. Careful selection of these values is a painstaking task which can be aided by evolutionary computation, as reported in [27].

Notice from the above expression that when all clusters are crisp (all boundary regions are empty) the RCM algorithm boils down to the traditional C-Means method. Encouraging, interpretable results were obtained after a comparison of both approaches in the context of mining three students' web sites [28]. It was observed that there were many similarities and a few differences between the characteristics of conventional and interval clusters for the three web sites. The interval set representation of clusters made it easier to identify these subtle differences between the three courses than the conventional C-means approach.

As to the former rough set genome approach, RCM proved to consume much less computational resources than its predecessor, this measured in the number of times the most frequent instruction (invoking the distance function) was executed [8]. With RCM, the 8,442 visits recorded at the students' web log were processed only in a few minutes. This speaks highly about its scalability properties. Although some improvements to the method are always a plausible and very welcomed option [29], this algorithm preserves good knowledge discovery capabilities and is an appealing choice to be regarded as one of the benchmarks in the field.

3.3 Rough Fuzzy C-Means

A fuzzy version of the RCM algorithm was designed by Mitra et al [10]. Now, each cluster receives fuzzy inputs in the form of membership grades u_{ij}, which translates into a greater robustness of the clustering with respect to different choices of parameters.

The partition matrix U borrowed from FCM plays a pivotal role in determining the membership of patterns to clusters. A pattern belongs to two or more upper approximations if the difference in its associated membership grades is less than some predefined threshold, otherwise the pattern is assigned to the lower approximation of the cluster with maximal membership value. The steps of the RFCM approach are displayed in Algorithm 2.

Algorithm 2. Rough Fuzzy C-Means

procedure RFCM(c, fuzzifier m, distance d, parameters $w_{low}, w_{up}, threshold$)
  repeat
    Assign initial centroids v_i for the c clusters
    Compute u_{ik} by (5) for c clusters and N patterns.
    for each object (pattern) x_k in the dataset **do**
      $u_{ik} \leftarrow$ the maximal membership grade for pattern k
      for each $j \in \{1, \ldots, c\}$, $j \neq i$ **do**
        if $u_{ik} - u_{jk} < threshold$ **then**
          Assign x_k to both upper approximations $x_k \in \overline{B}X_i, x_k \in \overline{B}X_j$
          and x_k can not be a member of any lower approximation
        else
          Assign x_k to the lower approximation $x_k \in \underline{B}X_i$
        end if
      end for
    end for
    Compute new cluster centroids v_i according to (6)
  until there are no more new assignments of objects
  Output $\underline{B}X_i, \overline{B}X_i$ for each cluster $i \in \{1, \ldots, c\}$
end procedure

$$u_{ik} = \frac{1}{\sum_{j=1}^{c} \left(\frac{d_{ik}}{d_{jk}} \right)^{\frac{2}{m-1}}} \tag{5}$$

$$v_i = \begin{cases} w_{low} \frac{\sum_{x_k \in \underline{B}X_i} u_{ik}^m x_k}{\sum_{x_k \in \underline{B}X_i} u_{ik}^m} + w_{up} \frac{\sum_{x_k \in BNX_i} u_{ik}^m x_k}{\sum_{x_k \in BNX_i} u_{ik}^m}, & \text{if } \underline{B}X_i \neq \emptyset \wedge BNX_i \neq \emptyset \\ \frac{\sum_{x_k \in BNX_i} u_{ik}^m x_k}{\sum_{x_k \in BNX_i} u_{ik}^m}, & \text{if } \underline{B}X_i = \emptyset \wedge BNX_i \neq \emptyset \\ \frac{\sum_{x_k \in \underline{B}X_i} u_{ik}^m x_k}{\sum_{x_k \in \underline{B}X_i} u_{ik}^m}, & \text{otherwise} \end{cases} \tag{6}$$

A recommended parameter configuration is $m = 2$, $w_{up} = 1 - w_{low}$, $0.5 < w_{low} < 1$ and $0 < threshold < 0.5$ although the optimal set of values for a given problem will be reached after undergoing a tuning stage.

4 Partially Supervised Rough Clustering

The introduction of some core concepts from rough set theory, like set approximations, into the conventional clustering machinery allows to provide a finer degree of control over the way patterns are assigned into clusters, thus continuing the departure from the original crisp allocation model which, truth be told, falls short of providing a coherent paint about the data at hand in many real-life scenarios.

Although this departure was successfully initiated with fuzzy membership grades at the heart of the clustering algorithms, as one can see along the domain of fuzzy clustering, when it comes to partial supervision mechanisms in this realm, there is an underlying need to quantify the degree of membership of a pattern to all clusters, viz it is necessary to fill out a column of the partition matrix in advance. This is seldom an straightforward task for the expert in charge of providing the knowledge-based tips which will have a direct bearing in the overall clustering result, especially if he is dealing with a borderline pattern. Actually, the trend of describing vague concepts (i.e. memberships of objects to groups) by means of precise, numerical values has raised many concerns in the past and a quest for more qualitative models has been undertaken long time ago.

In daily practice, however, it is more common to tag an object as definitely pertaining to a predefined class or, the other way around, to be confident about excluding the object from some group. This, in the language of rough set theory, is equivalent to locate the object inside the lower approximation (positive region) of the concept or to place it outside its upper approximation (i.e., within the negative region), respectively.

In light of this, one easily realizes that partial supervision mechanisms in rough clustering algorithms must be formulated in terms of memberships (and non-memberships) of patterns to classes. While foreign evidence about the inclusion of a pattern into two or more upper approximations could eventually be considered, the inherent ambiguity associated with this information could lead us to an undesirable outcome. This is why we confine ourselves to regard only precise domain knowledge hints, even though if they come in the fashion of fuzzy membership degrees, as happens with fuzzy clustering.

Mathematically put, we define a set $POS(X)$ where $X = \{X_1, \ldots, X_m\}$ is the set of m classes to be discovered during clustering and $POS(X) = \{POS(X_1), \ldots, POS(X_m)\}$ in which every $POS(X_i)$ contains the set of objects known to belong to the X_i class. Likewise, we define the set $NEG(X) = \{NEG(X_1), \ldots, NEG(X_m)\}$ with each $NEG(X_i)$ holding the set of objects clearly not members of the X_i group.

Contrary to what one encounters in partially supervised fuzzy clustering approaches, where the membership grades of a pattern to all possible clusters is to be provided in advance, the limited knowledge the expert might have on a certain object (for instance, the physician is sure that patient P_j doesn't suffer from sclerosis but can not say anything about the patient being or not being hit by arthritis) still proves useful as it is embedded into the rough clustering machinery.

We can equip the existing rough clustering algorithms with mechanisms of partial supervision by making slight modifications to their workflow of activities which, generally speaking, take few computational effort. In particular, we have assumed that the RCM approach will be able to assimilate external information coming in the form of positive and negative regions of clusters, whereas for the RFCM algorithm, a subset of labeled patterns has been supplied (i.e., some columns of the partition matrix are known beforehand). These, of course, are not the only ways in which foreign guidance schemes can be devised and incorporated into rough clustering approaches, but serve well to illustrate this point.

4.1 Partially Supervised Rough C-Means

The sequence of detailed activities for the Partially Supervised Rough C-Means (PS-RCM) algorithm is as follows:

$$v_i = \begin{cases} \dfrac{\sum\limits_{x_k \in POS(X_i)} x_k}{|POS(X_i)|}, & POS(X_i) \neq \emptyset; \\ \text{random pattern}, & \text{otherwise} \end{cases} \tag{7}$$

$$v_i = \begin{cases} w_{low} \times \varphi_i + w_{up} \times \phi_i, & \text{if } \underline{B}X_i \neq \emptyset \wedge BNX_i \neq \emptyset \\ \phi_i, & \text{if } \underline{B}X_i = \emptyset \wedge BNX_i \neq \emptyset \\ \varphi_i, & \text{otherwise} \end{cases} \tag{8}$$

Algorithm 3. Partially Supervised Rough C-Means

1: **procedure** PS-RCM(clusters c, distance measure d, parameters $w_{low}, w_{up}, threshold$, positive regions $POS(X)$, negative regions $NEG(X)$)
2:   **repeat**
3:     Compute initial centroids v_i by using (7)
4:     **for** each object (pattern) x_k in the dataset **do**
5:       **if** $x_k \in POS(X_i),\ i = \{1, \ldots, c\}$ **then**
6:         Assign $x_k \in \underline{B}X_i$
7:       **else**
8:         Compute distance to all cluster centroids $D = \{d_{jk}\}, j \in \{1, \ldots, c\}$
9:         $d_{ik} \leftarrow$ the minimal distance in D such that $x_k \notin NEG(X_i)$
10:         **for** each $j \in \{1, \ldots, c\}$, $j \neq i$ and $x_k \notin NEG(X_j)$ **do**
11:           **if** $d_{jk} - d_{ik} < threshold$ **then**
12:             Assign x_k to both upper approxs $x_k \in \overline{B}X_i, x_k \in \overline{B}X_j$
13:             and x_k can not be a member of any lower approximation
14:           **else**
15:             Assign x_k to the lower approximation $x_k \in \underline{B}X_i$
16:           **end if**
17:         **end for**
18:       **end if**
19:     **end for**
20:     Compute new cluster centroids v_i according to (8)
21:   **until** there are no more new assignments of objects
22:   Output $\underline{B}X_i, \overline{B}X_i$ for each cluster $i \in \{1, \ldots, c\}$
23: **end procedure**

where:

$$\varphi_i = \frac{\sum\limits_{x_k \in POS(X_i)} w_k x_k + \sum\limits_{x_k \in \underline{B}X_i - POS(X_i)} x_k}{\sum\limits_{x_k \in POS(X_i)} w_k + |\underline{B}X_i - POS(X_i)|}$$

$$\phi_i = \frac{\sum\limits_{x_k \in BNX_i} x_k}{|BNX_i|} \tag{9}$$

The first observation here is that the initial set of cluster prototypes (line 3), which was composed of randomly chosen patterns in the original RCM algorithm, now exploits the supervised hints in an active way. That is to say, if some patterns are known to belong to a given class, then its initial cluster centroid may be set up as the mean vector of those lying within its positive region. This allows us to "anchor" the cluster prototype by relying upon solid domain knowledge clues available, which will have a great importance later, as we will see during the experiments. All other classes having no a priori information of this type are assigned a random pattern in the dataset as their initial cluster prototype.

Since the partial supervision in this scenario arises as positive and negative regions of the clusters, we must prioritize the assignment of any pattern to the corresponding cluster's lower approximation (lines 5–6). We take for granted that the external information is consistent, so an object can only belong to at most one lower approximation.

When a pattern has not been a priori assigned to a certain class, the assignment must be done in the traditional way, i.e. considering the nearness of the pattern to the cluster prototypes (lines 8–17). Notice that we must carefully check that no negative-region-based tip is violated during the process, so the pattern cannot be allocated neither to the lower nor to the upper approximation of any class whose negative region the pattern belongs to.

The last significant modification in the proposed approach lies in the way cluster prototypes are recalculated (line 20). This time we purposefully make a distinction between those patterns definitely belonging to a concept according to the expert's consideration and those allocated there as a consequence of the algorithm's computations, thus highlighting the importance of the former ones. Each labeled pattern x_k has its own weight w_k which denotes its worth within the data set. What weight must accompany each labeled pattern is something that must be determined in view of the topological features of the data set we are working with. Further on we will illustrate this point by using a synthetic data repository.

4.2 Partially Supervised Rough Fuzzy C-Means

Although the RFCM algorithm [10] is not substantially different from its predecessor, the way patterns are assigned to classes is now governed by fuzzy membership grades. Labeling a share of the underlying partition matrix takes us into the realm of partially supervised fuzzy clustering but still in presence of rough clusters. In light of this, we can borrow some ideas from [14] and come up with the semi-supervised version of RFCM, which is fully outlined below.

Let us make clear some notation first. Given that some objects have been labeled, the set of patterns X can now be described as $X = X^d \cup X^u$ where X^d is the subset of labeled patterns whereas X^u contains the unlabeled ones, $n_d = |X^d|, n_u = |X^u|$. A similar situation occurs with the partition matrix, that is $U = [U^d|U^u]$ and U^d remains unaltered throughout the algorithm's execution.

$$v_i = \begin{cases} \dfrac{\sum\limits_{k=1}^{n_d} (u_{ik}^d)^m x_k^d}{\sum\limits_{k=1}^{n_d} (u_{ik}^d)^m}, & 1 \leq i \leq c, \qquad X^d \neq \emptyset; \\ \text{random pattern}, & \text{otherwise} \end{cases} \tag{10}$$

Algorithm 4. Partially Supervised Rough Fuzzy C-Means

1: **procedure** PS-RFCM(nr. of clusters c, fuzzifier m, distance measure d, parameters $w_{low}, w_{up}, threshold$, labeled subset of patterns X^d, labeled partition matrix U^d)
2: **repeat**
3: Compute initial centroids v_i by using (10)
4: Compute u^u_{ik} by (11) for c clusters and n_u patterns.
5: **for** each object (pattern) x_k in the dataset **do**
6: $u_{ik} \leftarrow$ the maximal membership grade for pattern k
7: **for** each $j \in \{1, \ldots, c\}$, $j \neq i$ **do**
8: **if** $u_{ik} - u_{jk} < threshold$ **then**
9: Assign x_k to both upper approximations $x_k \in \overline{B}X_i, x_k \in \overline{B}X_j$
10: and x_k can not be a member of any lower approximation
11: **else**
12: Assign x_k to the lower approximation $x_k \in \underline{B}X_i$
13: **end if**
14: **end for**
15: **end for**
16: Compute new cluster centroids v_i according to (12)
17: **until** there are no more new assignments of objects
18: Output $\underline{B}X_i, \overline{B}X_i$ for each cluster $i \in \{1, \ldots, c\}$
19: **end procedure**

$$u^u_{ik} = \frac{1}{\sum_{j=1}^{c}\left(\frac{d_{ik}}{d_{jk}}\right)^{\frac{2}{m-1}}}, \qquad 1 \leq i \leq c,\ 1 \leq k \leq n_u \tag{11}$$

$$v_i = \begin{cases} w_{low} \times \chi_i + w_{up} \times \psi_i, & \text{if } \underline{B}X_i \neq \emptyset \wedge BNX_i \neq \emptyset \\ \psi_i, & \text{if } \underline{B}X_i = \emptyset \wedge BNX_i \neq \emptyset \\ \chi_i, & \text{otherwise} \end{cases} \tag{12}$$

where:

$$\chi_i = \frac{\sum\limits_{x^d_k \in \underline{B}X_i} w_k (u^d_{ik})^m x^d_k + \sum\limits_{x^u_k \in \underline{B}X_i} (u^u_{ik})^m x^u_k}{\sum\limits_{x^d_k \in \underline{B}X_i} w_k (u^d_{ik})^m + \sum\limits_{x^u_k \in \underline{B}X_i} (u^u_{ik})^m}$$

$$\psi_i = \frac{\sum\limits_{x^d_k \in BNX_i} w_k (u^d_{ik})^m x^d_k + \sum\limits_{x^u_k \in BNX_i} (u^u_{ik})^m x^u_k}{\sum\limits_{x^d_k \in BNX_i} w_k (u^d_{ik})^m + \sum\limits_{x^u_k \in BNX_i} (u^u_{ik})^m} \tag{13}$$

Just like in the previous algorithm, the calculation of the initial cluster prototypes has been modified (line 3) so as to allow only labeled patterns exercise influence over the early class representatives, as expression (10) clearly states. There is no need to pick a random pattern as a cluster centroid, as in PS-RCM, as long as some labeled data are provided, since the auxiliary information available includes membership degrees for all clusters.

Afterwards, only the unlabeled entries of the partition matrix want to be calculated (line 4) due to the fact that the columns of U^d are regarded as reliable and must therefore be fixed. This computation involves only unlabeled objects from X^u, so when we read d_{ij} in expression (11) we mean $\|x_k^u - v_i\|$.

The last major change, like in PS-RCM, appears when updating the cluster prototypes (see line 16). It happens again that some distinction must be made between labeled and unlabeled patterns, no matter whether they have been allocated to the lower or to the upper approximation of any cluster. Expression (12) boils down to (6) when there is no supervised information at all.

5 Experimental Studies

In order to estimate the effects of the injected knowledge-based hints into some previously described clustering methods, it was necessary to conduct some experiments using both synthetic and real-world data. There is a need to shed light on how the traditional behavior of the clustering algorithms might be affected by partial supervision for data sets exhibiting different degrees of overlap between their classes. Also important is to determine how many patterns are to be labeled per class and what their corresponding weights should be so that the number of overall misclassifications gets reduced.

We have employed four rough clustering techniques, i.e. RCM, RFCM and their partially supervised variants (PS-RCM and PS-RFCM) which were presented in the previous Sect. The experiments were realized against three data sets having different characteristics. The first one (SD) is comprised of 43 synthetic, two-dimensional patterns which are unevenly distributed into two classes (40 for the bigger class and 3 for the smaller class). It is important to stress that these classes are well defined in the feature space and do not overlap. The whole data can be found in Appendix A of reference [14].

The second dataset (Iris) comes from the UCI Machine Learning Repository [30] and has been extensively used as a standard benchmark for testing both supervised and unsupervised learning algorithms. Patterns are equally distributed over the three classes (50 each) although two of them are difficult to discern due to the substantial degree of overlap.

Finally, a third data set, which we will denote by "AC" (anticancer), is concerned with 961 patterns representing drugs whose anticancer properties are described in terms of 11 medical features. The two classes stand for the (relative) success or failure in battling the disease and consist of 298 and 663 cases, respectively. This is the most imbricated data set since most of the patterns from both classes form an homogeneous mesh around the center of the 2D feature space. With the exception of a handful of patterns belonging to the first class (which resemble a large comet tail when plotted) and a bunch of objects

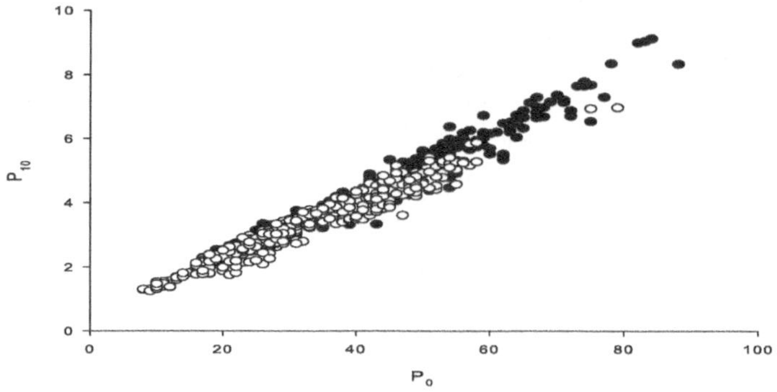

Fig. 2. A 2-D plot of the anticancer database (P_0 vs P_{10})

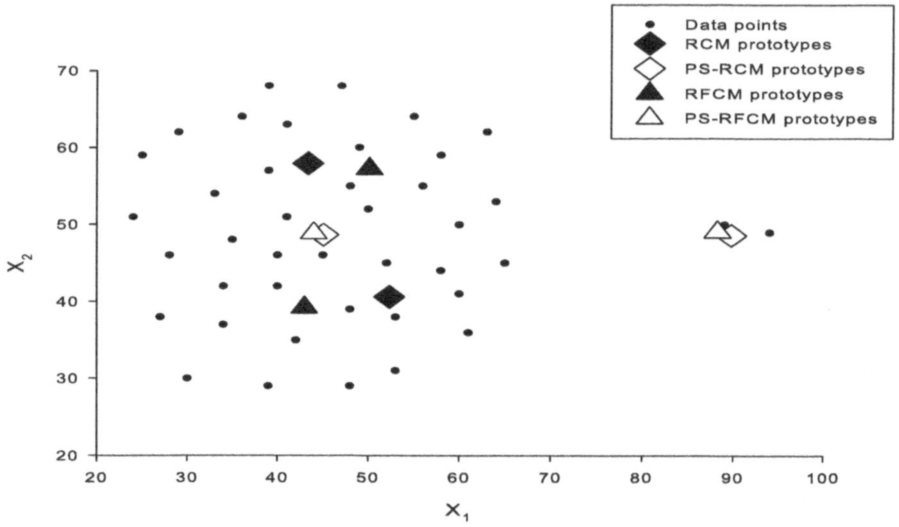

Fig. 3. Prototypes for the synthetic data set

located near the coordinate center (belonging to the second class), all remaining patterns are very difficult to segregate. Figure 2 displays a schematic view of the Anticancer data set.

Experiment 1. Largely used fuzzy clustering algorithms such as FCM and FCMED are prone to equalize the cluster populations, due to the minimization of the underlying objective function (the sum of the squared errors of the distances from patterns to cluster centroids). This is particularly detrimental in imbalanced data sets, i.e. data which are unevenly distributed into the existing classes, giving rise to poor representation of some natural groupings. With the SD data set, the observed behavior [14] is that the centroid of the smaller class is drifted toward the centroid of the larger class, thus leading to the incorrect assignment of some patterns and ending up with a more or less equalized population.

This very tendency has been corroborated with the unsupervised rough clustering algorithms. From Figure 3 one can notice that both RCM and RFCM locate the two centroids within the cloud of patterns actually belonging to the first class, for the influence of the three foreign data patterns in the middle right of the feature space is underestimated. Though RCM and its fuzzy version are not governed by any objective function to be optimized, the way the cluster prototypes are updated bestows the same importance to all objects lying in the lower approximation of each cluster and naturally arrives to such a result.

On the contrary, the partially supervised clustering algorithms (PS-RCM and PS-RFCM) clearly place the prototypes in their fair location, owing to the foreign guidance received by two labeled patterns which were a priori flagged as certain members of each lower approximation, i.e. $x_1 \in \underline{B}X_1$ and $x_{43} \in \underline{B}X_2$. The increase in the weight of the training pattern corresponding to the smallest class leads to a steeper impact of the knowledge-based hints and a more stable picture of the ensuing knowledge structures, as a direct outcome of expressions (9) and (13).

Figure 4 schematically portrays how the rough clusters were shaped after the execution of every algorithm. One can readily observe that unsupervised rough clustering techniques are incapable of capturing the topological aspects of the synthetic data and end up nearly balancing the number of patterns assigned to each cluster. This is not the behavior of their supervised versions, which significantly reduce the size of the boundary regions and mold the clusters in a much more accurate way, owing to the displacement of the cluster prototypes after being drawn by the two labeled data points. Moreover, the degree of fuzziness present in PS-RFCM approach led to a perfect description of the original data without any borderline pattern.

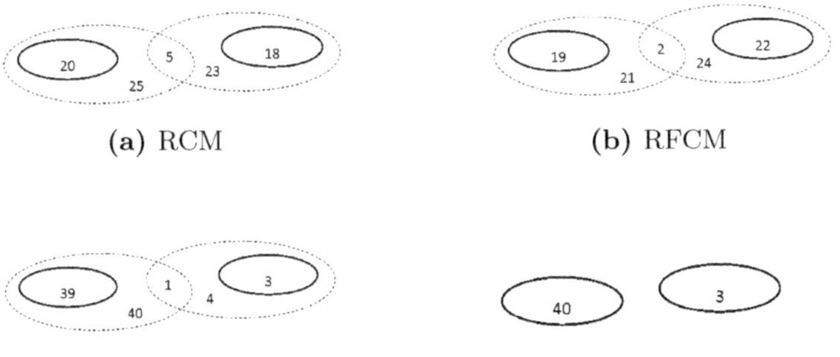

(a) RCM

(b) RFCM

(c) PS-RCM

(d) PS-RFCM

Fig. 4. Approximations for the synthetic data set. Dotted lines represent upper approximations whereas solid lines are lower approximations.

Table 1. Number of misclassifications reported by the algorithms with no auxiliary information provided

	RCM	RFCM	PS-RCM	PS-RFCM
Synthetic data	22 (51.2%)	21 (48.9%)	22 (51.2%)	18 (41.9%)
Iris	30 (20%)	28 (18.7%)	30 (20%)	29 (19.3%)
Anticancer	380 (39.5%)	385 (40.1%)	377 (39.2%)	384 (40%)

Table 2. Weight values and number of wrongly classified examples in the synthetic data set

	PS-RCM	PS-RFCM
w=1	21	4
w=2	10	0
w=3	8	2
w=4	1	2
⋮		
w=10	1	2

One encounters the same poor performance exhibited by RCM across all data repositories, as depicted in Table 1. Fully unsupervised pursuit of accurate knowledge structures doesn't lead to good results in presence of the three data sets under consideration.

Table 2 displays the improvements accomplished through the use of partial information with the SD repository. The same vectors x_1 and x_{43} were labeled as shown above. The values of **w** used in each run of the semi-supervised algorithms and the number of errors incurred during classification are reported as well. For both methods, parameters w_{low} and *threshold* were set to 0.9 and 0.3, respectively. While it is crystal clear that the supplied auxiliary tips had a greater impact on PS-RFCM (reflected in its accurate classification rate), even reaching the optimal classification for $\mathbf{w} = 2$, increasing values of **w** for all training patterns do not eventually lessen the number of misclassifications, as observed with PS-RCM.

Therefore some guidelines for setting the weights will become a useful tool if we are to achieve an acceptable classification rate. An appealing discussion of this topic, originally confined to the domain of fuzzy clustering, can be found in [14] and some of its ensuing considerations may be taken into account when dealing with rough clustering techniques. We can anticipate that, although some guiding principles in this sense can be derived after pondering, among other factors, the number and quality (reliability) of the training patterns per class, the intrinsic characteristics of the particular problem (degree of overlap between the classes, population density per class, etc.) will be better managed by an automated, heuristic-driven approach that proposes the optimal weight vector. This, of course, lies far beyond the scope of this chapter.

Experiment 2. One interesting point when dealing with rough clustering schemes under partial supervision arises when we own some knowledge about the domain but are hesitant on how to describe it, i.e. given the multiple choices of foreign guidance available as rough memberships (or non-memberships) of a pattern to a cluster and fuzzy membership grades of objects to classes, we must pick a convenient way for representing what we know about the environment so that it leads to the highest performance among all other options.

Let us shed some light on this by means of straightforward experiments considering the Iris database. Suppose we are fully certain about two exemplars belonging to the Iris-versicolor category, one being tagged as Iris-setosa and a last flower which is known to be Iris-virginica. Let us denote by x_{53}, x_{85}, x_{25} and x_{130} such known patterns and let be C_1, C_2 and C_3 the subsets of patterns corresponding to the Iris-setosa, Iris-versicolor and Iris-virginica classes, respectively. The subscript of the previously labeled objects refers to the zero-based index they occupy within the Iris repository.

The knowledge we possess about the four flowers can be put in various manners. The simplest of them is advising the algorithms that they are full members of the lower approximation of the corresponding clusters, i.e. $x_{25} \in \underline{B}C_1$, $x_{53} \in \underline{B}C_2$ and $x_{130} \in \underline{B}C_3$. With this type of auxiliary information, the PS-RCM method achieved a classification rate of 81% (see Table 3) which, although very close to the behavior of the algorithm without any supervised tips for this

Table 3. Behavior of the semi-supervised rough clustering approaches when fed with different types of knowledge-based hints

Supplied information	Performance	Comments
All positive hints	81%	No misclassified hints
All negative (accurate) hints	78%	$x_{25} \in C_2$, $x_{130} \in C_1$
All negative (imprecise) hints	50%	$x_{85} \in C_1$. C_2 and C_3 nearly blend
Positive and negative hints	81%	No misclassified hints
Fuzzy membership grades	67%	Only $x_{130} \in C_3$. C_2 and C_3 nearly blend

data set, has the advantage that the early prototypes for all clusters have been "anchored", thus allowing the subsequent assignments of patterns to groups to be driven by this positive guidance. The quest for the optimal number of training patterns per class remains a challenge and some underlying tendencies in this sense will be disclosed in the next experiment. We must, however, focus now on the effect provoked by changing the way the supervised information is wired into the clustering mechanisms.

If we choose not to employ lower approximations but negative regions for describing the information about the environment, we will seldom be in position of tagging a pattern as not pertaining to the majority of the clusters, for our knowledge about all categories will very likely be permeated to some extent by uncertainty and vagueness. Notice that the previous situation (i.e. all known patterns belong to some positive region) boils down to assigning those objects to the negative regions of all clusters except the one they are members of. If this is the case, we can talk about an "accurate" knowledge communicated to the clustering engines solely in the form of non-memberships of patterns to classes (2nd row of Table 3). We write accurate within quotation marks because we don't know in advance which cluster belongs to which class in practice. Put in other words, claiming that $x_{25} \in NEG(C_2)$ and $x_{25} \in NEG(C_3)$ will not be enough to classify this pattern as member of Iris-setosa unless the PS-RCM algorithm has grouped all of its related objects into the first cluster. This is deeply rooted in the manner in which initial cluster prototypes are selected in PS-RCM. If no external information is available in the form of positive memberships (as in the previous case), random patterns give rise to the first set of cluster centroids and the problem of cluster mismatch emerges. After this rationale, it is easy to understand why this second scenario, though structurally equivalent to the former one, may lead to some misclassified labeled patterns and, consequently, to a lower performance.

Most of the times, however, we will only be able to state that an object is not indeed a member of some predefined group due to our limited knowledge of the environment. The third row of Table 3 shows that leaning upon a few external hints formulated as non-memberships of patterns to classes takes one to the worst possible scenario in terms of classification performance. The problem of cluster mismatch becomes sharpened because we have less reliable information at hand. In the experiments it was demonstrated that clusters corresponding to Iris-versicolor and Iris-virginica almost completely amalgamated.

The good news is that we can combine both types of supervised information (positive and negative regions) in order to overcome the lack of assurance about all patterns fully belonging to a definite group, which might be hard to get in practice. Both types of auxiliary information are to be seamlessly integrated into the clustering algorithms, reaching its more profoundly visible effect after nourishing the formation of all early cluster prototypes with some already known patterns being located into the respective groups.

Finally, expressing this type of available information as fuzzy membership grades (Table 3, 5th row) allowed only to correctly classify 2/3 of the available

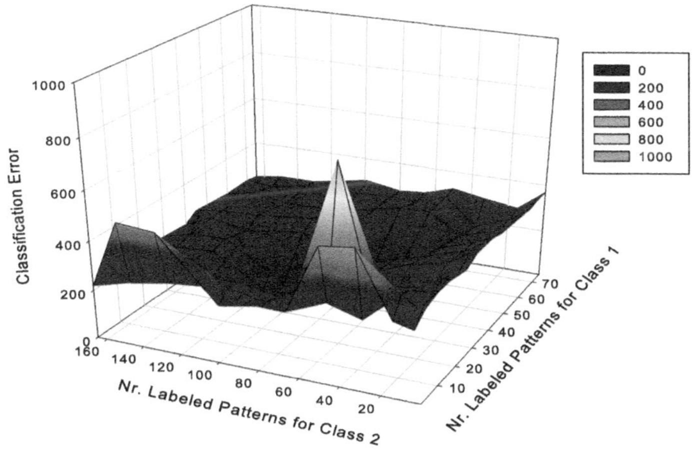

Fig. 5. Labeling a varying number of objects in the 'Anticancer' repository

patterns. Contrary to what one would expect, the PS-RFCM method behaved poorly after being driven by the foreign tips, even worse than the 80.7% classification rate reached with no supervised information at all, resulting in a very similar scenario to the one encountered in the 3rd row of the table. From this observation one can derive that fuzzy membership grades might need to be supplied abundantly to the partially supervised rough clustering approaches before an acceptable classification rate can be accomplished, although this observation may dramatically vary in presence of other data sets exhibiting different characteristics.

Experiment 3. We are going to finish this sect. taking a glimpse at the influence of the number of training patterns over the semi-supervised rough clustering techniques. Let us choose the 'Anticancer' data set for this experiment given that it consists of two largely overlapping clusters, as previously shown in Figure 2.

For each class, a different number of randomly chosen patterns was labeled in each run of the partially supervised rough c-means algorithm. The amount of patterns labeled a priori as sure members of each category varied from a single pattern up to 25% of the total class population. The results are portrayed in Figure 5. Axes x and y of the 3D plot correspond to the number of patterns flagged as full members of the lower approximation of classes 1 and 2, respectively, whereas axis z displays the error rate obtained. The PS-RCM was executed with the following parameter configuration: $w_{low} = 0.9$, $threshold = 0.3$ and $\mathbf{w} = 20$.

The 3D mesh clearly shows that the steep cone near the center of the feature space (whose peak almost reaches 400 wrongly assigned objects) only involves a limited subset of the possible numbers of labeled patterns per class, viz between

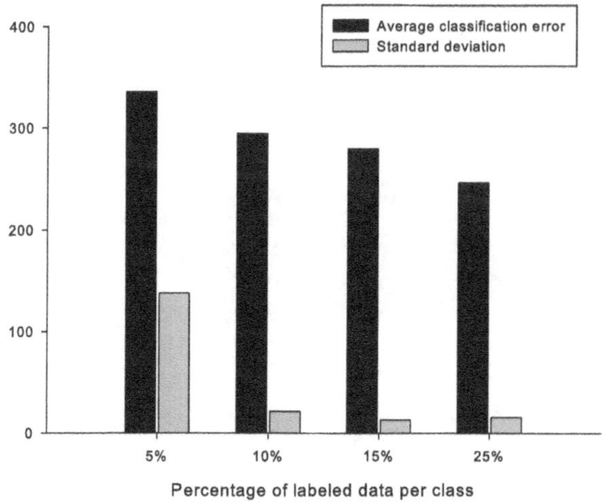

(a) PS-RCM

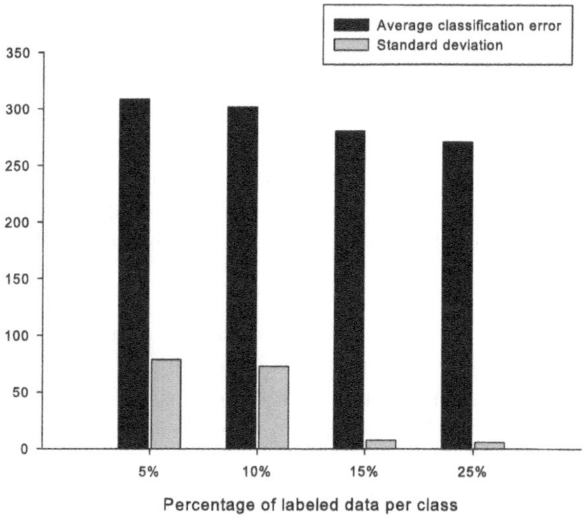

(b) PS-RFCM

Fig. 6. Means and standard deviations of the error rates incurred over 20 iterations with the 'Anticancer' repository for the two semi-supervised rough clustering approaches

30 and 40 patterns of class 1 and 60 - 100 patterns of the second cluster. From the chart one can additionally realize that high classification errors are also the direct outcome of labeling too few patterns in either group, although this effect is more visible regarding the less crowded cluster.

Notice that in no case the effect of partial supervision allows to drive the number of misclassifications below than 100 patterns, being 118 the lowest error rate obtained after randomly picking 24 objects from the first class and 99 objects from the second one and furnishing the algorithm with these knowledge-based tips. Nonetheless, this can be considered as a good, positive impact of the domain knowledge hints over the data set, given the highly interwoven nature of its underlying knowledge structures.

Moreover, it seems that when an acceptable number of objects (say, above 20%) for each class is supplied in advance to the PS-RCM method, there is an evident benefit which translates into an increase in the classification rate, as one can remark by looking at the roughly plain, valley-like region at the rear part of the graph.

This conclusion is empirically supported by the evidence collected after 20 iterations of every partially supervised rough clustering method over the repository under consideration. This information has been summarized and presented to the reader in Figure 6 in terms of averages and standard deviations of the error rates. The common denominator is that a greater subset of training patterns leads to a more accurate classification, even when there are two or more very overlapping clusters. The impact is more remarkable in the case of the PS-RFCM algorithm, given the inherent advantage that represents having fuzzy sets modeling the assignment of patterns to clusters. Another interesting issue is that the standard deviation of the error rates dramatically decreases as more patterns are labeled beforehand, thus becoming a reliable indicator of the pivotal role played by the foreign guidance throughout the clustering schemes.

6 Concluding Remarks

The problem of feeding clustering algorithms with some knowledge-based hints has been extended in this chapter to the realm of rough clustering techniques. More specifically, the rough c-means and rough-fuzzy c-means approaches were endowed with mechanisms of partial supervision leaning upon the observer's assurance of the membership (or non-membership) of some patterns to existing categories, this being expressed either as traditional fuzzy membership grades or simply by allocating objects to positive or negative regions of the underlying concepts under construction.

After examining the evidence collected throughout several conducted experiments, we witness an immediate, superior behavior in the way clustering schemes look for knowledge structures in the data set. While fuzzy entries of the partition matrix continue to be a major source of improvement of the algorithms' performance when associated to a labeled set of patterns, it has been empirically demonstrated that having early cluster prototypes being anchored by patterns known to lie within the confidence region (lower approximation) of some classes exercises a crucial influence over the ensuing optimization activities and leads to more stable, meaningful results.

We acknowledge there exists an array of multiple manners in which foreign guidance could be incorporated into rough clustering approaches. Existing and yet-to-come computational models falling under this umbrella will very likely be able to seamlessly assimilate the knowledge provided by the experts with little further computational effort.

Acknowledgment

We would like to thank the referees for their positive remarks and helpful insights about the chapter. Rafael Falcón and Rafael Bello gratefully acknowledge the support provided by the VLIR-UCLV cooperation programme. Gwanggil Jeon and Jechang Jeong were supported by the Korea Research Foundation Grant funded by the Korean Government (MOEHRD) (KRF-2004-005-D00164).

References

1. Jain, A.K., Murty, M.N., Flynn, P.J.: ACM Computing Surveys 31(3), 264–323 (1999)
2. Sato-Ilic, M., Jain, L.: Introduction to fuzzy clustering. In: Sato-Ilic, M., Jain, L. (eds.). Studies in Fuzziness and Soft Computing, vol. 205. Springer, Heidelberg (2006)
3. Pedrycz, W., Waletzky, J.: Fuzzy Clustering with Partial Supervision. IEEE Trans. on Systems, Man and Cybernetics Part B 27(5), 787–795 (1997)
4. Pawlak, Z.: Rough sets: theoretical aspects of reasoning about data. Kluwer, Dordrecht (1991)
5. Pawlak, Z.: Rough sets. International Journal of Information and Computer Sciences 11, 145–172 (1982)
6. Lingras, P., West, C.: Interval set clustering of Web users with rough k-means. Technical Report No. 2002-002, Saint Mary University, Halifax, Canada (2002)
7. Lingras, P.: Unsupervised rough set classification using GAs. Journal of Intelligent Information Systems 16(3), 215–228 (2001)
8. Lingras, P., Yao, Y.: Time complexity of rough clustering: GAs versus K-means. In: Alpigini, J.J., Peters, J.F., Skowron, A., Zhong, N. (eds.) RSCTC 2002. LNCS, vol. 2475, p. 263. Springer, Heidelberg (2002)
9. Lingras, P.: Rough set clustering for web mining. In: Proc. of 2002 IEEE International Conference on Fuzzy Systems (2002)
10. Mitra, S., Banka, H., Pedrycz, W.: Rough-Fuzzy Collaborative Clustering. IEEE Trans. on Systems, Man and Cybernetics 36(4), 795–805 (2006)
11. Pedrycz, W.: Knowledge-based clustering: from data to information granules. John Wiley & Sons, New York (2005)
12. Pedrycz, W.: Algorithms of fuzzy clustering with partial supervision. Pattern Recognition Letters 3, 13–20 (1985)
13. Kersten, P.R.: Including auxiliary information in fuzzy clustering. In: Proc. Biennial Conf. of the North American Fuzzy Information Processing Society, NAFIPS, pp. 221–224 (1996)
14. Bensaid, A.M., Hall, L.O., Bezdek, J.C., Clarke, L.P.: Partially supervised clustering for image segmentation. Pattern Recognition 29(5), 859–871 (1996)

15. Liu, H., Huang, S.T.: Evolutionary semi-supervised fuzzy clustering. Pattern Recognition Letters 24(16), 3105–3113 (2003)
16. Timm, H., Klawonn, F., Kruse, R.: An extension of partially supervised fuzzy cluster analysis. In: Proc. of the Annual Meeting of the North American Fuzzy Information Processing Society, pp. 63–68 (2002)
17. Pedrycz, W., Waletzky, J.: Fuzzy clustering with partial supervision. IEEE Trans. on Systems, Man and Cybernetics 5, 787–795 (1997)
18. Abonyi, J., Szeifert, F.: Supervised fuzzy clustering for the identification of fuzzy classifiers. Pattern Recognition Letters 24(14), 2195–2207 (2003)
19. Bezdek, J.C.: Pattern recognition with fuzzy objective function algorithms. Plenum Press, New York (1981)
20. Yao, Y.Y., Li, X., Lin, T.Y., Liu, Q.: Representation and classification of rough set models. In: Proc. of Third International Workshop on Rough Sets and Soft Computing, pp. 630–637 (1994)
21. Skowron, A., Stepaniuk, J.: Information granules in distributed environment. In: Zhong, N., Skowron, A., Ohsuga, S. (eds.) RSFDGrC 1999. LNCS, vol. 1711, pp. 357–366. Springer, Heidelberg (1999)
22. Buckles, B.P., Petry, F.E.: Genetic Algorithms. IEEE Computer Press, Los Alamitos (1994)
23. Wall, M.: GALib, a C++ library of genetic components (1993), `http://lancet.mit.edu/ga/`
24. Sharma, S.C., Werner, A.: Improved method of grouping provincewide permanent traffic counters. Transportation Research Record 815, Transportation Research Board 13–18 (1981)
25. Bezdek, J.C., Pal, N.R.: Some new indices for cluster validity. IEEE Trans. on Systems, Man and Cybernetics B 28(3), 301–315 (1998)
26. Lingras, P., Huang, X.: Statistical, Evolutionary and Neurocomputing Clustering Techniques: Cluster-Based vs Object-Based Approaches. Artificial Intelligence Review 23, 3–29 (2005)
27. Mitra, S.: An evolutionary rough partitive algorithm. Pattern Recognition Letters 25(12), 1439–1449 (2004)
28. Lingras, P., Yan, R., West, C.: Comparison of Conventional and Rough K-Means Clustering. In: Wang, G., Liu, Q., Yao, Y., Skowron, A. (eds.) RSFDGrC 2003. LNCS, vol. 2639. Springer, Heidelberg (2003)
29. Peters, G.: Some refinements of rough k-means clustering. Pattern Recognition Letters 39, 1481–1491 (2006)
30. Murphy, P.M., Aha, D.W.: Repository of Machine Learning Databases, `http://www.ics.uci.edu/~mlearn/MLRepository.html`

A Generic Scheme for Generating Prediction Rules Using Rough Sets

Hameed Al-Qaheri[1], Aboul Ella Hassanien[1,2], and Ajith Abraham[3]

[1] Department of Quantitative Methods and Information Systems
College of Business Administration, Kuwait University, Kuwait
[2] Information Technology Department, FCI, Cairo University
5 Ahamed Zewal Street, Orman, Giza, Egypt
alqaheri@cba.edu.kw, a.hassanien@fci-cu.edu.eg, abo@cba.edu.kw
[3] Center for Quantifiable Quality of Service
Norwegian University of Science and Technology
O.S. Bragstads plass 2E, NO-7491 Trondheim, Norway
ajith.abraham@ieee.org, abraham.ajith@acm.org

Abstract. This chapter presents a generic scheme for generating prediction rules based on rough set approach for stock market prediction. To increase the efficiency of the prediction process, rough sets with Boolean reasoning discretization algorithm is used to discretize the data. Rough set reduction technique is applied to find all the reducts of the data, which contains the minimal subset of attributes that are associated with a class label for prediction. Finally, rough sets dependency rules are generated directly from all generated reducts. Rough confusion matrix is used to evaluate the performance of the predicted reducts and classes. For comparison, the results obtained using rough set approach were compared to that of artificial neural networks and decision trees. Empirical results illustrate that rough set approach achieves a higher overall prediction accuracy reaching over 97% and generates more compact and fewer rules than neural networks and decision tree algorithm.

1 Introduction

Over the last few decades statistical techniques such as regression and Bayesian models and econometric techniques have dominated the research activities in prediction. Data mining [10] and computational intelligence techniques such as neural networks, fuzzy set, evolutionary algorithms, rough set theory, machine learning, multi-criteria decision aid (MCDA), etc., emerged as alternative techniques to the conventional statistical and econometric models and techniques that have dominated this field since the 1930s [56] and have paved the road for the increased usage of these techniques in various areas of economics and finance[45, 26, 21]. Examples of the utilization of these techniques are the applications of genetic algorithms and genetic programming [22] for portfolio optimization [5], neural network in stocks selection [33] and predicting the S&P 100 index using rough sets [46] and various types of intelligent systems for making trading decisions [1, 2, 3, 8, 16, 27, 28, 29, 34, 50, 51]. Other real world

A. Abraham, R. Falcón, and R. Bello (Eds.): Rough Set Theory, SCI 174, pp. 163–186.
springerlink.com

applications in the field of finance such as credit cards assessment, country risk evaluation, credit risk assessment, corporate acquisitions[56], business failure prediction, [32, 56, 11], prediction of the financial health of the dot.com firms.[7]and bankruptcy prediction[35], customer segmentation [9] are but few examples showing the diversity of the coverage of these new techniques.

In recent years, and since its inception, rough set theory has gained momentum and has been widely used as a viable intelligent data mining and knowledge discovery technique in many applications including economic, financial and investment areas. Applications of rough sets in economic and financial prediction can be divided into three main areas: database marketing, business failure prediction and financial investment [17, 4].

Database marketing is a method of analyzing customer data to look for patterns among existing preferences and to use these patterns for a more targeted selection of the customers [17, 15]. It is based on the principle that through collecting and organizing information about a business, one can reduce the cost of the businessŠs marketing efforts and increase profit. Database marketing is characterized by enormous amounts of data at the level of the individual consumer. However, these data have to be turned into information in order to be useful. To this end, several different problem specifications can be investigated. These include market segmentation, cross-sell prediction, response modelling, customer valuation and market basket analysis. Building successful solutions for these tasks requires applying advanced data mining and machine learning techniques to find relationships and patterns in historical data and using this knowledge to predict each prospect's reaction to future situations. The rough set model has been applied in this domain (see [41, 25]).

Business failure prediction [32, 44, 56, 11], of the financial health of the dot.com firms [7] and bankruptcy prediction[35], are examples of an important and challenging issue that has served as the impetus for many academic studies over the past three decades[32]. Recently, there has been a significant increase in interest in business failure prediction, from both industry and academia. Financial organizations, such as banks, credit institutes, clients, etc. need these predictions for evaluating firms in which they have an interest[17]. Accurate business failure prediction models would be extremely valuable to many industry sectors, particularly in financial investment and lending institutes. Despite the fact that Discriminant analysis has been the most popular approach, there are also a large number of alternative techniques available such as rough sets [12, 52].

Many financial analysis applications [45] such as *financial investment* employ predictive modeling techniques, for example, statistical regression, Bayesian approach and neural networks [45, 26, 21], to create and optimize portfolios and to build trading systems. Building trading systems using the rough set model was studied by several researchers. Ziarko et al. [54], Golan and Edwards [20] applied the rough set model to discover strong trading rules from the historical database of the Toronto stock exchange. Reader may refer to [17] for a detailed review of applications of rough sets in financial domain.

Despite the many prediction attempts using rough set models, prediction still remains a challenging and difficult task to perform specially within complicated, dynamic and often stochastic areas such as economic and finance. In response to this challenge, this chapter presents a generic scheme for generating prediction rules using rough set. The scheme, which could be applied in various areas of economic and finance such as stock price movement prediction, etc., is expected to extract knowledge in the form rules to guide the decision maker in making the right decision, say buy, hold or sell in the area of stock trading and portfolio management. To increase the efficiency of the prediction process, rough sets with Boolean reasoning discretization algorithm is used to discretize the data. Rough set reduction technique is, then, applied to find all reducts of the data which contains the minimal subset of attributes that are associated with a class used label for prediction. Finally, rough set dependency rules are generated directly from all generated reducts. Rough confusion matrix is used to evaluate the performance of the predicted reducts and classes.

This chapter is organized as follows. Sect. 2 gives a brief introduction to rough sets. Sect. 3 discusses the proposed rough set prediction model in detail. Experimentation is covered in Sect. 4 including data preparation and its characteristic, analysis, results and discussion of the results and finally, conclusions are provided in Sect. 5.

2 Rough Sets: Foundations

Rough set theory , a new intelligent mathematical tool proposed by Pawlak [37, 38, 39], is based on the concept of approximation spaces and models of sets and concepts. The data in rough set theory is collected in a table called a decision table. Rows of the decision table correspond to objects, and columns correspond to features. In the data set, we also assume that a set of examples with a class label to indicate the class to which each example belongs are given. We call the class label a decision feature, the rest of the features are conditional. Let $\mathcal{O}, \mathcal{F}$ denote a set of sample objects and a set of functions representing object features, respectively. Assume that $B \subseteq \mathcal{F}, x \in \mathcal{O}$. Further, let $[x]_B$ denote:

$$[x]_B = \{y : \ x \sim_B \ y\} .$$

Rough set theory defines three regions based on the equivalent classes induced by the feature values: lower approximation $\underline{B}X$, upper approximation $\overline{B}X$ and boundary $BND_B(X)$. A lower approximation of a set X contains all equivalence classes $[x]_B$ that are subsets of X, and upper approximation $\overline{B}X$ contains all equivalence classes $[x]_B$ that have objects in common with X, while the boundary $BND_B(X)$ is the set $\overline{B}X \setminus \underline{B}X$, *i.e.*, the set of all objects in $\overline{B}X$ that are not contained in $\underline{B}X$. So, we can define a rough set as any set with a non-empty boundary.

The indiscernibility relation $\sim_B$ (or by Ind_B) is a fundamental principle of rough set theory. Informally, $\sim_B$ is a set of all objects that have matching descriptions. Based on the selection of B, $\sim_B$ is an equivalence relation partitions a set of objects $\mathcal{O}$ into equivalence classes. The set of all classes in a partition is denoted by $\mathcal{O}/\sim_B$ (also by $\mathcal{O}/Ind_B$). The set $\mathcal{O}/Ind_B$ is called the quotient set. Affinities between objects of interest in the set $X \subseteq \mathcal{O}$ and classes in a partition can be discovered by identifying those classes that have objects in common with X. Approximation of the set X begins by determining which elementary sets $[x]_B \in \mathcal{O}/\sim_B$ are subsets of X.

In the following subsections, we provide a brief explanation of the basic framework of rough set theory, along with some of the key definitions. For a detailed review of the basic material, reader may consult sources such as [37, 38, 39].

2.1 Information System and Approximation

Definition 1. *(Information System) Information system is a tuple (U, A), where U consists of objects and A consists of features. Every $a \in A$ corresponds to the function $a : U \to V_a$ where V_a is a's value set. In applications, we often distinguish between conditional features C and decision features D, where $C \cap D = \emptyset$. In such cases, we define decision systems (U, C, D).*

Definition 2. *(Indiscernibility Relation) Every subset of features $B \subseteq A$ induces indiscernibility relation*

$$Ind_B = \{(x, y) \in U \times U : \forall_{a \in B}\, a(x) = a(y)\}$$

For every $x \in U$, there is an equivalence class $[x]_B$ in the partition of U defined by Ind_B.

Due to the imprecision, which exists in real world data, there are sometimes conflicting classification of objects contained in a decision table. The conflicting classification occurs whenever two objects have matching descriptions, but are deemed to belong to different decision classes. In such cases, the decision table is said to contain inconsistencies.

Definition 3. *(Lower and Upper Approximation)*
In rough set theory, approximations of sets are introduced to deal with inconsistency. A rough set approximates traditional sets using a pair of sets named the lower and upper approximation of the set. Given a set $B \subseteq A$, the lower and upper approximations of a set $Y \subseteq U$, are defined by equations (1) and (2), respectively.

$$\underline{B}Y = \bigcup_{x:[x]_B \subseteq X} [x]_B. \tag{1}$$

$$\overline{B}Y = \bigcup_{x:[x]_B \cap X \neq \emptyset} [x]_B. \tag{2}$$

Definition 4. *(Lower Approximation and positive region) The positive region $POS_C(D)$ is defined by*

$$POS_C(D) = \bigcup_{X:X \in U/Ind_D} \underline{C}X.$$

$POS_C(D)$ is called the positive region of the partition U/Ind_D with respect to $C \subseteq A$, i.e., *the set of all objects in U that can be uniquely classified by elementary sets in the partition U/Ind_D by means of C [40].*

Definition 5. *(Upper Approximation and Negative Region) The negative region $NEG_C(D)$ is defined by*

$$NEG_C(D) = U - \bigcup_{X:X \in U/Ind_D} \overline{C}X,$$

i.e., *the set of all all objects that can be definitely ruled out as members of X.*

Definition 6. *(Boundary region) The boundary region is the difference between upper and lower approximation of a set X that consists of equivalence classes having one or more elements in common with X. It is given as follows:*

$$BND_B(X) = \underline{B}X - \bar{B}X \tag{3}$$

2.2 Reduct and Core

Often we wonder whether there are features in the information system, which are more important to the knowledge represented in the equivalence class structure than other features and whether there is a subset of features which by itself can fully characterize the knowledge in the database. Such a feature set is called a reduct. Calculation of reducts of an information system is a key issue in RS theory [38, 39, 42] and we use reducts of an information system in order to extract rule-like knowledge from an information system.

Definition 7. *(Reduct) Given a classification task related to the mapping $C \rightarrow D$, a reduct is a subset $R \subseteq C$ such that*

$$\gamma(C, D) = \gamma(R, D)$$

and none of proper subsets of R satisfies analogous equality.

Definition 8. *(Reduct Set) Given a classification task mapping a set of variables C to a set of labeling D, a reduct set is defined with respect to the power set $P(C)$ as the set $R \subseteq P(C)$ such that $Red = \{A \in P(C) : \gamma(A, D) = \gamma(C, D)\}$. That is, the reduct set is the set of all possible reducts of the equivalence relation denoted by C and D.*

Definition 9. *(Minimal Reduct) A minimal reduct $R_{minimal}$ is the reduct such that $\|R\| \leq \|A\|, \forall A \in R$. That is, the minimal reduct is the reduct of least cardinality for the equivalence relation denoted by C and D.*

Definition 10. *(Core) Attribute $c \in C$ is a core feature with respect to D, if and only if it belongs to all the reducts. We denote the set of all core features by $Core(C)$. If we denote by $R(C)$ the set of all reducts, we can put:*

$$Core(C) = \bigcap_{R \in R(C)} R \tag{4}$$

The computation of the reducts and the core of the condition features from a decision table is a way of selecting relevant features. It is a global method in the sense that the resultant reduct represents the minimal set of features which are necessary to maintain the same classification power given by the original and complete set of features. A straight forward method for selecting relevant features is to assign a measure of relevance to each feature and then select the features with higher values. And based on the generated reduct system, we generate a list of rules that will be used for building the classifier model which will be able to identify new objects and assign them the correct class label corresponding decision class in the reduced decision table (i.e. the reduct system). Needless to say, the calculation of all the reducts is fairly complex (see [47, 23, 48]).

2.3 Significance of the Attribute

The significance of features enables us to evaluate features by assigning a real number from the closed interval [0,1], expressing the important a feature in an information table. Significance of a feature a in a decision table DT can be evaluated by measuring the effect of removing of the feature a in C from feature set C on a positive region defined by the table DT. As shown in definition 2.3, the number $\gamma(C, D)$ express the degree of dependency between feature C and D or accuracy of approximation of U/D by C.. The formal definition of the significant is given as follows:

Definition 11. *(Significance) For any feature $a \in C$, we define its significance ζ with respect to D as follows:*

$$\zeta(a, C, D) = \frac{|POS_{C \setminus \{a\}}(D)|}{|POS_C(D)|} \tag{5}$$

Definitions 7-11 are used to express the importance of particular features in building the classification model. For a comprehensive study, reader may consult [49]. An important measure is to use frequency of occurrence of features in reducts. One can also consider various modifications of Definition 7, for example approximate reducts, which preserve information about decisions only to some degree [47]. Further more, positive region in Definition 4 can be modified by allowing for the approximate satisfaction of inclusion $[x]_C \subseteq [x]_D$, as proposed, e.g., in VPRS model [53]. Finally, in Definition 2, the meaning of $IND(B)$ and $[x]_B$ can be changed by replacing equivalence relation with similarity relation, especially useful when considering numeric features. For further reading, see [38, 42].

2.4 Decision Rules

In the context of supervised learning, an important task is the discovery of classification rules from the data provided in the decision tables. These decision rules not only capture patterns hidden in the data but also can be used to classify new unseen objects. Rules represent dependencies in the dataset, and represent extracted knowledge, which can be used when classifying new objects not present in the original information system. Once reducts were found, the job of creating definite rules for the value of the decision feature of the information system is practically done. To transform a reduct into a rule, one has to bind the condition feature values of the object class from which the reduct originated to the corresponding features of the reduct. To complete the rule, a decision part comprising the resulting part of the rule is added. This is done in the same way as for the condition features. To classify objects, which has never been seen before, rules generated from a training set are used. These rules represent the actual classifier. This classifier is used to predict classes to which new objects are attached. The nearest matching rule is determined as the one whose condition part differs from the feature vector of re-object by the minimum number of features. When there is more than one matching rule, a voting mechanism is used to choose the decision value. Every matched rule contributes votes to its decision value, which are equal to the number of times objects are matched by the rule. The votes are added and the decision with the largest number of votes is chosen as the correct class. Quality measures associated with decision rules can be used to eliminate some of the decision rules.

3 Rough Set Prediction Model (RSPM)

Figure 1 illustrates the overall steps in the proposed Rough Set Prediction Model(RSPM) using a UML Activity Diagram where a square or rectangular represents a data object, a rounded rectangular represents an activity, solid and dashed directed lines indicate control flow and data object flow respectively. Functionally, RSPM can be partitioned into three distinct phases:

- *Pre-processing phase(Activities in Dark Gray).* This phase includes tasks such as extra variables addition and computation, decision classes assignments, data cleansing, completeness, correctness, attribute creation, attribute selection and discretization.
- *Analysis and Rule Generating Phase(Activities in Light Gray).* This phase includes the generation of preliminary knowledge, such as computation of object reducts from data, derivation of rules from reducts, rule evaluation and prediction processes.
- *Classification and Prediction phase (Activities in Lighter Gray).* This phase utilize the rules generated from the previous phase to predict the stock price movement

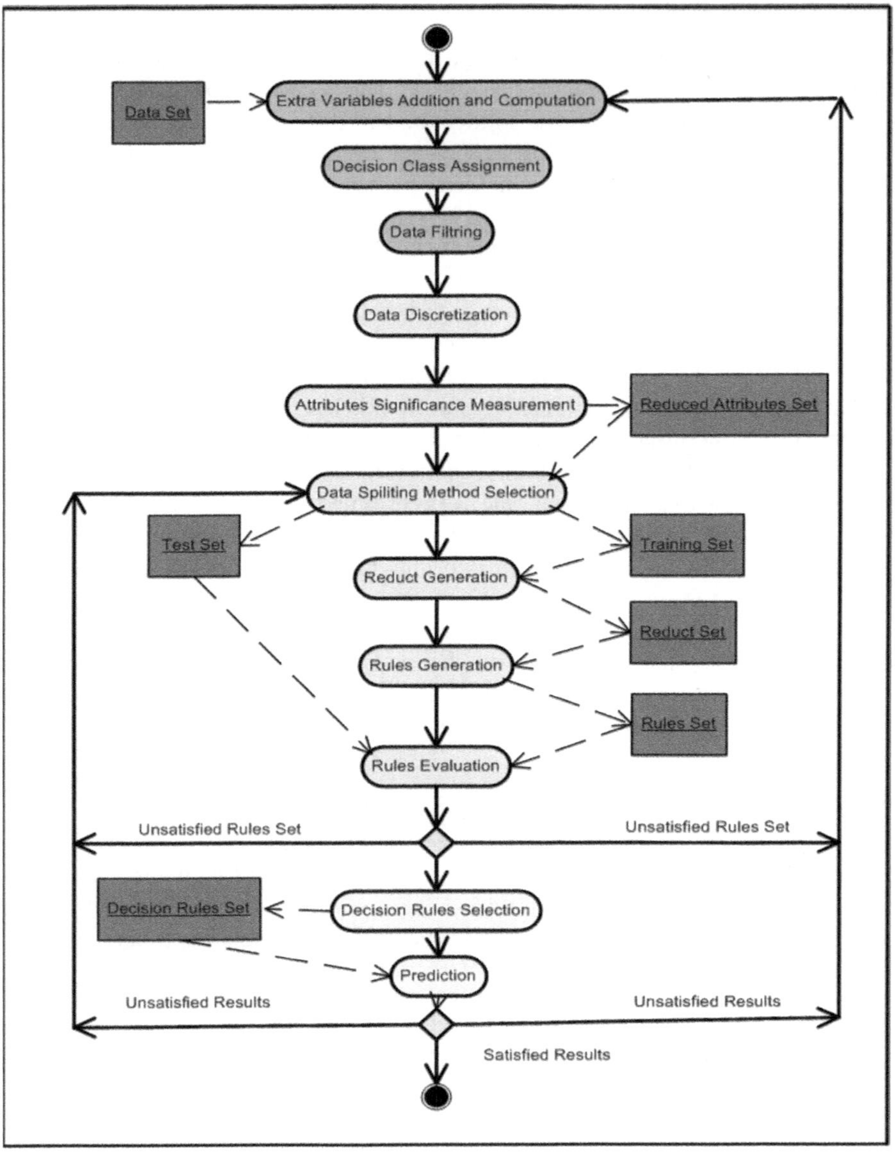

Fig. 1. General overview of rough set prediction model

3.1 Pre-processing Phase

In this phase, the decision table required for rough set analysis is created. In doing so, a number of data preparation tasks such as data conversion, data cleansing, data completion checks, conditional attribute creation, decision attribute generation, discretization of attributes are performed. Data splitting is also performed which created two randomly generated subsets, one subset for analysis containing 75% of the objects in the data set and one validation containing the remainder 25% of the objects. It must be emphasized that data conversion performed on the initial data must generate a form in which specific rough set tools can be applied.

Data Completion and Discretization Processes

Data Completion

Often, real world data contain missing values. Since rough set classification involves mining for rules from the data, objects with missing values in the data set may have undesirable effects on the rules that are constructed. The aim of the data completion procedure is to remove all objects that have one or more missing values. Incomplete data or information systems exist broadly in practical data analysis, and approaches to complete the incomplete information system through various completion methods in the preprocessing stage are normal in data mining and knowledge discovery. However, these methods may result in distorting the original data and knowledge, and can even render the original data to be un-minable. To overcome these shortcomings inherent in the traditional methods, we used the decomposition approach for incomplete information system (i.e. decision table)proposed in [43].

Data Discretization

When dealing with attributes in concept classification and prediction, it is obvious that they may have varying importance in the problem being considered. Their importance can be pre-assumed using auxiliary knowledge about the problem and expressed by properly chosen weights. However, in the case of using the rough set approach to concept classification and prediction, it avoids any additional information aside from what is included in the information table itself. Basically, the rough set approach tries to determine from the data available in the information table whether all the attributes are of the same strength and, if not, how they differ in respect of the classifier power.

Therefore, some strategies for discretization of real valued features must be used when we need to apply learning strategies for data classification (*e.g.*, equal width and equal frequency intervals). It has been shown that the quality of learning algorithm is dependent on this strategy, which has been used for real-valued data discretization [14]. It uses data transformation procedure which involves finding cuts in the data sets that divide the data into intervals. Values lying within an interval are then mapped to the same value. Performing this process leads to reducing the size of the attributes value set and ensures that the rules that are mined are not too specific. For the discretization of continuous-valued attributes, we adopt, in this chapter, rough sets with boolean reasoning (RSBR) algorithm proposed by Zhong et al. [43] The main advantage of RSBR is that it combines discretization of real-valued attributes and classification. For the main steps of the RSBR discretization algorithm, reader may consult [4].

3.2 Analysis and Rule Generating Phase

Analysis and Rule Generating Phase includes generating preliminary knowledge, such as computation of object reducts from data, derivation of rules from reducts, and prediction processes. These stages lead towards the final goal of generating rules from information system or decision table.

Relevant Attribute Extraction and Reduction

One of the important aspects in the analysis of decision tables is the extraction and elimination of redundant attributes and also the identification of the most important attributes from the data set. Redundant attributes are attributes that could be eliminated without affecting the degree of dependency between the remaining attributes and the decision. The degree of dependency is a measure used to convey the ability to discern objects from each other. The minimum subset of attributes preserving the dependency degree is called reduct. The computation of the core and reducts from a decision table is, in a way, selecting the relevant attributes [6, 48].

In decision tables, there often exist conditional attributes that do not provide (almost) any additional information about the objects. These attributes need to be removed in order to reduce the complexity and cost of decision process [6, 18, 42, 48]. A decision table may have more than one reduct. Any of these reducts could be used to replace the original table. However, finding all the reducts from a decision table is NP-complete but fortunately, in applications, it is usually not necessary to find all of them – one or a few of them are sufficient. Selecting the best reduct is important. The selection depends on the optimality criterion associated with the attributes. If a cost function could be assigned to attributes, then the selection can be based on the combined minimum cost criteria. But in the absence of such cost function, the only source of information to select the reduct from is the contents of the table. In this chapter, we adopt the criteria that the best reducts are the those with minimal number of attributes and – if there are more such reducts – with the least number of combinations of values of its attributes cf. [6, 36].

In general, rough set theory provides useful techniques to reduce irrelevant and redundant attributes from a large database with a lot of attributes. The dependency degree (or approximation quality, classification quality) and the information entropy are two most common attribute reduction measures in rough set theory. In this chapter, we use the dependency degree measure to compute the significant features and measuring the effect of removing a feature from the feature sets. [24].

Computation of the Reducts

A reduced table can be seen as a rule set where each rule corresponds to one object of the table. The rule set can be generalized further by applying rough set value reduction method. The main idea behind this method is to drop those redundant condition values of rules and to unite those rules in the same class. Unlike most value reduction methods, which neglect the difference among the classification capabilities of condition attributes, we first remove values of those attributes that have less discrimination factors. Thus more redundant values can be reduced from decision table and more concise rules can be generated. The main steps of the Rule Generation and classification algorithm are outlined in Algorithm-1:

Algorithm 1. Reduct Generation algorithm

Input: information table (ST) with discretized real valued attribute.
Output: reduct sets $R_{final} = \{r_1 \cup r_2 \cup .... \cup r_n\}$

1: **for** each condition attribute $c \in C$ **do**
2: Compute the correlation factor between c and the decisions attributes D
3: **if** the correlation factor > 0 **then**
4: Set c as relevant attributes.
5: **end if**
6: **end for**
7: Divide the set of relevant attribute into different variable sets.
8: **for** each variable sets **do**
9: Compute the dependency degree and compute the classification quality
10: Let the set with high classification accuracy and high dependency as an initial reduct set.
11: **end for**
12: **for** each attribute in the reduct set **do**
13: Calculate the degree of dependencies between the decisions attribute and that attribute.
14: Merge the attributes produced in previous step with the rest of conditional attributes
15: Calculate the discrimination factors for each combination to find the highest discrimination factors
16: Add the highest discrimination factors combination to the final reduct set.
17: **end for**
18: **repeat**
19: statements 12
20: **until** all attributes in initial reduct set are processed

Rule Generation from a Reduced Table

The generated reducts are used to generate decision rules. The decision rule, at its left side, is a combination of values of attributes such that the set of (almost) all objects matching this combination have the decision value given at the rule's right side. The rule derived from reducts can be used to classify the data. The set of rules is referred to as a classifier and can be used to classify new and unseen data. The main steps of the Rule Generation and classification algorithm are outlined as Algorithm-2):

Algorithm 2. Rule Generation

Input: reduct sets $R_{final} = \{r_1 \cup r_2 \cup .... \cup r_n\}$
Output: Set of rules

for each reduct r **do**
  for each corresponding object x **do**
    Contract the decision rule $(c_1 = v_1 \wedge c_2 = v_2 \wedge .... \wedge c_n = v_n) \longrightarrow d = u$
    Scan the reduct r over an object x
    Construct $(c_i, 1 \leq i \leq n)$
    for every $c \in C$ **do**
      Assign the value v to the corresponding attribute a
    end for
    Construct a decision attribute d
    Assign the value u to the corresponding decision attribute d
  end for
end for

The quality of rules is related to the corresponding reduct(s). We are especially interested in generating rules which cover largest parts of the universe U. Covering U with more general rules implies smaller size rule set.

3.3 Classification and Prediction Phase

Classification and prediction is the last phase of our proposed approach. We present a classification and prediction scheme based on the methods and techniques described in the previous sections. Figure 2 illustrates the classification scheme for a construction of particular classification and prediction algorithm. To transform a reduct into a rule, one only has to bind the condition feature values of the object class from which the reduct originated to the corresponding features of the reduct. Then, to complete the rule, a decision part comprising the resulting part of the rule is added. This is done in the same way as for the condition features. To classify objects, which has never been seen before, rules generated from a training set will be used. These rules represent the actual classifier. This classifier is used to predict to which classes new objects are attached. The nearest matching rule is determined as the one whose condition part differs from the feature vector of re-object by the minimum number of features. When there is more than one matching rule, we use a voting mechanism to choose the decision value. Every matched rule contributes votes to its decision value, which are equal to the t times number of objects matched by the rule. The votes are added and the decision with the largest number of votes is chosen as the correct class. Quality measures associated with decision rules can be used to eliminate some of the decision rules.

The global strength defined in [6] for rule negotiation is a rational number in [0, 1] representing the importance of the sets of decision rules relative to the considered tested object. Let us assume that $T = (U, A \bigcup (d))$ is a given decision table, u_t is a test object, $Rul(X_j)$ is the set of all calculated basic decision rules

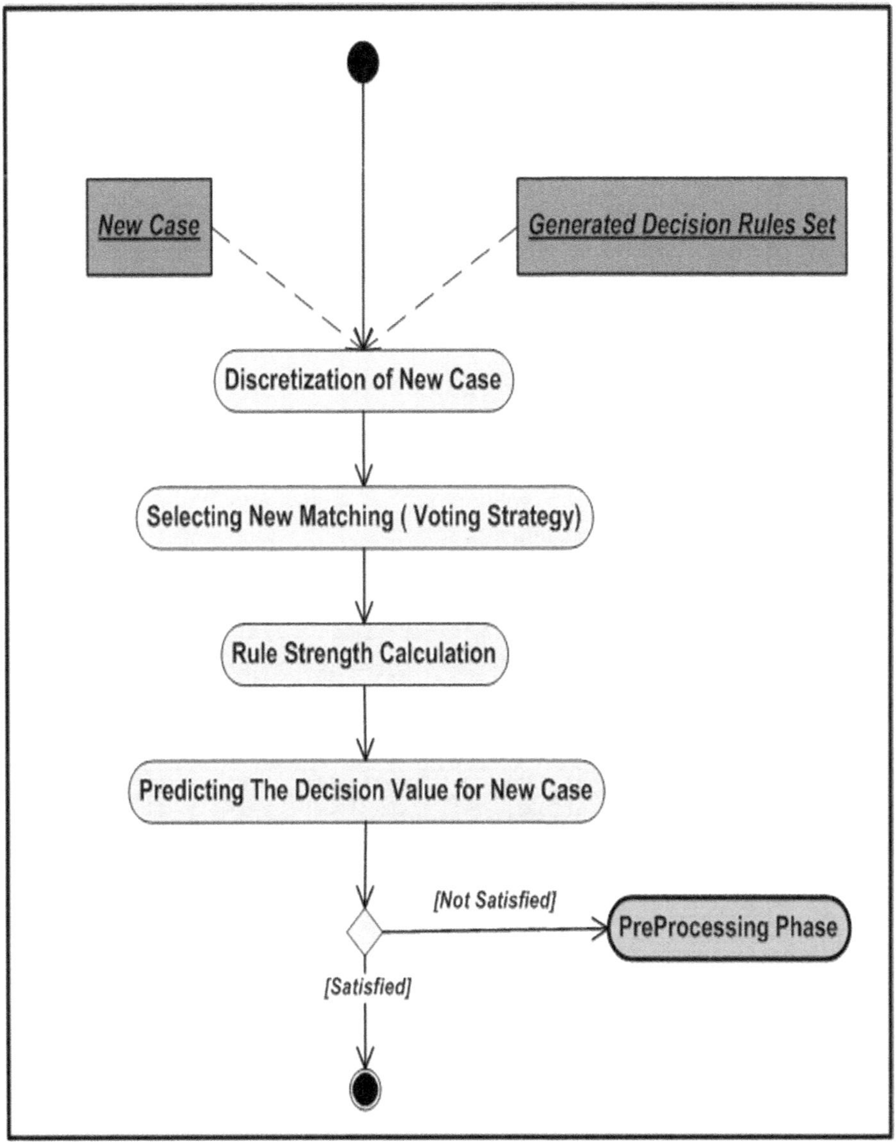

Fig. 2. Rough set classification and prediction scheme

for T, classifying objects to the decision class $X_j(v_d^j = v_d)$, $MRul(X_j, u_t) \subseteq Rul(X_j)$ is the set of all decision rules from $Rul(X_j)$ matching tested object u_t. The global strength of decision rule set $MRul(X_j, u_t)$ is defined by the following form [6]:

$$MRul(X_j, u_t) = \frac{\left| \bigcup_{r \subset MRul(X_j,u_t)} |Pred(r)|_A \cap |d = v_d^j|_A \right|}{\left| |d = v_d^j|A \right|}.$$

Measure of strengths of rules defined above is applied in constructing classification algorithm. To classify a new case, rules are first selected matching the new object. The strength of the selected rule sets is calculated for any decision class, and then the decision class with maximal strength is selected, with the new object being classified to this class.

4 Experimental Results

4.1 Data Set and Its Characteristics

To test and verify the prediction capability of the proposed RSPM, the daily stock movement of a banking stock traded in Kuwait Stock Exchange and spanning over a period of 7 years (2000-2006), were captured. Figure 3 depicts a sample of the stock's daily movements.

Sector	stk	Ticker	Date	Last	High	Low	Vol	Trade	Value
Banking	102	GBK	01/02/2000	410	410	400	150000	9	60800
Banking	102	GBK	01/03/2000	405	405	405	140000	4	56700
Banking	102	GBK	01/04/2000	405	405	405	1010000	31	409050
Banking	102	GBK	01/05/2000	405	405	405	370000	7	149850
Banking	102	GBK	01/11/2000	400	400	400	130000	5	52000
Banking	102	GBK	01/12/2000	400	400	400	10410000	14	4164000
Banking	102	GBK	15/01/2000	400	400	400	990000	16	396000
Banking	102	GBK	16/01/2000	400	400	400	1450000	27	580000
Banking	102	GBK	17/01/2000	400	400	400	1740000	30	696000
Banking	102	GBK	18/01/2000	400	400	400	1550000	22	620000

Fig. 3. A sample of the stock daily movement

Table 1 shows the attributes used in the creation of the rough set decision table, where MA: Moving average of price,UP: Upward price change, Dw:Downward price change; P_i: closing price. The first five attributes in the Table, i.e. Last(or Closing Price), High, Low, Trade, and Value) were extracted from the stock daily movement. The other important attributes in the table were compiled from the literature [31] along with the formula for their computation. The decision attributed, D, in Table 1, which indicates the future direction of the the data set, is constructed using the following formula:

$$Dec_{att} = \frac{\sum_{i=1}^{i=n} ((n+1) - i).sign[close(i) - close(0)]}{\sum_{i=1}^{n} i} \quad (6)$$

where close (0) is today's closing price and close (i) is the ith closing price in the future. Equation (1) specifies a range -1 to +1 for Dec_{att} A value of +1 indicate

Table 1. Stock price movement decision table

Attribute	Attribute description
Last	closing price
High	High price
Low	Low price
Trade	
Value	
Lag_i, $i = 1..6$	An event occurring at time $t + k$ $(k > 0)$
	is said to lag behind event occurring at time t,
$Aver_5$	moving average of 5 days for close price
Momentum	$P_i - P_{i-4}$
Disparity in 5 days	$\frac{P_i}{MA_5} * 100$
Price Osculiator	$OSCP = 100 - \frac{100}{1+\frac{\sum_{i=0}^{n-1} UP_{i-1/n}}{\sum_{i=0}^{n-1} DW_{i-1/n}}}$
RSI (relative strength index)	$= 100 - \frac{100}{\sum_{i=0^{n-1}} UP_i/n}$
ROC	rate of change $\frac{P_i - P_{i-n}}{P_i} * 100$
D	Decision attribute

that every day up to n days in the future, the market closed higher than today. Similarly, -1 indicates that every day up to n days in the future, the market closed lower than today.

Figure 4 presents a snapshot of the 21 index for the period covering from Jan. 1st 2000 to Jan. 31th 2000, and the fluctuation of the Dec_att. Figure 5 illustrates part of the calculated daily stock movement time series data set according the attributes described in Table 1.

4.2 Analysis, Results and Discussion

For many data mining tasks, it is useful to learn about the general characteristics of the given data set and to identify the outliers - samples that are not consistent with the general behavior of the data model. Outlier detection is important because it may affect the classifier accuracy. As such we performed several descriptive statistical analysis, such as measures of central tendency and data dispersion. In our statistical analysis, we used the mean and the median to detect the outliers in our data set. Table 2 represents the statistical analysis and essential distribution of attributes, respectively.

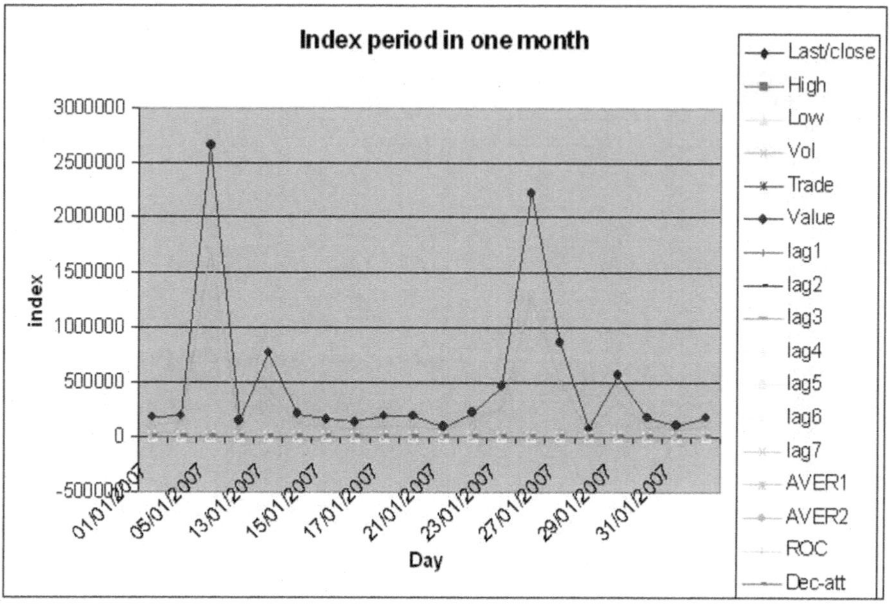

Fig. 4. Snapshot of 21 index for the period covering January 2000

Date	Last/close	High	Low	Vol	Trade	Value	lag1	lag2	lag3	lag4	lag5	
01/01/2007	1720	1720	1700	105000	10	178700	1720	1720	1720	1720	1720	
04/01/2007	1500	1500	1480	130000	12	194000	1480	1480	1520	1520	1520	
05/01/2007	1580	1620	1560	1685000	76	2657300	1580	1620	1580	1600	1600	
07/01/2007	1740	1780	1740	82500	8	143950	1780	1740	1740	1740	1700	
13/01/2007	1800	1820	1760	435000	24	773000	1800	1840	1860	1840	1780	
14/01/2007	1800	1800	1780	117500	9	209200	1800	1800	1840	1860	1840	
15/01/2007	1780	1780	1760	95000	25	168050	1800	1800	1800	1840	1860	
16/01/2007	1780	1800	1760	72500	20	128750	1780	1800	1800	1800	1840	
17/01/2007	1760	1760	1740	110000	11	191600	1780	1780	1800	1800	1800	
20/01/2007	1760	1760	1740	110000	11	191600	1760	1780	1780	1800	1800	
21/01/2007	1760	1760	1700	55000	6	94900	1760	1760	1780	1780	1800	
22/01/2007	1740	1740	1700	132500	24	226450	1760	1760	1760	1780	1780	
23/01/2007	1740	1740	1700	270000	24	464050	1740	1760	1760	1760	1780	
24/01/2007	1760	1780	1720	1280000	91	2216600	1740	1740	1760	1760	1760	
27/01/2007	1740	1760	1720	502500	42	869100	1760	1740	1740	1760	1760	
28/01/2007	1740	1740	1720	40000	3	69200	1740	1760	1740	1740	1760	

Fig. 5. Samples of the banking sector data - after post processing

We reach the minimal number of reducts that contains a combination of attributes which has the same discrimination factor. The final generated reduct sets, which are used to generate the list of rules for the classification are:

{high, low, last, momentum, disparity in 5 days, Roc}

A natural use of a set of rules is to measure how well the ensemble of rules is able to classify new and unseen objects. To measure the performance of the

Table 2. Statistical results of the attributes

Attribute	Mean	Std. Dv	Median	Correlation with decision class
Last-Close	497.8	145.17	490.0	0.255
High	498.9	145.6	490	0.2500
Low	493.7	143.5	485.0	0.24
Vol	626189.3	1314775.6	240000	0.097
Trade	13.3	15.12	8.0	0.185
Value	322489.3	674862.3	118900.0	0.1065
Lag1	522.25	94.5	490.0	-0.0422
Lag2	493.8	0.4828	490.0	0.0055
Lag3	496.4	148.5	490.0	0.092
Aver5	501.5	103.6	488.0	0.075
Momentum	2.44	163.1	0.0	0.266
Disparity in 5 days	99.0	25.2	100.3	0.28
Price Osculator	.0002	0.095	0.006	0.156
RSI	49.8	1.4.36	49.8	-0.035
ROC	-4.7	21.5	0.0	-0.365

rules is to assess how well the rules perform in classifying new cases. So we apply the rules produced from the training set data to the test set data.

The following present the rules in a more readable format:

R1: IF Closing Price(Last) = (403 **OR** 408) **AND**
High = (403 **OR** 408) **AND**
Low = (3 **OR** 8) **AND**
momentum = (403 **OR** 408) **AND**
disparityin5dayes = (100.48700 **OR** 100.60700) **AND**
ROC = (−0.50505 **OR** 0.51021)
THEN Decision Class is 0.0

Table 3 shows a partial set of the generated rules. These obtained rules are used to build the prediction system.

Several runs were conducted using different setting with strength rule threshold. Rule importance and rule strength measures are used to obtain a sense of the quality of the extracted rules. These measures are chosen according to the number of times a rule appears in all reducts, number of generated reducts, and

Table 3. A partial set of the generated rules

Rule number	Rule form
R1	Last/close=(403 or 408) AND High=(403 RO 408) AND Low=(403 or 408) AND momentum=(3 OR 8) AND disparityin5dayes=(100.48700 or 100.60700) AND ROC=(-0.50505 or 0.51021) $\Longrightarrow d = 0$
R2	Last/close=(398 or 403) AND High=(398 or 403) AND Low=(393 or 398) AND momentum=(-2 or 3) AND disparityin5dayes=(125.19600 or 125.43000) AND ROC=(-0.50505 or 0.51021) $\Longrightarrow d = 0$
R3	Last/close=(403 or 408)) AND High(403 or 408) AND Low=(398 or 403) AND momentum(3 or 8) AND disparityin5dayes=(100.93900 or 101.01500) AND ROC=(0.51021) $\Longrightarrow d = 1.0$
R4	Last/close=(378 or 385) AND High(378 or 385) AND Low=(378 or 385)) AND momentum=(-25 or -17) AND disparityin5dayes=(97.70110) AND ROC=(-0.50505) $\Longrightarrow d = -1.0$
R5	Last/close=(183 or 370) AND High=(368, 373) AND Low=(183, 368) AND momentum=(-37, -32) AND disparityin5dayes=(113.76700 or 120.81700) AND ROC=(-0.50505) $\Longrightarrow d = 1.0$
R6	Last/close=(403, 408) AND High=(403 or 408) AND Low=([398 or 403)) AND momentum=(-2 or 3) AND disparityin5dayes=(100.24500 or 100.27300) AND ROC=(0.51021) $\Longrightarrow d = 1.0$

Table 4. Number of generated rules

Method	Generated rule number
Neural networks	630
Rough sets	371

Table 5. Model prediction performance (confusion matrix)

Actual	Predict Class1	Predict Class2	Predict Class3	Accuracy
Class1 (-1)	39	1	0	0.975 %
Class2 (0)	0	76	0	1.0 %
Class3 (+1)	0	2	34	0.94%
	1.0	.962	1.0	0.9802 %

the support the strength of a rule. The rule importance and Rule Strength are given by the following forms:

Rule Importance. Rule Importance measures ($Importance_{rule}$) is used to assess the quality of the generated rule and it is defined as follows:

$$Importance_{rule} = \frac{\tau_r}{\rho_r}, \quad (7)$$

where τ_r is the number of times a rule appears in all reducts and ρ_r is the number of reduct sets.

Rule Strength. The strength of a rule, $Strength_{rule}$, states how well the rule covers or represent the data set and can be calculated as follows:

$$Strength_{rule} = \frac{Support_{rule}}{|U|}, \quad (8)$$

where $|U|$ denotes the number of all objects in the training data or objects in the universe in general. The strength of a rule states how well the rule covers or represents the data set.

Table 4 shows the number of generated rules using rough sets and for the sake of comparison we have also generated rules using neural network. Table 4 indicates that the number of rules generated using neural networks is much larger than that of the rough set approach.

Measuring the performance of the rules generated from the training data set in terms of their ability to classify new and unseen objects is also important. Our measuring criteria were Rule Strength and Rule Importance [30] and to check

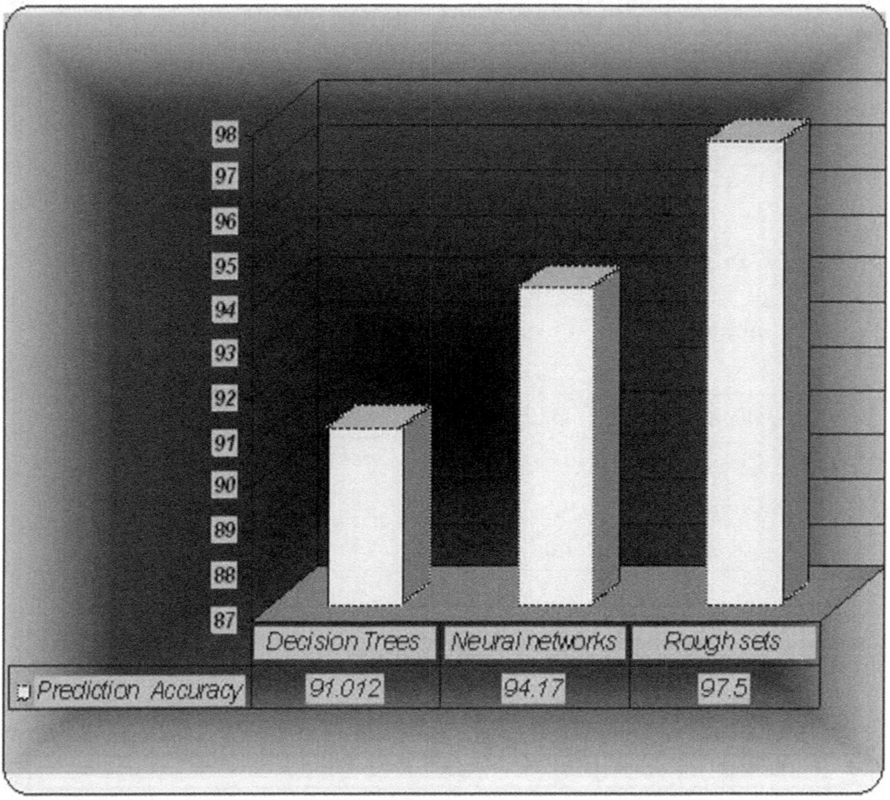

Fig. 6. Comparative analysis in terms of the prediction accuracy

the performance of our method, we calculated the confusion matrix between the predicted classes and the actual classes as shown in Table 5. The confusion matrix is a table summarizing the number of true positives, true negatives, false positives, and false negatives when using classifiers to classify the different test objects.

Figure 6 shows the overall prediction accuracy of well known two approaches compared with the proposed rough set approach. Empirical results reveal that the rough set approach is much better than neural networks and ID3 decision tree. Moreover, for the neural networks and the decision tree classifiers, more robust features are required to improve their performance.

5 Conclusions and Future Research

This chapter presented a generic stock price prediction model using rough set theory. The model was able to extract knowledge in the form of rules from daily stock movements. These rules then could be used to guide investors whether to buy, sell or hold a stock. To increase the efficiency of the prediction process, rough sets with Boolean reasoning discretization algorithm is used to discretize the data. Rough set reduction technique is, then, applied to find all reducts of the data which contains the minimal subset of attributes that are associated with a class used label for prediction. Finally, rough set dependency rules are generated directly from all generated reducts. Rough confusion matrix is used to evaluate the performance of the predicted reducts and classes.

Using a data set consisting of daily movements of a stock traded in Kuwait Stock Exchange, a preliminary assessment showed that performance of the rough set based stock price prediction model, given the limited scoped of the data set, was highly accurate and as such this investigation could lead to further research using a much larger data set consisting of the entire Kuwait Stock Exchange, which would in turn prove the model's generalizability that the model is accurate and sufficiently robust and reliable as a forecasting and prediction model. For comparison purposes, the results obtained using rough sets were compared to those generated by neural networks and decision tree algorithms. It was shown, using the same constrained data set, that rough set approach has a higher overall accuracy rates and generate more compact and fewer rules than neural networks. A future research, based on this finding, could be to implement a hybrid approach using rough sets as reducts generator and neural networks for knowledge discovery and rule generator utilizing the rough set reducts.

References

1. Abraham, A., Nath, B., Mahanti, P.K.: Hybrid intelligent systems for stock market analysis. In: Alexandrov, V.N., Dongarra, J., Juliano, B.A., Renner, R.S., Tan, C.J.K. (eds.) ICCS-ComputSci 2001. LNCS, vol. 2074, pp. 337–345. Springer, Heidelberg (2001)
2. Abraham, A., Philip, N.S., Nath, B., Saratchandran, P.: Performance analysis of connectionist paradigms for modeling chaotic behavior of stock indices. In: Second International Workshop on Intelligent Systems Design and Applications, Computational Intelligence and Applications, pp. 181–186. Dynamic Publishers Inc., USA (2002)
3. Abraham, A., Philip, N.S., Saratchandran, P.: Modeling chaotic behavior of stock indices using intelligent paradigms. Int. J. Neural Parallel Sci. Comput. 11(12), 143–160 (2003)
4. Al-Qaheri, H., Hassanien, A.E., Abraham, A.: Discovering Stock Price Prediction Rules Using Rough Sets. Neural Network World Journal (in press, 2008)
5. Bauer, R.J.: Genetic Algorithms and Investment Strategies. Wiley, New York (1994)
6. Bazan, J., Nguyen, H.S., Nguyen, S.H., Synak, P., Wróblewski, J.: Rough Set Algorithms in Classification Problem. In: Polkowski, L., Tsumoto, S., Lin, T.Y. (eds.) ICCV-WS 1999, pp. 49–88. Physica Verlag (2000)
7. Indranil, B.: Deciding the financial health of dot-coms using rough sets. Information and Management 43(7), 835–846 (2006)
8. Chan, W.S., Liu, W.N.: Diagnosing shocks in stock markets of southeast Asia, Australia, and New Zealand. Math. Comput. Simulation 59(13), 223–232 (2002)

9. Cheng, C., Chen, Y.: Classifying the segmentation of customer value via RFM model and RS theory. Expert Systems with Applications (article in progress) (2008)
10. Cios, K., Pedrycz, W., Swiniarski, R.: Data Mining Methods for Knowledge Discovery. Kluwer Academic Publishers, Dordrecht (1998)
11. Dimitras, A.I., Slowinski, R., Susmaga, R., Zopounidis, C.: Business failure prediction using rough sets. European Journal of Operational Research 114(2), 263–281 (1999)
12. Dimitras, A.I., Slowinski, R., Susmagab, R., Zopounidisa, C.: Business failure prediction using rough sets. European Journal of Operational Research 114(2), 263–280 (1999)
13. Dougherty, J., Kohavi, R., Sahami, M.: Supervised and Unsupervised Discritization of Continuous Features. In: Proc. of the XII International conference on Machine Learning, San Francisco, pp. 294–301 (1995)
14. Dougherty, J., Kohavi, R., Sahami, M.: Supervised and Unsupervised Discritization of Continuous Features. In: Proc. of the XII International conference on Machine Learning, pp. 294–301 (1995)
15. Fayyad, U.M., Piatetsky-Shapiro, G., Smyth, P.: From data mining to knowledge discovery: An overview. In: Fayyad, U.M., et al. (eds.) Advances in Knowledge Discovery and Data Mining, p. 129. AAAI, New York (1996)
16. Francis, E.S., Cao, L.J.: Modified support vector machines in financial time series forecasting. Neurocomputing 48(14), 847–861 (2002)
17. Francis, E.H., Lixiang, S.: Economic and financial prediction using rough set model. European Journal of Operational Research 141(3), 641–659 (2002)
18. Gediga, G., Duentsch, I.: Statistical Techniques for Rough Set Data Analysis. In: Rough Sets: New Developments, pp. 545–565. Physica Verlag (2000)
19. Golan, R.: Stock market analysis utilizing rough set theory. Ph.D. Thesis, Department of Computer Science, University of Regina, Canada (1995)
20. Golan, R., Ziarko, W.: A methodology for stock market analysis utilizing rough set theory. In: Proceedings of the IEEE/IAFE Computational Intelligence for Financial Engineering, pp. 32–40 (1995)
21. Graflund, A.: Bayesian Inference Approach to Testing Mean Reversion in the Swedish Stock Market (2002) SSRN: `http://ssrn.com/abstract=250089`
22. Grosan, C., Abraham, A.: Stock Market Modeling Using Genetic Programming Ensembles, Genetic Systems Programming. In: Nedjah, N., et al. (eds.) Studies in Computational Intelligence, pp. 131–146. Springer, Germany (2006)
23. Guoyin, W.: Rough reduction in algebra view and information view. International Journal of Intelligent Systems 18(6), 679–688 (2003)
24. Hassanien, A.E., Abdelhafez, M.E., Own, H.S.: Rough Set Data Analysis in Knowledge Discovery: A Case of Kuwaiti Diabetic Children Patients, Advances in Fuzzy System (in press, 2008)
25. Jiajin, H., Chunnian, L., Chuangxin, O., Yao, Y.Y., Zhong, N.: Attribute reduction of rough sets in mining market value functions. In: IEEE/WIC International Conference Proceedings on Web Intelligence, WI 2003, pp. 470–473 (2003)
26. Jovina, R., Akhtar, J.: Backpropagation and Recurrent Neural Networks in Financial Analysis of Multiple Stock Market Returns. In: Proceedings of the 29th Hawaii International Conference on System Sciences (HICSS), Decision Support and Knowledge-Based Systems, vol. 2, p. 454 (1996)
27. Kim, K.J., Han, I.: Genetic algorithms approach to feature discretization in artificial neural networks for the prediction of stock price index. Expert Systems Applications 19(2), 125–132 (2000)

28. Leigh, W., Modani, N., Purvis, S., Roberts, T.: Stock market trading rule discovery using technical charting heuristics. Expert Systems and Applications 23(2), 155–159 (2002)
29. Leigh, W., Purvis, R., Ragusa, J.M.: Forecasting the NYSE composite index with technical analysis, pattern recognizer, neural network and genetic algorithm: a case study in romantic decision support. Decision Support Syst. 32(4), 361–377 (2002)
30. Li, J., Cercone, N.: A rough set based model to rank the importance of association rules. In: Ślęzak, D., Yao, J., Peters, J.F., Ziarko, W.P., Hu, X. (eds.) RSFDGrC 2005. LNCS, vol. 3642, pp. 109–118. Springer, Heidelberg (2005)
31. Lixiang, S., Loh, H.T.: Applying rough sets to market timing decisions. Decision support Systems journal 37, 583–597 (2004)
32. Lin, C., Wang, Y., Wu, C., Chuang, C.: Developing a business failure prediction model via RST, GRA and CBR. Expert Systems with Applications (2008) (article in press)
33. Mani, G., Quah, K.K., Mahfoud, S., Barr, D.: An analysis of neural-network forecasts from a large-scale, real-world stock selection system. In: Proceedings of the IEEE/IAFE 1995 Conference on Computational Intelligence for Financial Engineering (CIFER 1995), pp. 72–78. IEEE, Los Alamitos (1995)
34. Maqsood, M.R., Abraham, A.: Neural network ensemble method for weather forecasting. Neural Computing and Applications 13(2), 112–122 (2004)
35. McKee, T.E.: Rough Set Bankruptcy Prediction Models versus Auditor Signalling Rates. Journal of Forecasting 22(8), 569–586 (2003)
36. Ning, S., Ziarko, W., Hamilton, J., Cercone, N.: Using Rough Sets as Tools for Knowledge Discovery. In: Fayyad, U.M., Uthurusamy, R. (eds.) First International Conference on Knowledge Discovery and Data Mining KDD 1995, pp. 263–268. AAAI, Montreal (1995)
37. Pawlak, Z.: Rough Sets. Int. J. Computer and Information SCI. 11, 341–356 (1982)
38. Pawlak, Z.: Rough Sets- Theoretical aspects of Reasoning about Data. Kluwer Academic Publishers, Dordrecht (1991)
39. Pawlak, Z., Grzymala-Busse, J., Slowinski, R., Ziarko, W.: Rough sets. Communications of the ACM 38(11), 89–95 (1995)
40. Pawlak, Z., Skowron, A.: Rudiments of rough sets. Information Sciences 177, 3–27 (2007)
41. Poel, D.: Rough sets for database marketing. In: Polkowski, L., Skowron, A. (eds.) Rough Sets in Knowledge Discovery 2, pp. 324–335. Physica-Verlag (1998)
42. Polkowski, L.: Rough Sets: Mathematical Foundations. Physica-Verlag (2003)
43. Qizhong, Z.: An Approach to Rough Set Decomposition of Incomplete Information Systems. In: 2nd IEEE Conference on Industrial Electronics and Applications, ICIEA 2007, pp. 2455–2460 (2007)
44. Salcedo-Sanza, S., Fernandez-Villacanasa, J.L., Segovia-Vargasb, J., Bousono-Calzona, C.: Genetic programming for the prediction of insolvency in non-life insurance companies. Computers & Operations Research 32, 749–765 (2005)
45. Shuxiang, X., Ming, Z.: Data mining - an adaptive neural network model for financial analysis. In: Third International Conference on Information Technology and Applications, ICITA 2005, vol. 1(4-7), pp. 336–340 (2005)
46. Skalko, C.: Rough sets help time the OEX. Journal of Computational Intelligence in Finance 4(6), 20–27 (1996)
47. Ślęzak, D.: Various approaches to reasoning with frequency-based decision reducts: a survey. In: Polkowski, L., Tsumoto, S., Lin, T.Y. (eds.) Rough Sets in Soft Computing and Knowledge Discovery: New Developments, Physica Verlag (2000)

48. Starzyk, J.A., Dale, N., Sturtz, K.: A mathematical foundation for improved reduct generation in information systems. In: Knowledge and Information Systems Journal, pp. 131–147. Springer, Heidelberg (2000)
49. Swiniarski, R., Skowron, A.: Rough set methods in feature selection and recognition. Pattern Recognition Letters 24(6), 833–849 (2003)
50. Tsaih, R., Hsu, Y., Lai, C.C.: Forecasting S&P 500 stock index futures with a hybrid AI system. Decision Support Syst. 23(2), 161–174 (1998)
51. Van den Berg, J., Kaymak, U., Van den Bergh, W.M.: Financial markets analysis by using a probabilistic fuzzy modelling approach. Int. J. Approximate Reasoning 35(3), 291–305 (2004)
52. Wang, Z., Li, H., Deng, X.: Empirical Study of Business Failure Prediction Based on Rough Set. In: International Conference on Management Science and Engineering, ICMSE 2006, pp. 848–852 (2006)
53. Ziarko, W.: Variable Precision Rough Set Model. Journal of Computer and Systems. Sciences 46(1), 39–59 (1993)
54. Ziarko, W., Golan, R., Edwards, D.: An application of datalogic/R knowledge discovery tool to identify strong predictive rules in stock market data. In: Proceedings of AAAI Workshop on Knowledge Discovery in Databases, Washington, DC, pp. 89–101 (1993)
55. Zhong, N., Skowron, A.: Rough set based Knowledge Discovery Process. International Journal of Applied Maths and Computer Science 11(3), 603–619 (2001)
56. Zopounidis, C., Doumpos, M.: Multi-group discrimination using multi-criteria analysis: Illustrations from the field of finance. European Journal of Operational Research 139(2), 371–389 (2002)

Rough Web Caching

Sarina Sulaiman[1], Siti Mariyam Shamsuddin[1], and Ajith Abraham[2]

[1] Soft Computing Research Group, Faculty of Computer Science and Information System, Universiti Teknologi Malaysia, Johor, Malaysia
{sarina,mariyam}@utm.my

[2] Centre for Quantifiable Quality of Service in Communication Systems, Norwegian University of Science and Technology, Trondheim, Norway
ajith.abraham@ieee.org

Summary. The demand for Internet content rose dramatically in recent years. Servers became more and more powerful and the bandwidth of end user connections and backbones grew constantly during the last decade. Nevertheless users often experience poor performance when they access web sites or download files. Reasons for such problems are often performance problems, which occur directly on the servers (e.g. poor performance of server-side applications or during flash crowds) and problems concerning the network infrastructure (e.g. long geographical distances, network overloads, etc.). Web caching and prefetching have been recognized as the effective schemes to alleviate the service bottleneck and to minimize the user access latency and reduce the network traffic. In this chapter, we model the uncertainty in Web caching using the granularity of rough set (RS) and inductive learning. The proposed framework is illustrated using the trace-based experiments from Boston University Web trace data set.

1 Introduction

Good interactive response-time has long been known to be essential for user satisfaction and productivity [1, 2, 3]. This is also true for the Web [4, 5]. A widely-cited study from Zona Research [6] provides an evidence for the "eight second rule" in electronic commerce, *"if a Web site takes more than eight seconds to load, the user is much more likely to become frustrated and leave the site"*.

Lu *et al.*[7] has mentioned that most business organizations and government departments nowadays have developed and provided Internet based electronic services (e-services) that feature various intelligent functions. This form of e-services is commonly called *e-service intelligence* (ESI). ESI integrates intelligent technologies and methodologies into e-service systems for realizing intelligent Internet information searching, presentation, provision, recommendation, online system design, implementation, and assessment for Internet users. These intelligent technologies include machine learning, soft computing, intelligent languages, and data mining etc. ESI has been recently identified as a new direction for the future development stage of e-services. E-services offer great opportunities and challenges for many areas of services, such as government, education, tourism, commerce, marketing, finance, and logistics. They involve various online service

A. Abraham, R. Falcón, and R. Bello (Eds.): Rough Set Theory, SCI 174, pp. 187–211.
springerlink.com

providers, delivery systems and applications including e-government, e-learning, e-shopping, e-marketing, e-banking, and e-logistics.

A surprising fact is that many people tend to access the same piece of information repeatedly [7, 8] in any ESI. This could be weather related data, news, stock quotes, baseball scores, course notes, technical papers, exchange rate information and so on. If too many people attempt to access a Web site simultaneously, then they may experience problems in getting connected to the Web site. This is due to slow responses from the server as well as incapability of Web site in coping with the load.

An alternative way to tackle these problems is an implementation of Web caching in enhancing Web access [9, 8]. Web caching is beneficial to broad users including those who are relied on slow dial-up links as well as on faster broadband connections. The word caching refers to the process of saving data for future use. In other words, Web caching is the process of saving copies of content from the Web closer to the end user for quicker access. Web caching is a fairly new technology whose history is linked to that of the Web [10].

At the same time, Web prefetching is another well-known technique for reducing user web latency by preloading the web object that is not requested yet by the user [8, 9, 11, 12]. In other words, prefetching is a technique that downloads the probabilistic pages that are not requested by the user but could be requested again by the same user. Conventionally, there is some elapse time between two repeated requests by the same user. Prefetching usually performs the preloading operation within an elapse time and puts web objects into the local browser or proxy cache server to satisfy the next user's requests from its local cache.

However, the Web caching and prefetching technologies are the most popular software based solutions [11, 12]. Caching and prefetching can work individually or combined. The blending of caching and prefetching (called as pre-caching) enables doubling the performance compared to single caching [13]. These two techniques are very useful tools to reduce congestion, delays and latency problems. There are three most important features of web caching [14]:

- Caching that reduces network bandwidth usage
- Caching that also reduces user-perceived delays
- Caching that reduce loads on the original server

1.1 Problem in WWW Services

World Wide Web (WWW) has become the most ideal place for business and entertainment to enrich their presentation with interactive features. This has caused the evolution of Web growing and rising fast and drastically. Human interaction with objects or so called interactive features has leaded the Web to be more easily guided and capable to perform business task between distance places. These pages are linked and managed for certain purposes that perform as a Web application. These interactive Web pages consist of pages that are able to perform application logical task. The rising popularity of using Web applications in WWW causes tremendous demands on the Internet.

A key strategy for scaling the Internet to meet these increasing demands is to cache data near clients and thus improve access latency and reduce the network and server load [15, 16]. Mohamed *et. al* [17, 18, 19] has proposed an intelligent concept of *Smart Web Caching* with integrated modules of artificial neural networks (ANN), environment analysis and conventional caching procedure. The results are convincing in reducing the internet traffic flow and enhancing performances. However, implementing this integrated analyzer in Web caching environment causes highly computational cost [20, 17, 21] due to the complexity of the integrated process generation.

Caching is a technique used to store popular documents closer to the user. It uses algorithms to predict user's needs to specific documents and stores important documents. According to Curran and Duffy [22], caching can occur anywhere within a network, on the user's computer or mobile devices, at a server, or at an Internet Service Provider (ISP). Many companies employ web proxy caches to display frequently accessed pages to their employees, as such to reduce the bandwidth with lower costs [22, 23]. Web cache performance is directly proportional to the size of the client community [24, 22]. The bigger the client community, the greater the possibility of cached data being requested, hence, the better the cache's performance [22].

Moreover, caching a document can also cause other problems. Most documents on the Internet change over time as they are updated. Static and Dynamic Caching are two different technologies that widely used to reduce download time and congestion [20]. *Static Caching* stores the content of a web page which does not change. There is no need to request the same information repeatedly. This is an excellent approach to fight congestion. *Dynamic Caching* is slightly different. It determines whether the content of a page has been changed. If the contents have changed, it will store the updated version [23]. This unfortunately can lead to congestion and thus it is possibly not a very good approach as it does require verification on the source of the data prior to updating. If these two technologies are implemented simultaneously, then the latency and congestion can be diminished.

According to Davison [14] caching helps to bridge the performance gap between local activity and remote content. Caching assists improvement of Web performance by reducing the cost and end-user latency for Web access within a short term. However, in the long term, even as bandwidth costs continue to drop and higher end-user speeds become available; caching will continue to obtain benefits for the following reasons:

Bandwidth will always have some ***cost***. The cost of bandwidth will never reach zero, even though the competition is increasing, the market is growing, and the economies of scale will reduce end-user costs. The cost of bandwidth at the core has stayed relatively stable, requiring ISPs to implement methods such as caching to stay competitive and reduce core bandwidth usage so that edge bandwidth costs can be low.

Nonuniform bandwidth and ***latencies*** *will persist.* Because of physical limitations such as environment and location as well as financial constraints, there will always be variations in bandwidth and latencies. Caching can help to smooth these effects.

Network ***distances*** *are increasing.* Firewalls, other proxies for security and privacy, and virtual private networks for telecommuters have increased the number of hops for contents delivery, hence slow Web response time.

Bandwidth ***demands*** *continue to increase.* The growth of user base, the popularity of high-bandwidth media, and user expectations of faster performance have guaranteed the exponential increase in demand for bandwidth.

Hot spots *in the Web will continue.* Intelligent load balancing can alleviate problems when high user demand for a site is predictable. However, a Web site's popularity can also appear as a result of current events, desirable content, or gossips. Distributed Web caching can help alleviate these "hot spots" resulting from flash traffic loads.

Communication ***costs*** *exceed computational costs.* Communication is likely to always be more expensive (to some extent) than computation. The use of memory caches are preferred because CPUs are much faster than main memory. Likewise, the cache mechanisms will prolong as both computer systems and network connectivity become faster.

Furthermore, caching is the most relevant technique to improve storage system, network, and device performance. In mobile environments, caching can contribute to a greater reduction in the constraint of utilization resources such as network bandwidth, power, and allow disconnected operation [29]. A lot of studies are focused on developing a better caching algorithm to improve the choice of item to replace, and simultaneously, building up techniques to model access behavior and prefetch data. From 1990's until today, researchers on caching have produced different caching policies to optimize a specific performance and to automate policy parameter tuning. Prior to this, administrator or programmer had to select a particular parameter to observe workload changes. However, an adaptive and self-optimizing caching algorithm offer another advantage when considered mobile environments, where users of mobile devices should not expect to tune their devices to response the workload changes [29]. The workload depends on the current position of the mobile node in relation to other nodes and stations, and also depends on the current location and context of the mobile user.

Caching is effectively for data with infrequent changes. Besides, caching data locally to mobile nodes helps the ability to retrieve data from a nearby node, rather than from a more distant base station [28]. By simply retrieving data using multiple short-range transmissions in wireless environments provides a reduction in overall energy consumed. Santhanakrishnan *et al.* [29] illustrated on the demand-based retrieval of the Web documents in the mobile Web. They proposed caching scheme; Universal Mobile Caching which performed the most basic and general form of caching algorithms and largely emphasize the impact of the adaptive policy. This scheme is suitable for managing object caches in structurally varying environments. Ari *et al.* [30] proposed Adaptive Caching using Multiple Experts (ACME), which the individual experts were full replacement algorithms, applied to virtual caches, and their performance was estimated based on the observed performance of the virtual caches. The term expert refers to any mechanism for offering an answer to the question. For cache replacement, the answer they seek is the identity of the object in the cache with the least likelihood of subsequent future access.

Contrast to a single algorithm, there are not so many research works on integrated schemes. Aiming at integrating caching and prefetching, Yang and Zhang [26] employed a prediction model, whereas Teng *et al.* [31] presented a new cache replacement algorithm, considering the impact of prefetching engine located at Web server and a few cache parameters. Kobayashi and Yu [32] discussed the performance model for mobile Web caching and prefetching and provided the estimate of the total average latency, hit ratio, cache capacity and wireless bandwidth required.

Prefetching is an intelligent technique used to reduce perceived congestion, and to predict the subsequent page or document to be accessed [24, 12]. For example, if a user is on a page with many links, the prefetching algorithm will predict that the user may want to view associated links within that page. The prefetcher will then appeal the predicted pages, and stores them until the actual request is employed. This approach will display the page significantly faster compared to the page request without prefetching. The only drawback is that if the user does not request the pages, the prefetching algorithm will still implement the prediction of the subsequent pages, thus causes the network to be congested [25, 26, 27, 28].

In addition, Web prefetching method evolves from prefetching top-10 popular pages [33] or hyperlinks [34] into prefetching by user's access patterns. Statistical prefetching algorithms [35] make use of Markov modeling, and establish a Markov graph based on user's access histories and make prefetching predictions based on the graph which needs to be updated continuously while accessing Web. Prefetching strategies in [25, 36] used data mining technique, to decide whether to prefetch or not according to the probability of the pages accessed recently. But it is possible that the prefetched pages are far away from the current page sequence so that the cache hit ratio may not benefit from prefetching.

Hence, Web prefetching strategy need to achieve a balance between network loads and performance gains. Some research studies have found that too aggressive prefetching will increase Web access latency, since more prefetching will lead to replacement of more cache items even including the pages that will be accessed in near future. Under the wireless environment, Yin and Cao [37] proposed to dynamically adjust the number of prefetching according to power consumption for mobile data dissemination.

Wu *et al.* [38] introduced a rule-based modular framework for building self-adaptive applications in mobile environments. They developed techniques that combine static and dynamic analysis to uncover phase structure and data access semantics of a rule program. The semantic information is used to facilitate intelligent caching and prefetching for conserving limited bandwidth and reducing rule processing cost. As well, Komninos and Dunlop [39] found that calendars can really provide information that can be used to prefetch useful Internet content for mobile users. While it is expected that such an approach cannot fulfill the whole of Internet content needs for a user, the work presented provided evidence to the extent to which a mobile cache can be populated with relevant documents that the user could find of interest. However, a foreseeable problem with the current system is that the current adaptation algorithm adjusts the system gradually, and not immediately, to the needs of a user. Thus, if a dramatic change of circumstances was to occur, or if a user was to require information from a very specific and known source, it is likely the system would fail to provide the necessary information.

2 Why Web Caching?

Web caching is the temporary storage of Web objects (such as HTML documents) for later retrieval. There are three significant advantages to Web caching: reduced bandwidth consumption (fewer requests and responses that need to go over the network), reduced server load (fewer requests for a server to handle), and reduced latency (since responses for cached requests are available immediately, and closer to the client being served). Together, they make the Web less expensive and better performing.

Caching can be performed by the client application, and is built in to most Web browsers. There are a number of products that extend or replace the built-in caches with systems that contain larger storage, more features, or better performance. In any case, these systems cache net objects from many servers but all for a single user.

Caching can also be utilized in the middle, between the client and the server as part of a proxy. Proxy caches are often located near network gateways to reduce the bandwidth required over expensive dedicated Internet connections. These systems serve many users (clients) with cached objects from many servers. In fact, much of the usefulness (reportedly up to 80% for some installations) is in caching objects requested by one client for later retrieval by another client. For even greater performance, many proxy caches are part of cache hierarchies, in which a cache can inquire of neighboring caches for a requested document to reduce the need to fetch the object directly.

Finally, caches can be placed directly in front of a particular server, to reduce the number of requests that the server must handle. Most proxy caches can be used in this fashion, but this form has a different name (reverse cache, inverse cache, or sometimes httpd accelerator) to reflect the fact that it caches objects for many clients but from (usually) only one server [21].

2.1 How Web Caching Works?

All caches have a set of rules that they use to determine when to serve an object from the cache, if it's available. Some of these rules are set in the protocols (HTTP 1.0 and 1.1), and some are set by the administrator of the cache (either the user of the browser cache, or the proxy administrator).

Generally speaking, these are the most common rules that are followed for a particular request [21]:

1. If the object's headers notify the cache not to keep the object, then it will do so. Simultaneously, if there is no validation, then most caches will mark that as uncacheable item.
2. If the object is authenticated or secured, then it will not be cached.
3. A cached object is considered *fresh* (that is, able to be sent to a client without checking with the origin server) if:
   - It has an expiry time or other age-controlling directive set, and is still within the fresh period.
   - If a browser cache has already seen the object, and has been set to check once a session.
   - If a proxy cache has seen the object recently, and it was modified relatively long ago. Fresh documents are served directly from the cache, without checking with the origin server.
4. If an object is stale, the origin server will be executed to *validate* the object, or notify the cache whether the existing copy is still good.

Mutually **freshness** and **validation** are the most important mechanisms that make cache works with content. A fresh object will be available instantly from the cache, while a validated object will avoid sending the entire object all over again if it has not been changed.

3 Performance Measurement for Web Optimization

Performance measurement of Web caching is needed to establish the efficiency of a Web caching solution [9, 17, 32]. Some performance benchmarks or standards are required for a particular Web caching solution to be evaluated. Such benchmarks may assist in choosing the most suitable Web caching solution for the problem we encounter. In this situation, a possibility of a particular structure will beneficial for certain applications while other applications may require some other substitutes.

Some organizations may choose for proxy based caching solutions. They may try to overcome the problem of configuration Web browsers by forcing the use of browsers that provide auto-configuration. For massive organizations, network components such as routers and switches [9, 10] might be considered; otherwise, transparent caching can be employed. Some organizations may prefer highly scalable solutions for anticipating future needs. Besides, organizations which Web sites contain highly dynamic content might occupy Active Cache [41] or possibly will utilize Web server accelerators. Obviously, the subject of measurement of performance is controlled not just to find the competence of a given Web caching solution but also to cover evaluation of the performance of cache consistency protocols, cache replacement algorithms, the role of fundamental protocols such as HTTP and TCP and others.

3.1 Parameters for Measuring Web Performance

Several metrics are commonly used when evaluating Web caching policies [41]. These include [42]:

1. Hit rate is generally a percentage ratio of documents obtained by using the caching mechanism and total documents requested. If measurement focuses on byte transfer efficiency, then weighted hit rate is a better performance measurement [43].
2. Bandwidth utilization is an efficiency metric measurement. The reduction bandwidth consumption shows that the cache is better.
3. Response time/access time –response time is the time taken for a user to get a document.

The are various parameters such as user access patterns, cache removal policy, cache size and document size that can significantly affect cache performance. Other common metrics that are used to quantify the performance of Web caching solutions proposed by Mohamed [17] include hit ratio, byte hit ratio, response time, bandwidth saved, script size and current CPU usage.

Performance of Web caching solutions may be quantified by measuring parameters as follows [9]:

1. price
2. throughput (e.g. the number of HTTP requests per second generated by users, the rate at which a product delivers cache hits etc.)
3. cache hit ratio (the ratio of the number of requests met in the cache to the total number of requests)
4. byte hit ratio (the fraction of the number of bytes served by the cache divided by the total number of bytes sent to its clients)
5. the number of minutes until the first cache hit/miss after a breakdown
6. the cache age (the time after which the cache become full)
7. hit ratio/price (e.g. hits/second per thousand dollars)
8. downtime (e.g. time to recover from power outrages or cache failures)

Techniques for measuring the efficiency and usefulness of Web caching solutions have been evolving slowly since this field is relatively a new discipline; the theory of Web Caching has advanced much faster than practice [9].

Despite quantifying the performance of caching clarifications, other aspects such as client side latencies, server side latencies, aborted requests, DNS lookup latencies, cookies, different popularity characteristics among servers, the type of content, network packet losses should not be disregarded since there are some parameters are interrelated. For illustration, hit ratio is affected by inadequate disk space in a cache server, and these lacking in the object placement/replacement policies can cause the network to be overloaded. Hence, by maximizing a single parameter alone may not be adequate [9].

4 Uncertainty in Web Caching

Uncertainty, as well as evolution, is a part of nature. When humans describe complex environments, they use linguistic descriptors of cognized real-world circumstances that are often not precise, but rather "fuzzy". The theory of fuzzy sets [44] provides an effective method of describing the behavior of a system, which is too complex to be handling with the classical precise mathematical analysis. The theory of rough sets [61] emerged as another mathematical approach for dealing with uncertainty that arises from inexact, noisy or incomplete information. Fuzzy set theory assumes that the membership of the objects in some set is defined as a degree ranging over the interval [0,1]. Rough Set Theory (RST) focuses on the ambiguity caused by the limited distinction between objects in a given domain.

Uncertainty occurs in many real-life problems. It can cause the information used for problem solving being unavailable, incomplete, imprecise, unreliable, contradictory, and changing [46]. In computerized system, uncertainty is frequently managed by using quantitative approaches that are computationally intensive. For example, a binary that processes 'TRUE or FALSE', or 'YES' or 'NO' type of decisions, is likely to arrive at a conclusion or a solution faster than one that needs to handle uncertainty.

Organizing uncertainty is a big challenge for knowledge-processing systems [46]. In some problems, uncertainty can possibly be neglected, though at the risk of compromising the performance of a decision support system. However, in most cases, the management of uncertainty becomes necessary because of critical system requirements or more complete rules are needed. In these cases, eliminating inconsistent or incomplete information when extracting knowledge from an information system may introduce inaccurate or even false results, especially when the available source information is limited. Ordinarily, the nature of uncertainty comes from the following three sources: incomplete data, inconsistent data, and noisy data.

Thus, in a proxy cache, the superfluous of logs dataset with the huge number of records, the frequency of errors (incomplete data), and the diversity of log formats (inconsistent data) [10] will ground the practical challenges to analyze it either to cache or not cache objects in the popular documents. Table 1 depicts the sample of Web log data from Boston University Web Trace [47].

4.1 How Rough Sets Boost Up Web Caching Performance?

Another approach to represent uncertainty is using Rough Set (RS). RS are based on equivalence relations and set approximations, and the algorithms for computing RS properties are combinatorial in nature. The main advantages of RST are as follows [48]:

- It does not need any preliminary or additional information about data;
- It is easy to handle mathematically;
- Its algorithms are relatively simple.

Table 1. Sample Web log data

bugs 791131220 682449 "http://cs-www.bu.edu/" 2009 0.518815
bugs 791131221 620556 "http://cs-www.bu.edu/lib/pics/bu-logo.gif" 1805 0.320793
bugs 791131222 312837 "http://cs-www.bu.edu/lib/pics/bu-label.gif" 717 0.268006
bugs 791131266 55484 "http://cs-www.bu.edu/courses/Home.html" 3279 0.515020
bugs 791131266 676413 "http://cs-www.bu.edu/lib/pics/bu-logo.gif" 0 0.0
bugs 791131266 678045 "http://cs-www.bu.edu/lib/pics/bu-label.gif" 0 0.0
bugs 791131291 183914 "http://cs-www.bu.edu/students/grads/tahir/CS111/" 738 0.292915
bugs 791131303 477482 "http://cs-www.bu.edu/students/grads/tahir/CS111/hw2.ps" 41374 0.319514
bugs 791131413 265831 "http://cs-www.bu.edu/students/grads/tahir/CS111/if-stat.ps" 10202 0.380549
bunsen 791477692 218136 "http://cs-www.bu.edu/" 2087 0.509628
bunsen 791477693 134805 "http://cs-www.bu.edu/lib/pics/bu-logo.gif" 1803 0.286981
bunsen 791477693 819743 "http://cs-www.bu.edu/lib/pics/bu-label.gif" 715 0.355871
bunsen 791477719 107934 "http://cs-www.bu.edu/techreports/Home.html" 960 0.335809
bunsen 791477719 518262 "http://cs-www.bu.edu/lib/pics/bu-logo.gif" 0 0.0
bunsen 791477719 520770 "http://cs-www.bu.edu/lib/pics/bu-label.gif" 0 0.0

Wakaki *et al.* [48] used the combination of the RS-aided feature selection method and the support vector machine with the linear kernel in classifying Web pages into multiple categories. The proposed method gave acceptable accuracy and high dimensionality reduction without prior searching of better feature selection. Liang *et al.* [49] used RS and RS based inductive learning to assist students and instructors with WebCT learning. Decision rules were obtained using RS based inductive learning to give the reasons for the student failure. Consequently, RS based WebCT Learning improves the state-of-the-art of Web learning by providing virtual student/teacher feedback and making the WebCT system much more powerful.

Ngo and Nguyen [50] proposed an approach to search results clustering based on tolerance RS model following the work on document clustering. The application of tolerance RS model in document clustering was proposed as a way to enrich document and cluster representation to increase clustering performance. Furthermore, Chimphlee *et al.* [51] present a RS clustering to cluster web transactions from web access logs and using Markov model for next access prediction. Users can effectively mine web log records to discover and predict access patterns while using this approach. They perform experiments using real web trace logs collected from www.dusit.ac.th servers. In order to improve its prediction ration, the model includes a rough sets scheme in which search similarity measure to compute the similarity between two sequences using upper approximation.

In [52], the authors employed RS based learning program for predicting the web usage. In their approach, web usage patterns are represented as rules generated by the inductive learning program, BLEM2. Inputs to BLEM2 are clusters generated by a hierarchical clustering algorithm that are applied to preprocess web log records. Their empirical results showed that the prediction accuracy of rules induced by the learning program is better than a centroid-based method, and the learning program can generate shorter cluster descriptions.

In general, the basic problems in data analysis that can be undertaken by using RS approach is as follows [46]:

- Characterization of a set of objects in terms of attribute values;
- Finding the dependencies (total or partial) between attributes;
- Reduction of superfluous attributes (data);
- Finding the most significant attributes;
- Generation of decision rules.

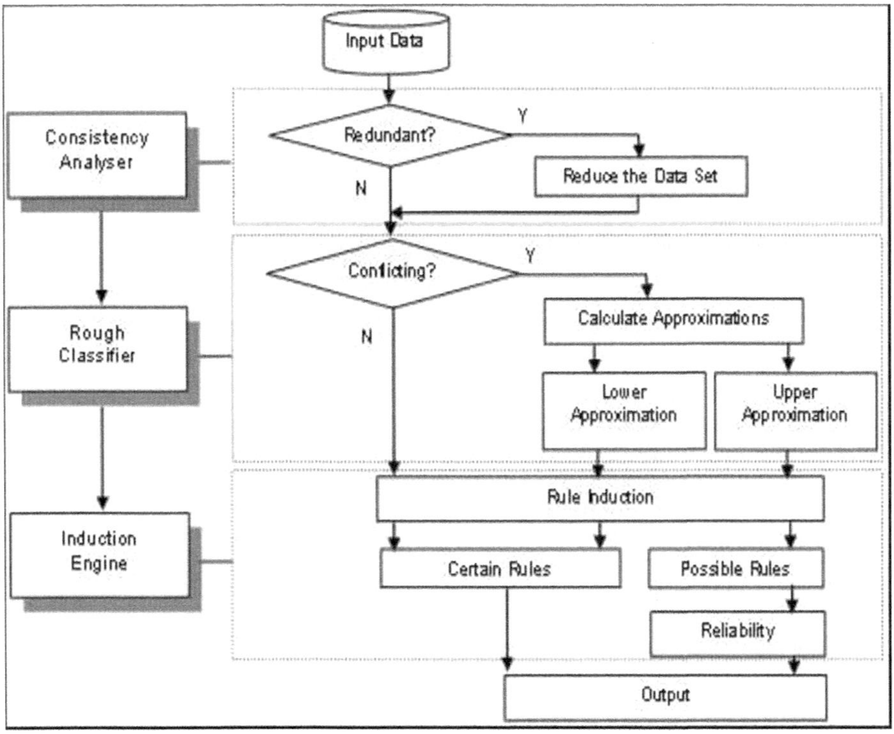

Fig. 1. Framework of the RClass System [46]

4.2 A Framework of Rough Sets

The RClass system integrates RST with an ID3-like learning algorithm [46] as shown in Figure 1. It includes three main modules; a consistency analyzer, a rough classifier and an induction engine. The consistency analyzer analyses the training data and performs two tasks; elimination of redundant data items, and identification of conflicting training data. The rough classifier has two approximators; the upper approximator and the lower approximator. The rough classifier is employed to treat inconsistent training data. The induction engine module has an ID3-like learning algorithm based on the minimum-entropy principle. The concept of entropy is used to measure how informative an attribute is.

5 Rough Sets and Inductive Learning

Rough Set Theory [53] was introduced by Zdzislaw Pawlak as a tool to solve problems with ambiguity and uncertainty [46]. Typically, data to be analyzed consists of a set of *objects* whose properties can be described by multi-valued *attributes*. The objects are described by the data that can be represented by a structure called the *information system* (S) [54]. An information system can be viewed as information table with its rows and columns consequent to objects and attributes.

Given a set E of examples described by an information table T, we classify objects in two different ways: by a subset C of the condition attributes and by a decision attribute D in the information table to find equivalence classes called indiscernibility classes $\Omega = \{\Omega_1,...,\Omega_n\}$ [55]. Objects within a given indiscernibility class are indistinguishable from each other on the basis of those attribute values. Each equivalence class based on the decision attribute defines a concept. We use $Des(\Omega_i)$ [49] to denote the description, i.e., the set of attribute values, of the equivalence class Ω_i. RS theory allows a concept to be described in terms of a pair of sets, lower approximation and upper approximation of the class. Let Y be a concept. The lower approximation $\underline{Y}$ and the upper approximation $\overline{Y}$ of Y are defined as [49]:

$$\underline{Y} = \{e \in E | e \in \Omega_i and X_i \subseteq Y\} \tag{1}$$

$$\overline{Y} = \{e \in E | e \in \Omega_i and X_i \cap Y = \emptyset\} \tag{2}$$

Lower approximation is the intersection of all those elementary sets that are contained by Y and upper approximation is the union of elementary sets that are contained by Y.

Inductive Learning is a well-known area in artificial intelligence. It is used to model the knowledge of human experts by using a carefully chosen sample of expert decisions and inferring decision rules automatically, independent of the subject of interest [56]. RS based Inductive Learning uses RS theory to find general decision rules [57, 58]. These two techniques are nearness to determine the relationship between the set of attributes and the concept.

5.1 Rough Set Granularity in Web Caching

In our research, BU Web trace dataset from Oceans Research Group at Boston University are used [47]. We considered 20 sample objects only, i.e., January 1995 records. In our previous research, we used the same dataset with implementation of RS [59] and integration of Neurocomputing and Particle Swarm Optimization (PSO) algorithm [60] to optimize the Web caching performance. Three conditional attributes are taken into consideration; request time (*Timestamp, TS*) in seconds and microseconds, a current CPU usage (*Sizedocument, SD*) in bytes and response time (*Objectretrievaltime, RT*) in seconds. Consequently, a cache, CA is chosen as a decision for the information table; 1 for cache and 0 for not cache. Decision rules are obtained using RS based Inductive Learning [57] for

Table 2. Sample of log files dataset information table

Object	Attributes			Decision
	TS	***SD***	***RT***	***CA***
S_1	790358517	367	0.436018	0
S_2	790358517	514	0.416329	0
S_3	790358520	297	0.572204	0
S_4	790358527	0	0	1
S_5	790358529	0	0	1
S_6	790358530	0	0	1
S_7	790358530	0	0	1
S_8	790358538	14051	0.685318	0
S_9	790362535	1935	1.021313	0
S_{10}	790362536	1804	0.284184	0
S_{11}	790362537	716	0.65038	0
S_{12}	790363268	1935	0.76284	0
S_{13}	790363270	716	1.050344	0
S_{14}	790363270	1804	0.447391	0
S_{15}	790363329	1935	0.553885	0
S_{16}	790363330	716	0.331864	0
S_{17}	790363330	1804	0.342798	0
S_{18}	790363700	0	0	1
S_{19}	790363700	0	0	1
S_{20}	790363700	1136	0.428784	0

Web caching. Table 2 depicts the structure of the study: 20 objects, 3 attributes, and a decision.

Detailed description and analysis are given in Table 3. The domain E and two concepts Y_{cache} and $Y_{notcache}$ from the decision attribute (CA) are obtained as follows:

$E=\{e_1,e_2,e_3,e_4,e_5,e_6,e_7,e_8,e_9,e_{10},e_{11},e_{12},e_{13},e_{14},e_{15},e_{16},e_{17},e_{18},e_{19},e_{20}\}$
$Y_{cache} = \{e_4,e_5,e_6,e_{17}\}$
$Y_{notcache} = \{e_1,e_2,e_3,e_7,e_8,e_9,e_{10},e_{11},e_{12},e_{13},e_{14},e_{15},e_{16},e_{18}\}$

Initially we find the indiscernibility classes based on TS that are $\{e_1,e_2\}$, $\{e_{12}, e_{13}\}$, $\{e_{15},e_{16}\}$,$\{e_{17},e_{18}\}$and$\{e_3\}$, $\{e_4\}$,$\{e_5\}$,$\{e_6\}$,$\{e_7\}$, $\{e_8\}$,$\{e_9\}$,$\{e_{10}\}$, $\{e_{11}\}$, $\{e_{14}\}$.

The discriminant index of a concept Y is defined using the following formula:

$$\alpha_{C_i}(Y) = 1 - |\overline{Y} - \underline{Y}|/|E| \tag{3}$$

Consequently, the discriminant index of TS is $\alpha_{C1}(Y) = 1 - |\overline{Y} - \underline{Y}|/|E| =$ 1-(9-0)/20 = 0.55 determines the effectiveness of the singleton set of attributes consisting of TS in specifying the membership in Y (the cache concept). Subsequently, the indiscernibility classes of SD is conducted and the results are $\{e_4,e_5,e_6,e_{17}\}$,$\{e_{10},e_{12},e_{15}\}$, $\{e_9,e_{13},e_{16}\}$,$\{e_8,e_{11},e_{14}\}$ and $\{e_1\}$,$\{e_2\}$,$\{e_3\}$,$\{e_7\}$, $\{e_{18}\}$.

The lower approximation is illustrated as $\underline{Y}=\cup_{\Omega_i\subseteq Y}\Omega_i=\{e_1\}$,$\{e_2\}$,$\{e_3\}$, $\{e_7\}$,$\{e_{18}\}$. The upper approximation is given as $\overline{Y}=\cup_{\Omega_i\cap Y\neq\emptyset}\Omega_i=\{e_4,e_5,e_6,e_{17},e_{10},e_{12},e_{15},e_9,e_{13},e_{16},e_8,e_{11},e_{14}\}$. Hence, the discriminant index of SD is $\alpha_{C2}(Y) = 1 - |\overline{Y} - \underline{Y}|/|E| = 1 - (15-5)/20 = 0.5$.

The indiscernibility classes based on RT are $\{e_4,e_5,e_6,e_{17}\}$ and $\{e_1\}$,$\{e_2\}$, $\{e_3\}$,$\{e_7\}$,$\{e_8\}$,$\{e_9\}$,$\{e_{10}\}$,$\{e_{11}\}$,$\{e_{12}\}$,$\{e_{13}\}$,$\{e_{14}\}$,$\{e_{15}\}$,$\{e_{16}\}$,$\{e_{18}\}$. The lower approximation is given as $\underline{Y}=\cup_{\Omega_i\subseteq Y}\Omega_i=\emptyset$. The upper approximation is $\overline{Y}=\cup_{\Omega_i\cap Y\neq\emptyset}\Omega_i=\{e_4,e_5,e_6,e_{17}\}$. The discriminant index of RT is $\alpha_{C3}(Y) = 1 - |\overline{Y} - \underline{Y}|/|E| =$ 1 - (6 - 0)/20 = 0.7.

By comparing the discriminant indices of all attributes, we identify that the discriminant index of RT has the highest value, $\alpha_{Condition3}(Y) = 0.7$. This value determines better membership in Y. Hence, the first rule is obtained as:

$$R_1 : \{Objectretrievaltime = 0\} \Rightarrow \{Cache = 1\}$$

Since RT is the most important condition attribute, we merge this condition attribute with other condition attributes to produce a new domain and to execute new rules (refer to Table 3).

To discover the new domain, initially, the following equation is used to remove unnecessary elements. $(E - \overline{Y}) \cup (\underline{Y}) = \{e_1,e_2,e_3,e_7,e_8,e_9,e_{10},e_{11},e_{12},e_{13},e_{14}, e_{15},e_{16}\} \cup \emptyset$. The new element set are given as, $(E - [(E - \overline{Y}) \cup (\underline{Y})] = (E - \{e_1,e_2,e_3,e_7,e_8,e_9,e_{10},e_{11},e_{12},e_{13},e_{14},e_{15},e_{16}\} \cup \emptyset) = \{e_4,e_5,e_6,e_{17}\}$

Table 3. Collapsed log files dataset information table

Object	Attributes			Decision	Total
	TS	***SD***	***RT***	***CA***	
e_1	790358517	367	0.436018	0	1
e_2	790358517	514	0.416329	0	1
e_3	790358520	297	0.572204	0	1
e_4	790358527	0	0	1	1
e_5	790358529	0	0	1	1
e_6	790358530	0	0	1	2
e_7	790358538	14051	0.685318	0	1
e_8	790362535	1935	1.021313	0	1
e_9	790362536	1804	0.284184	0	1
e_{10}	790362537	716	0.65038	0	1
e_{11}	790363268	1935	0.76284	0	1
e_{12}	790363270	716	1.050344	0	1
e_{13}	790363270	1804	0.447391	0	1
e_{14}	790363329	1935	0.553885	0	1
e_{15}	790363330	716	0.331864	0	1
e_{16}	790363330	1804	0.342798	0	1
e_{17}	790363700	0	0	1	2
e_{18}	790363700	1136	0.428784	0	1

Table 4. Horizontal selection of collapsed table

Object	Attributes			Decision	Total
	TS	***SD***	***RT***	***CA***	
e_4	790358527	0	0	1	1
e_5	790358529	0	0	1	1
e_6	790358530	0	0	1	2
e_{17}	790363700	0	0	1	2

Table 5. Further horizontally collapsed reduction table

Object	Attributes *TS*	Decision *CA*	Total
e_4	790358527	1	1
e_5	790358529	1	1
e_6	790358530	1	2
e_{17}	790363700	1	2

Subsequently, the horizontal selection of the collapsed information table is obtained (Table 4). The total number of objects becomes 6.

The illustrations of this selected information table are given as $Y_{cache} = \{e_4, e_5, e_6, e_{17}\}$ and $Y_{notcache} = \emptyset$, and the domain is $E = \{e_4, e_5, e_6, e_{17}\}$. We locate the indiscernibility classes based on SD and RT as $\emptyset$. The lower approximation is $\underline{Y} = \cup_{\Omega_i \subseteq Y} \Omega_i = \emptyset$ and the upper approximation is $\overline{Y} = \cup_{\Omega_i \cap Y \neq \emptyset} \Omega_i = \{e_4, e_5, e_6, e_{17}\}$. The discriminant index of SD and RT is $\alpha_{C2,C3}(Y) = 1 - |\overline{Y} - \underline{Y}|/|E| = 1 - (6-0)/6 = 0$.

The indiscernibility classes based on TS and RT is $\{e_4, e_5, e_6, e_{17}\}$. The lower approximation is $\underline{Y} = \cup_{\Omega_i \subseteq Y} \Omega_i = \{e_4, e_5, e_6, e_{17}\}$ and the upper approximation is $\overline{Y} = \cup_{\Omega_i \cap Y \neq \emptyset} \Omega_i = \{e_4, e_5, e_6, e_{17}\}$. The discriminant index of TS and RT is $\alpha_{C1,C3}(Y) = 1 - |\overline{Y} - \underline{Y}|/|E| = 1 - (6 - 6)/6 = 1$.

By comparing the discriminant indices, we discover that $\alpha_{C1,C3}(Y) = 1$ best determines the membership in Y. Thus, we attain the sample of second rule:

$\boldsymbol{R_2}$: **{*Timestamp* = 790358527, *Objectretrievaltime* = *0*} ⇒{*Cache* = *1*}**

Two rules have been found. If new domain is uncovered and new rules are computed using the same method as previous, then the irrelevant elements can be removed as $(E - \overline{Y}) \cup (\underline{Y}) = \emptyset \cup \{e_4, e_5, e_6, e_{17}\}$.

By referring to Table 3, we can see that the first set is empty and the second set has been handled by rule 2. Hence, the new set of elements becomes $(E - [(E - \overline{Y}) \cup (\underline{Y})] = \{e_4, e_5, e_6, e_{17}\}$.

Based on this assumption, we obtain supplementary collapsed information table in which SD and RT are omitted due to superfluous attributes (see Table 5).

The rules are fruitfully induced. A question that rises is how much we can believe in these rules. Therefore, we need to evaluate the strength of the rules as follows [61, 54]:

$$\frac{\#\text{ of positive objects covered by the rule}}{\#\text{ of objects covered by the rule (including both positive and negative)}}$$

Based on this equation, the first rule has strength of 6/20. It shows that 30% Classes of e_4, e_5, e_6, ande_{17} (Table 3) are positive examples covered by the rule. Class e_1 is a negative example covered by the first rule. The second rule has the strength of 6/6, that is, 100%. In applying the first rule to this object, there is a 30% chance that the reason for cache the object is exclusively the cache of RT. However, there is a higher probability that the reason for cache is due to extra timing of TS and RT, due to 100% strength of the second rule. Algorithm 1 illustrates the algorithm of rules induction using RS [57].

Algorithm 1. Rough set algorithm [57]

```
for each decision class do
    Initialise universe of objects
    Select decision class
    Find class relation
    repeat
        for each attribute do
            Select attribute
            Find equivalence relation
            Find lower subset
            Find upper subset
            Calculate discriminant index
        end for
        Select attribute with highest discriminant index
        Generate rules
        Reduce universe of objects
        Reduce class relation
    until no objects with selected decision class
end for
```

This part presents substantial RS analysis based on Inductive Learning methods to optimize Web caching performance to probe significant attributes and generate the decision rules. RS granularity in Web caching allows decision rules to be induced. These rules are important in optimizing user storage by executing caching strategy in specifying the most relevant condition attributes. This approach provides guidance to the administrator in Web caching regarding to selection of the best parameters to be cached. Based on this analysis, the administrator may reorganize the parameter of log data set in proxy caching accordingly.

6 Experimental Results

In this part describes experimental results of dataset for HTTP requests and user behavior of a set of Mosaic clients running in the Boston University (BU), Computer Science Department [47].

6.1 BU Log Dataset

In this experiment, BU Web Trace collected by Oceans Research Group at Boston University is employed. BU traces records consist of 9,633 files with a population of 762 different users, and recording 1,143,839 requests for data transfer. The data for January 1995 comprises of 11 to 220 users with 33,804 records. However, after data cleaning, only 10,727 dataset is left.

Moreover, in this research RS is exploited to reduce the rules of a log file and simultaneously to enhance the prediction performance of user behavior. RS is beneficial in probing the most significant attributes with crucial decision rules to facilitate intelligent caching and prefetching to safeguard limited bandwidth and minimize the processing cost.

The dataset is split in two; 70% (7,187 objects) for training and 30% (3,540 objects) for testing. To simplify data representation, a Naïve Discretization Algorithm (NA) is exploited and Genetic Algorithm (GA) is chosen to generate the object rules. Next, Standard Voting Classifier (SVC) is selected to classify the log file dataset. The derived rules from the training are used to test the effectiveness of the unseen data. In addition, 3-Fold Cross Validation is implemented for validation of our experiment. First fold (K1) the testing data from 1 to 3540, second fold (K2) from 3541 to 7081 and third fold (K3) from 7082 to 10622. Data are stored in decision table. Columns represent *attributes*, rows represent *objects* whereas every cell contains *attribute value* for corresponding objects and attributes. A set of attributes are *URL, Machinename, Timestamp, Useridno, Sizedocument, Objectretrievaltime*, and *Cache* as a decision.

6.2 Data Discretization and Reduction

Training data is discretized using NA. This discretization technique is implemented a very straightforward and simple heuristic that may result in very many cuts, probably far more than are desired. In the worst case, each observed value is assigned its own interval. GA is used for reduct generation [63] as it provides more exhaustive search of the search space. Reducts generation have two options [64]; full object reduction and object related reduction. Full object reduction produces set of minimal attributes subset that defines functional dependencies, while reduct with object related produce a set of decision rules or general pattern through minimal attributes subset that discern on a per object basis. The reduct with object related is preferred due to its capability in generating reduct based on discernibility function of each object.

Table 6 illustrates the comparison results of generation of a log file dataset in different K-fold (K1, K2 and K3). The highest testing accuracy is 98.46% achieved through NA discretization method and GA with full reduct method. Number of reducts for K1, K2 and K3 are equivalent. Object related reduct, 22 and full reduct, 6. In our observation, the highest number of rules are GA with full reduct, 63311 for K1, K2 and K3 and the highest testing accuracy is GA with full reduct for K1, 98.46%.

Table 6. Comparison reduct for K1, K2 and K3

Discretize Method	Reduct Method	K-fold	No.of Reduct	No.of Rules	Testing Accuracy (%)
NA	GA (object related)	K1	22	26758	96.8644
		K2	22	26496	96.8644
		K3	22	26496	96.8079
	GA (full object)	K1	6	63311	98.4618
		K2	6	63311	5.76271
		K3	6	63311	5.79096

6.3 Rule Derivation

A unique feature of the RS method is its generation of rules that played an important role in predicting the output. ROSETTA tool has listed the rules and provides some statistics for the rules which are support, accuracy, coverage, stability and length. Below is the definition of the rule statistics [64]:

- The rule LHS support is defined as the number of records in the training data that fully exhibit property described by the IF condition.
- The rule RHS support is defined as the number of records in the training data that fully exhibit the property described by the THEN condition.
- The rule RHS accuracy is defined as the number of RHS support divided by the number of LHS support.
- The rule LHS coverage is the fraction of the records that satisfied the IF conditions of the rule. It is obtained by dividing the support of the rule by the total number of records in the training sample.
- The rule RHS coverage is the fraction of the training records that satisfied the THEN conditions. It is obtained by dividing the support of the rule by the number of records in the training that satisfied the THEN condition.

The rule length is defined as the number of conditional elements in the IF part. Table 7 shows the sample of most significant rules. These rules are sorted according to their support value. The highest support value is resulted as the most significant rules. From the Table 7, the generated rule of {Sizedocument(0) $\Rightarrow$ Cache(1)} is considered the most significant rules with the outcome of not cache (output=0) and with cache (output=1). This is supported by 3806 for LHS support and RHS support value. Subsequently, the impact of rules length on testing accuracy are evaluated based on rules set from Table 7. Consequently, the same rules are divided into two groups; $1\leq$ rules of length ≤ 2. It seems that the rules with length ≥ 1 contribute better classification compared to the rules with length ≤ 2.

Table 7. Sample for sorted of highest rule support values from data decision table for K1, K2 and K3

Rule	LHS Support	RHS Support	LHS Length	RHS Length
K1				
Sizedocument(0) ⇒ Cache(1)	3806	3806	1	1
Objectretrievaltime(0.000000) ⇒ Cache(1)	3805	3805	1	1
Sizedocument(2009) ⇒ Cache(0)	233	233	1	1
Sizedocument(717) ⇒ Cache(0)	128	128	1	1
K2				
URL(http://cs-www.bu.edu/lib/pics/bu-logo.gif) AND Sizedocument(0) ⇒ Cache(1)	1009	1009	2	1
URL(http://cs-www.bu.edu/lib/pics/bu-logo.gif) AND Objectretrievaltime(0.00000) ⇒ Cache(1)	1009	1009	2	1
Machinename(beaker) AND Sizedocument(0) ⇒ Cache(1)	308	308	2	1
Machinename(beaker) AND Objectretrievaltime(0.00000) ⇒ Cache(1)	308	308	2	1
K3				
URL(http://cs-www.bu.edu/lib/pics/bu-logo.gif) AND Objectretrievaltime(0.00000) ⇒ Cache(1)	989	989	2	1
URL(http://cs-www.bu.edu/lib/pics/bu-logo.gif) AND Sizedocument(0) ⇒ Cache(1)	989	989	2	1
Machinename(beaker) AND Sizedocument(0) ⇒ Cache(1)	306	306	2	1
Machinename(beaker) AND Objectretrievaltime(0.00000) ⇒ Cache(1)	306	306	2	1

6.4 Classification

From the analysis, it shows that the classification is better. Furthermore, the core attributes and the significant rules can improve the accuracy of classification. Table 8 shows the result of classification performance of K1, K2 and K3 for the original table and the new decision table of log file dataset. Hence, Figure 2 depicts an overall accuracy for log file, 36.67% for all rules in original decision table and 96.85% for selected rules in new decision table. This result shows a different of overall accuracy up to 60.18% between the original decision table and new decision table.

Table 8. Classification performance of K1, K2 and K3 for both original decision table and new decision table of log file dataset

Decision Table	Rule Set	K-fold	Accuracy (%)	Overall Accuracy (%)
New decision table	Selected rules	K1	96.8644	96.85
		K2	96.8644	
		K3	96.8079	
Orig. decision table	All rules	K1	98.4618	36.67
		K2	5.76271	
		K3	5.79096	

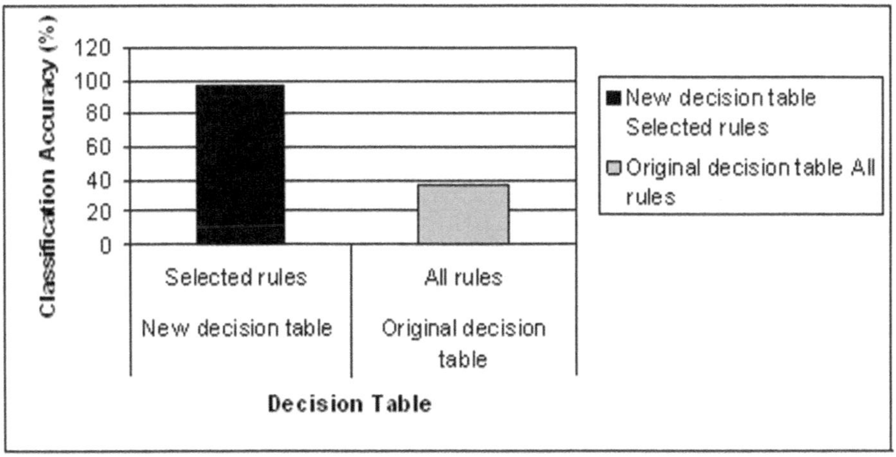

Fig. 2. Overall classification accuracy for both original decision table and new decision table of log file dataset

7 Conclusions

This chapter illustrated the usage of rough set theory for performance enhancement of Web caching. The RClass System framework [46] is used as a knowledge representation scheme for uncertainty in data for optimizing the performance of proxy caching that use to store the knowledge discovery of user behaviors in log format.

Furthermore, substantial RS analysis based on Inductive Learning methods is presented to optimize Web caching performance to probe significant attributes and generate the decision rules. RS granularity in Web caching allows decision rules to be induced. These rules are important in optimizing users' storage by executing caching strategy in specifying the most relevant condition attributes. This approach provides guidance to the administrator in Web caching regarding to selection of the best parameters to be cached. Based on this analysis, the administrator may reorganize the parameter of log data set in proxy caching accordingly.

Moreover, an empirical study has been conducted for searching optimal classification. A RS framework for log dataset is illustrated mutually with an analysis of reduced and derived rules, with entrenchment of their implicit properties for better classification outcomes.

In the future, more experiments on huge data will be conducted on hybridization of RS and evolutionary computation to deal with multiple knowledge of Web caching in reducing network latency.

Acknowledgements

This work is supported by MOSTI and RMC, Universiti Teknologi Malaysia, MALAYSIA. Authors would like to thank Soft Computing Research Group, Faculty of Computer Science and Information System, UTM for their continuous support and fondness in making this work an accomplishment.

References

1. Walter, J.D., Ahrvind, J.T.: The economic value of rapid response time. Technical Report GE20-0752-0, IBM, White Plains, NY (November 1982)
2. James, T.B.: A theory of productivity in the creative process. IEEE Computer Graphics and Applications 6(5), 25–34 (1986)
3. Chris, R.: Designing for delay in interactive information retrieval. Interacting with Computers 10, 87–104 (1998)
4. Judith, R., Alessandro, B., Jenny, P.: A psychological investigation of long retrieval times on the World Wide Web. Interacting with Computers 10, 77–86 (1998)
5. Nina, B., Anna, B., Allan, K. (2000) Integrating user perceived quality into Web server design. In: Proceedings of the Ninth International World Wide Web Conference, Amsterdam (May 2000)
6. Zona, The economic impacts of unacceptable Web site download speeds. White paper, Zona Research (1999), `http://www.zonaresearch.com/deliverables/whitepapers/wp17/index.htm`

7. Lu, J., Ruan, D., Zhang, G.: E-Service Intelligence Methodologies. In: Technologies and Applications. Springer, Heidelberg (2006)
8. Davison, B.D.: The Design and Evaluation of Web Prefetching and Caching Techniques. Doctor of Philosophy thesis, Graduate School of New Brunswick Rutgers, The State University of New Jersey, United State (2002)
9. Nagaraj, S.V.: Web Caching and Its Applications. Kluwer Academic Publishers, Dordrecht (2004)
10. Krishnamurthy, B., Rexford, J.: Web Protocols and Practice: HTTP 1.1, Networking Protocols. Caching and Traffic Measurement. Addison Wesley, Reading (2001)
11. Acharjee, U.: Personalized and Intelligence Web Caching and Prefetching. Master thesis, Faculty of Graduate and Postdoctoral Studies, University of Ottawa, Canada (2006)
12. Garg, A.: Reduction of Latency in the Web Using Prefetching and Caching. Doctor of Philosophy thesis, University of California, Los Angeles, United State (2003)
13. Kroeger, T.M., Long, D.D.E., Mogul, J.C.: Exploring The Bounds of Web Latency Reduction from Caching and Prefetching. In: Proceedings of the USENIX Symposium on Internet Technology and Systems, pp. 13–22 (1997)
14. Davison, B.D.: A Web Caching Primer. IEEE Internet Computing, pp. 38–45 (2001), http://computer.org/internet
15. Wong, K.Y., Yeung, K.H.: Site-Based Approach in Web Caching Design. IEEE Internet Comp. 5(5), 28–34 (2001)
16. Wong, K.Y.: Web Cache Replacement Policies: A Pragmatic Approach. IEEE Network (January/Feburary 2006)
17. Mohamed, F.: Intelligent Web Caching Architecture. Master thesis, Faculty of Computer Science and Information System, Universiti Teknologi Malaysia, Malaysia (2007)
18. Mohamed, F., Shamsuddin, S.M.: Smart Web Caching with Structured Neural Networks. In: Proc. Of The 2nd National Conf. on Computer Graphics and Multimedia, Selangor, pp. 348–353 (2004)
19. Mohamed, F., Shamsuddin, S.M.: Smart Web Cache Prefetching with MLP Network. In: Proc. Of The 1st IMT-GT Regional Conf. On Mathematics, Statistics and their Applications, pp. 555–567 (2005)
20. Curran, K., Duffy, C.: Understanding and Reducing Web Delays. Int. J. Network Mgmt. 15, 89–102 (2005)
21. Web Caching, Caching Tutorial for Web Authors (2008), http://www.web-caching.com/mnot_tutorial/intro.html
22. Saiedian, M., Naeem, M.: Understanding and Reducing Web Delays. IEEE Computer Journal 34(12) (December 2001)
23. Foedero.com, Dynamic Caching (2006), http://www.foedero.com/dynamicCaching.html
24. Fan, L., Jacobson, Q., Cao, P., Lin, W.: Web prefetching between low-bandwidth clients and proxies: potential and performance. In: Proceedings of the 1999 ACM SIGMETRICS International Conference on Measurement and Modelling of Computer Systems, Atlanta, Georgia, USA, pp. 178–187 (1999)
25. Yang, Q., Zhang, Z.: Model Based Predictive Prefetching. IEEE, 1529-4188/01: 291–295 (2001)
26. Yang, W., Zhang, H.H.: Integrating Web Prefetching and Caching Using Prediction Models. In: Proceedings of the 10th international conference on World Wide Web (2001)
27. Jiang, Z., Kleinrock, L.: Web Prefetching in a Mobile Environment. IEEE Personal Communications, 1070-9916/98: 25–34 (1998a)

28. Jiang, Z., Kleinrock, L.: An Adaptive Network Prefetch Scheme. IEEE Journal on Selected Areas in Communications 16(3), 1–11 (1998b)
29. Santhanakrishnan, G., Amer, A., Chrysanthis, P.K.: Towards Universal Mobile Caching. In: Proceedings of MobiDE 2005, Baltimore, Maryland, USA, pp. 73–80 (2005)
30. Ari, I., Amer, A., Gramacy, R., Miller, E.L., Brandt, S., Long, D.D.E.: Adaptive Caching using Multiple Experts. In: Proc. of the Workshop on Distributed Data and Structures (2002)
31. Teng, W.-G., Chang, C.-Y., Chen, M.-S.: Integrating Web Caching and Web Prefetching in Client-Side Proxies. IEEE Transaction on Parallel and Distributed Systems 16(5) (May 2005)
32. Kobayashi, H., Yu, S.-Z.: Performance Models of Web Caching and Prefetching for Wireless Internet Access. In: International Conference on Performance Evaluation: Theory, Techniques and Applications (PerETTA 2000) (2000)
33. Markatos, E.P., Chironaki, C.E.: A Top 10 Approach for Prefetching the Web. In: Proc. of INET 1998:Internet Global Summit (1998)
34. Duchamp, D.: Prefetching Hyperlinks. In: Proceedings of the 2nd USENIX Symposium on Internet Technologies & Systems(USITS 1999), Boulder, Colorado, USA (1999)
35. Deshpande, M., Karypis, G.: Selective Markov Models for Predicting Web-Page Accesses. In: Proceedings SIAM International Conference on Data Mining (2001)
36. Song, H., Cao, G.: Cache-Miss-Initiated Prefetch in Mobile Environments. Computer Communications 28(7) (2005)
37. Yin, L., Cao, G.: Adaptive Power-Aware Prefetch in Mobile Networks. IEEE Transactions on Wireless Communication 3(5) (September 2004)
38. Wu, S., Chang, C., Ho, S., Chao, H.: Rule-based intelligent adaptation in mobile information systems. Expert Syst. Appl. 34(2), 1078–1092 (2008)
39. Komninos, A., Dunlop, M.D.: A calendar based Internet content pre-caching agent for small computing devices. Pers Ubiquit Comput. (2007), DOI 10.1007/s00779-007-0153-4
40. Cao, P., Zhang, J., Beach, K.: Active Cache:Caching Dynamic Contents on The Web. Distributed Systems Engineering 6(1), 43–50 (1999)
41. Shi, Y., Watson, E., Chen, Y.-S.: Model-Driven Simulation of World-Wide-Web Cache Policies. In: Proceedings of the 1997 Winter Simulation Conference, pp. 1045–1052 (1997)
42. Abrams, M.: WWW:Beyond the Basics (1997), `http://ei.cs.vt.edu/~wwwbtb/fall.96/bppk/chap25/index.html`
43. Abrams, M., Standridge, C.R., Abdulla, G., Williams, S., Fox, E.A.: Caching proxies: Limitations and Potentials. In: Proceedings of the 4th International WWW Conference, Boston, MA (December 1995), `http://www.w3.org/pub/Conferences/WWW4/Papers/155/`
44. Zadeh, L.: Fuzzy sets. Information and Control 8, 338–353 (1965)
45. Pawlak, Z.: Rough Sets - Theoretical Aspects of Reasoning about Data. Kluwer Academic Publishers, Dordrecht (1991)
46. Triantaphyllou, E., Felici, G.: Data Mining and Knowledge Discovery Approaches Based on Rule Induction Techniques. Massive Computing Series, pp. 359–394. Springer, Heidelberg (2006)
47. BU Web Trace, `http://ita.ee.lbl.gov/html/contrib/BU-Web-Client.html`
48. Wakaki, T., Itakura, H., Tamura, M., Motoda, H., Washio, T.: A Study on Rough Set-Aided Feature Selection for Automatic Web-page Classification. Web Intelligence and Agent Systems: An International Journal 4, 431–441 (2006)

49. Liang, A.H., Maguire, B., Johnson, J.: Rough Set WebCT Learning, pp. 425–436. Springer, Heidelberg (2000)
50. Ngo, C.L., Nguyen, H.S.: A Tolerence Rough Set Approach to Clustering Web Search Results, pp. 515–517. Springer, Heidelberg (2004)
51. Chimphlee, S., Salim, N., Ngadiman, M.S., Chimphlee, W., Srinoy, S.: Rough Sets Clustering and Markov model for Web Access Prediction. In: Proceedings of the Postgraduate Annual Research Seminar, Malaysia, pp. 470–475 (2006)
52. Khasawneh, N., Chan, C.-C.: Web Usage Mining Using Rough Sets. In: Annual Meeting of the North American Fuzzy Information Processing Society (NAFIPS 2005), pp. 580–585 (2005)
53. Pawlak, Z.: Rough Sets, pp. 3–8. Kluwer Academic Publishers, Dordrecht (1997)
54. Johnson, J., Liu, M.: Rough Sets for Informative Question Answering. In: Proceedings of the International Conference on Computing and Information (ICCI 1998), pp. 53–60 (1996)
55. Liang, A.H., Maguire, B., Johnson, J.: Rough set based webCT learning. In: Lu, H., Zhou, A. (eds.) WAIM 2000. LNCS, vol. 1846, pp. 425–436. Springer, Heidelberg (2000)
56. Johnson, J.A., Johnson, G.M.: Student Characteristics and Computer Programming Competency: A Correlational Analysis. Journal of Studies in Technical Careers 14, 23–92 (1992)
57. Tsaptsinos, D., Bell, M.G.: Medical Knowledge Mining Using Rough Set Theory, `http://citeseer.ist.psu.edu/86438.html`
58. Shan, N., Ziarko, W., Hamilton, H.J., Cercone, N.: Using rough sets as tools for knowledge discovery. In: Proceedings of the International Conference on Knowledge Discovery and Data Mining (KDD 1995), pp. 263–268. AAAI Press, Menlo Park (1995)
59. Sulaiman, S., Shamsuddin, S.M., Abraham, A.: An Implementation of Rough Set in Optimizing Mobile Web Caching Performance. In: Tenth International Conference on Computer Modeling and Simulation, UKSiM/EUROSiM 2008, pp. 655–660. IEEE Computer Society Press, Los Alamitos (2008)
60. Sulaiman, S., Shamsuddin, S.M., Forkan, F., Abraham, A.: Intelligent Web Caching Using Neurocomputing and Particle Swarm Optimization Algorithm. In: Second Asia International Conference on Modeling and Simulation, AMS 2008, pp. 642–647. IEEE Computer Society Press, Los Alamitos (2008)
61. Pawlak, Z., Grzymala-Busse, J., Slowinski, R., Ziarko, W.: Rough sets. Communications of the ACM 38(11), 89–95 (1995)
62. Johnson, J., Liu, M.: Rough Sets for Informative Question Answering. In: Proceedings of the International Conference on Computing and Information (ICCI 1998), pp. 53–60 (1996)
63. Wróblewski, J.: Finding minimal reducts using genetic algorithms. In: Proceedings of Second International Joint Conference on Information Science, pp. 186–189 (1995)
64. Sulaiman, N.S.: Generation of Rough Set (RS) Significant Reducts and Rules for Cardiac Dataset Classification. Master thesis, Faculty of Computer Science and Information System, Universiti Teknologi Malaysia, Malaysia (2007)

Software Defect Classification: A Comparative Study of Rough-Neuro-fuzzy Hybrid Approaches with Linear and Non-linear SVMs

Rajen Bhatt[1], Sheela Ramanna[2], and James F. Peters[3]

[1] Samsung India Software Center,
Noida-201305, Uttar Pradesh, India
rajen.bhatt@gmail.com

[2] Department of Applied Computer Science
University of Winnipeg,
Winnipeg, Manitoba R3B 2E9 Canada
s.ramanna@uwinnipeg.ca

[3] Computational Intelligence Laboratory
Electrical and Computer Engineering Department
University of Manitoba,
Winnipeg, Manitoba R3T 5V6 Canada
jfpeters@ee.umanitoba.ca

Summary. This chapter is an extension of our earlier work in combining and comparing rough hybrid approaches with neuro-fuzzy and partial decision trees in classifying software defect data. The extension includes a comparison of our earlier results with linear and non-linear support vector machines (SVMs) in classifying defects. We compare SVM classification results with partial decision trees, neuro-fuzzy decision trees(NFDT), LEM2 algorithm based on rough sets, rough-neuro-fuzzy decision trees(R-NFDT), and fuzzy-rough classification trees(FRCT). The analyses of the results include statistical tests for classification accuracy. The experiments were aimed at not only comparing classification accuracy, but also collecting other useful software quality indicators such as number of rules, number of attributes (metrics) and the type of metrics (design vs. code level). The contribution of this chapter is a comprehensive comparative study of several computational intelligence methods in classifying software defect data. The different methods also point to the type of metrics data that ought to be collected and whether the rules generated by these methods can be easily interpreted.

Keywords: Classification, fuzzy-rough classification trees, neuro-fuzzy decision trees, rough sets, software defects, support vector machines.

1 Introduction

In the context of software defect classification, the term data mining refers to knowledge-discovery methods used to find relationships among defect data and the extraction of rules useful in making decisions about defective modules either during development or during post-deployment of a software system. A software

A. Abraham, R. Falcón, and R. Bello (Eds.): Rough Set Theory, SCI 174, pp. 213–231.
springerlink.com

defect is a product anomaly (e.g, omission of a required feature or imperfection in the software product) [19]. As a result, defects have a direct bearing on the quality of the software product and the allocation of project resources to program modules. Software metrics make it possible for software engineers to measure and predict quality of both the product and the process.

There have been several studies in applying computational intelligence techniques such as rough sets [18], fuzzy clustering [8, 29], neural networks [13] to software quality data. Statistical predictive models correlate quality metrics to number of changes to the software. The predicted value is a numeric value that gives the number of changes (or defects) to each module. However, in practice, it is more useful to have information about modules that are highly defective rather than knowing the exact number of defects for each module.

This chapter is an extension of our earlier work [20, 21] in comparing rough hybrid approaches in classifying software defect data. Significant enhancements include i) the comparison with linear and non-linear support vector machines (SVMs) ii) Using rough set based LEM2 algorithm [27] iii) preprocessing of data and experimenting with several iterations.

We compare SVM classification results with partial decision trees [25], neuro-fuzzy decision trees [1], rough-neuro-fuzzy decision trees, and fuzzy-rough classification trees [3, 4]. The analyses of the results include statistical tests for classification accuracy. In [20], the hybrid approach was limited to employing the strength of rough sets to attribute reduction as the first step in classification with neuro-fuzzy decision trees. In [21], Fuzzy-Rough Classification Trees that employ fuzzy-rough dependency degree [9, 2] for the induction of FRCT. Other data mining methods reported in this chapter are from rough set theory [17] and fuzzy decision trees [28].

In this work, the defect data consists of product metrics drawn from the PROMISE[1] Software Engineering Repository data set. The results are very promising in terms of how different methods point to the type of metrics data that ought to be collected and whether the rules generated by these methods can be easily interpreted.. In addition, we observed that the rule set with LEM2 was significantly smaller than our earlier reported result [20, 21]. The contribution of this chapter is a comprehensive comparative study of several computational intelligence methods in classifying software defect data.

This chapter is organized as follows. In Sect. 2, we give a brief overview of the various methods that were used in our experiments. The details of the defect data and classification methods are presented in Sect. 3. This is followed by an analysis of the classification results in Sect. 4.

2 Approaches

2.1 Neuro-Fuzzy Decision Trees

Fuzzy decision trees are powerful, top-down, hierarchical search methodology to extract easily interpretable classification rules [2]. However, they are often

[1] http://promise.site.uottawa.ca/SERepository

criticized for poor learning accuracy [26]. In [1] a Neuro-Fuzzy Decision Trees (NFDT) algorithm was proposed to improve the learning accuracy of fuzzy decision trees. In the forward cycle, NFDT constructs a fuzzy decision tree using the standard FDT induction algorithm fuzzy ID3 [28]. In the feedback cycle, parameters of fuzzy decision trees (FDT) have been adapted using stochastic gradient descent algorithm by traversing back from each leaf to root nodes. Forward cycle means construction of fuzzy decision tree by doing forward pass through data. Feedback cycle means tuning of FDT parameters using N-FDT algorithm. With the computation of mean-square-error, feedback regarding the classification performance of FDT is continuously available, which is being used to tune the FDT parameters. During the parameter adaptation stage, NFDT retains the hierarchical structure of fuzzy decision trees. A detailed discussion of NFDT algorithm with computational experiments using real-world datasets and analysis of results are available in [1].

We will now give a brief discussion about standard crisp decision trees and fuzzy decision trees. This will provide the useful reference for the study of neuro-fuzzy decision trees. The most important feature of decision trees is their capability to break down complex decision making method into a collection of locally optimal simple decisions through top-down greedy search technique. State-of-the-art survey of various crisp decision tree generation algorithms, including the most important and popular Quinlan's ID3 family and C4.5 [25], is given in [23, 24],and by Safavian and Landgrebe [22]. Although the decision trees generated by these methods are useful in building knowledge-based expert systems, they are often not capable of handling cognitive uncertainties consistent with human information processing, such as vagueness and ambiguity. In general, vagueness is related to the difficulty in making sharp classification boundaries. Ambiguity is associated with one-to-man mapping. To overcome these deficiencies, various researchers have developed fuzzy decision tree induction algorithms [28]. All fuzzy decision tree generation techniques evaluate classification abilities of fuzzified attributes using some suitable measure of uncertainty. Incorporating this measure in crisp decision tree generation algorithm like ID3, fuzzy decision trees can be constructed.

Figure 1 shows fuzzy decision tree using fuzzy ID3 algorithm for a toy dataset of two class classification problem. As shown in Fig. 1, fuzzy decision trees are composed of a set of internal nodes representing variables used in the solution of a classification problem, a set of branches representing fuzzy sets of corresponding node variables, and a set of leaf nodes representing the degree of certainty with which each class has been approximated. Patterns are classified by starting from the root node and then reaching to one or more leaf nodes by following the path of degree of memberships greater than zero. Each *path-m* is defined on the premise space composed of input features available in traversing from root node to m^{th}leaf node. In Fig. 1, *path-1*, *path-2*, and *path-3* are composed on the premise space $x1 = x2 = x3 = [x6, x2]$, where as *path-4* and *path-5* are composed on the premise space $x4 = x5 = [x6]$. Number of variables appearing on the path defines the length of that path. For example, in Fig. 1, length of

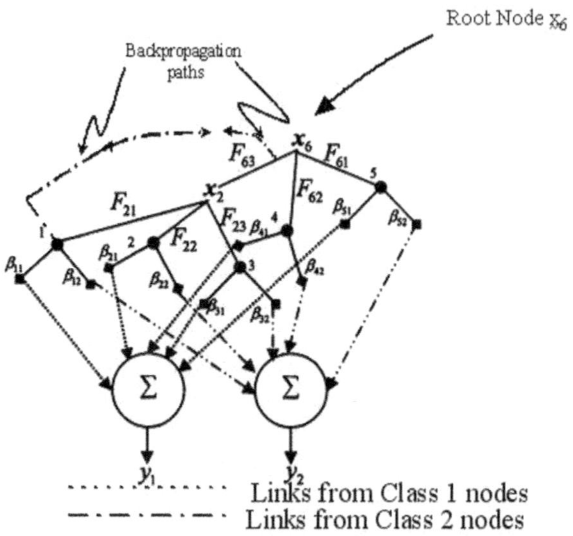

Fig. 1. Exemplary neuro-fuzzy decision tree

path-1, *path-2*, *path-3* = 2. Fuzzy decision tree being a hierarchical structure, share the membership functions along the paths leading to each leaf node from root node. In Fig. 1, membership function F_{63} has been shared by *path-1*, *path-2*, and *path-3*. With this preliminary discussion on fuzzy decision trees and its advantages over crisp decision trees, what follows is the formal notation and details of the neuro-fuzzy decision trees:

Figure 1 shows an exemplary NFDT with two summing nodes to carry out the inference process. There are five paths starting from root node to five leaf nodes. Root node is indicated by x_6. Leaf nodes are shown by dots and indexed as $m = 1, 2, , 5$. Training patterns are labeled as $\{x_i, y_i\}$ where $i = 1, ..., n$ and $y_i \in \{0, 1\}$ where where x_i represents the ith object(pattern) and y_i represents the prediction certainty of the decision class for the ith object. The input to NFDT is a decision table. Let β_{ml} be the certainty factor corresponding to m^{th} leaf node and decision *class-l*. From all the leaf nodes, certainty corresponding to decision *class-l* are summed up to calculate output y_l. Traversing path from root node x6 to second leaf node is represented by:

$$\begin{aligned} &path_2 = \mathrm{x}_6 \text{ is } \mathrm{F}_{63} \wedge \mathrm{x}_2 \text{ is } \mathrm{F}_{22} \\ &leaf_2 : y_1 = \mathrm{Class}_1(\beta_{21}), y_2 = \mathrm{Class}_2(\beta_{22}) \end{aligned} \tag{1}$$

The firing strength (FS) of *path-m* with respect to decision *class-l* for the ith object is defined by (2)

$$FS_m = \mu^i_{path_m} \times \beta_{ml}, \tag{2}$$

where $\mu^i_{path_m}$ is the membership degree of ith object for *path-m* and can be calculated as shown in (3)

$$\mu^i_{path_m} = \prod_j \mu_{F^m_j}\left(x^i_j\right). \tag{3}$$

where $\mu_{F^m_j}\left(x^i_j\right)$ is the degree of membership of the ith pattern of the jth input x_j into F^m_j. F^m_j is fuzzy membership function for j^{th} attribute on *path-m*. This way, $\mu^i_{path_m}$ is zero if for any of the input variable on m^{th} path the degree of membership of ith pattern to the fuzzy membership function F^m_j is zero.

Firing strengths of all the paths for a particular decision *class-l* are summed up to calculate the prediction certainty y^i_l of ith object(pattern) to l^{th} class through fuzzy decision tree as shown in (4)

$$y^i_l = \sum_{m=1}^{M} FS_m, \tag{4}$$

where $0 \leq y^i_l \leq 1$ and q is total number of classes. When classification to a unique class is desired, the class with the highest membership degree needs to be selected, i.e., classify given object(pattern) to class l_0, where

$$l_0 = \arg \max_{l=1,\dots,q} \left\{y^i_l\right\}. \tag{5}$$

To fuzzify input attributes, we have selected Gaussian membership functions out of many alternatives due to its differentiability property (i.e., existence of the differentiation). For i^{th} object(pattern), membership degree of *path-m* can be calculated as shown in (6)

$$\mu^i_{path_m} = \prod_j \mu_{F^m_j}\left(x^i_j\right) = \prod_j \exp\left(\frac{\left(x^i_j - c_{jm}\right)^2}{2\sigma^2_{jm}}\right), \tag{6}$$

where c_{jm} and σ_{jm} are center and standard deviation (width) of Gaussian membership function of jth attribute of input object(pattern) x_i on *path-m*, i.e., of F^m_j. We now briefly outline the strategy of NFDT by performing an adaptation of all types of parameters (centers, widths, and certainty factors) simultaneously on the structure shown in Fig. 1. We define as the error function of the fuzzy decision tree, the mean-square-error defined by (7)

$$MSE = \frac{1}{2n} \sum_{l=1}^{q} \sum_{i=1}^{n} \left(d^i_l - y^i_l\right)^2, \tag{7}$$

where n is the total number of training patterns and d^i_l and y^i_l are the desired prediction certainty and the actual prediction certainty of *class-l* for i^{th} object(pattern), respectively. At each epoch (iteration), the complete parameter $\mathrm{P} = \{c_{jm}, \sigma_{jm}, \beta_{ml} \mid m = 1, ..., M; l = 1, ..., q\}$ is moved by a small distance η in the direction in which MSE decreases most rapidly, i.e., in the direction of the negative gradient $-\frac{\partial E}{\partial \theta}$ where θ is the parameter vector constituted from the set P. This leads to the parameter update rule shown in (8)

$$\theta^{\tau+1} = \theta^{\tau} - \eta \frac{\partial E}{\partial \theta}, \tag{8}$$

where τ is the iteration index and η is the learning rate. The update equations for centers, widths, and certainty factors can be found in [1]. Parameter adaptation continues until error goes below certain small positive error goal ϵ or the specified number of training epochs has been completed. Neuro-fuzzy decision trees take as input the fuzzy decision tree structure and try to minimize the mean-square-error by tuning the centers and standard deviations of Gaussian membership functions along with certainty factors associated with each node. For example, if there is high degree of overlap (or very less overlap) between adjacent membership functions in the initial fuzzy decision tree structure if cluster centers are too close (or too far). Neuro-fuzzy decision tree algorithm tries to minimize the mean-square-error and in the process adjust the cluster centers and standard deviations.

What follows are details about the computational set-up for the experiments reported in this chapter with NFDT. All the attributes have been fuzzified using fuzzy c-means algorithm [5] into three fuzzy clusters. From the clustered row data, Gaussian membership functions have been approximated by introducing the width control parameter λ. The center of each gaussian membership function has been initialized by fuzzy cluster centers generated by the fuzzy c-means algorithm. To initialize standard deviations, we have used a value proportional to the minimum distance between centers of fuzzy clusters. For each numerical attribute x_j and for each gaussian membership function, the Euclidean distance between the center of F_{jk} and the center of any other membership function F_{jh} is given by $dc\left(c_{jk}, c_{jh}\right)$, where $h \neq k$. For each kth membership function, after calculating $dc_{\min}\left(c_{jk}, c_{jh}\right)$, the standard deviation σ_{jk} has been obtained by (9)

$$\sigma_{jk} = \lambda \times dc_{\min}\left(c_{jk}, c_{jh}\right); 0 < \lambda \leq 1, \tag{9}$$

where λ is the width control parameter. For the computational experiments reported here, we have selected various values of $\lambda \in (0, 1]$ to introduce variations in the standard deviations of initial fuzzy partitions.

2.2 Fuzzy-Rough Classification Trees

Fuzzy-Rough Classification Trees (FRCT) integrate rule generation technique of fuzzy decision trees and rough sets. The measure used for the induction of FRCT is fuzzy-rough dependency degree proposed in [3, 4]. The dependency degree measure in the context of rough set theory has been proposed by Pawlak [17]. Fuzzy-rough dependency degree measure is an extension of Pawlak's measure to accommodate fuzzy data. Pawlak's measure is only applicable to crisp partitions of feature space. In [3, 4], we have shown that our measure of fuzzy-rough dependency degree is more general one and covers Pawlak's measure as a limiting case when partitions are crisp rather than fuzzy. In this sect., we briefly outline the steps of computing the fuzzy-rough dependency degree.

Formally, a data (information) table IS is represented by a pair (X, A), where X is a non-empty, finite set of objects and A is a non-empty, finite set of attributes, where $a_j : X \rightarrow Va_j$ for every $a_j \in A$ where a_j is the jth input of attribute a and Va_j is the value set of a_j. A decision table is represented by a pair (X, C, D), where $C, D \subseteq A$. In other words, the attribute set A is partitioned into: condition attributes C and decision attribute D. Let $|C| = p$ i.e., the total number of input of variables are p. Let $|D| = q$, i.e., the total number of classes are q. In other words, each input pattern is classied into one of the q classes. Let F_{jk} be the kth fuzzy set of attribute a_j. The fuzzy set represents overlapping and non-empty partitions of real-valued attributes $a_j \in C$ where $(1 \leq j \leq p)$ on the set of training set $T \subseteq X$. For notational convenience, we will use j instead of a_j for attributes.

The membership function of the lower approximation of an arbitrary *class-l* of F_{jk} for the ith object(pattern) denoted by $x_j^i \in T$ with decision d^i is given by:

$$\mu_{\underline{l}}\left(F_{jk}\right) = \inf_{\forall i \in U} \max\left\{1 - \mu_{F_{jk}}\left(x_j^i\right), \mu_l\left(d^i\right)\right\}$$

The dependency degree $\gamma_{x_j}(d)$ for the ith object (pattern) and jth attribute can be calculated as follows:

- Calculate the lower approximation member function $\mu_{\underline{l}}\left(F_{jk}\right)$ using the above definition
- Calculate fuzzy positive region $\mu_{POS}\left(F_{jk}\right) = \sup\limits_{l=1,..,q} \left\{\mu_{\underline{l}}\left(F_{jk}\right)\right\}$
- Calculate the degree of membership of the ith pattern to the fuzzy positive region
$\mu_{POS}\left(x_j^i\right) = \sup\limits_{l=1,..,q} \min\left\{\mu_{F_{jk}}\left(x_j^i\right), \mu_{POS}\left(F_{jk}\right)\right\}$
- Calculate the dependency degree $\gamma_{x_j}(d) = \frac{\sum_{i=1}^{n} \mu_{POS}\left(x_j^i\right)}{n}$

Fuzzy-rough dependency degree of attribute x_j, denoted here as γ_{x_j} lies between 0 and 1, i.e., $0 \leq \gamma_{x_j}(d) \leq 1$. $d = 1$ indicates that decision attribute d completely depends on input attribute x_j, in other words, x_j alone is sufficient to approximate all the decisions given in decision attribute d. $d = 0$ indicates that decision attribute d is not completely dependent on on input attribute x_j. Any value of $\gamma_{x_j}(d)$ that is in (0,1) indicates partial dependency. Partial dependency means *addition* of other input attributes is required to completely approximate all the decisions given in decision attribute d. This property of fuzzy-rough dependency degree makes it a good choice as an attribute selection criterion for the induction of fuzzy decision trees. We call fuzzy decision trees wherein fuzzy-rough dependency degree is used as an attribute selection criterion, a fuzzy-rough classification trees.

Given fuzzy partitions of feature space, leaf selection threshold β_{th}, and fuzzy-rough dependency degree γ as expanded attribute (attribute to represent each node in fuzzy decision tree) selection criterion, the general procedure for generating fuzzy decision trees using FRCT algorithm is outlined in Alg. 1.

Algorithm 1. Algorithm for generating fuzzy decision trees using FRCT

Require: fuzzy partitions of feature space, β_{th}, γ
Ensure: fuzzy decision trees

```
while ∃ candidate nodes do
    Select node with highest γ;                ▷ dependency degree
    Generate its child nodes;    ▷ root node will contain attribute with highest γ
    if β_child-node >= β_th then
        child-node = leaf-node
    else
        Search continues with child-node as new root node
    end if
end while
```

Before training the initial data, the α cut is usually used for the initial data [4]. Usually, α is in the interval $(0, 0.5]$. A detailed description of fuzzy-rough dependency degree is available in [3]. The cut of a fuzzy set F is defined as:

$$\mu_{F_\alpha}(a) = \begin{cases} \mu_F(a)\,; \mu_F(a) \geq \alpha \\ 0; \mu_F(a) < \alpha \end{cases} .$$

In the case of FRCT experiments, fuzzy partitioning of the feature space has been generated by the following method. Fuzzy c-means [5] algorithm has been utilized to fuzzify continuous attributes into three fuzzy clusters. The triangular approximation of the clustered raw data is done in two steps. First, the convex hull of the original clustered data is determined through MATLAB® function "convhull", and then the convex hull is approximated by a triangular membership function. We mention here that three fuzzy clusters have been chosen only to report experimental results. Choosing different number of clusters may affect the result. In general, one should iterate from a few minimum to maximum number of clusters, construct fuzzy-rough classification trees, and choose one which gives best classification accuracy with acceptable number of rules.

2.3 Support Vector Machines

In this sect., we give a brief discussion of linear and nonlinear Support Vector Machines (SVMs) used in our computational experiments [6]. Linear and nonlinear SVMs trained on non separable (and separable) data results in a quadratic programming problem.

Linear SVM

Let training patterns are labeled as $\{x_i, y_i\}$, where $i = 1, ..., n$, $y_i \in \{-1, +1\}, x_i \in \Re$. Let there exist some separating hyper plane which separates the positive from the negative patterns. The points x, which lie on the hyper plane satisfy

$$x_i \cdot \mathbf{w} + b = 0 \tag{10}$$

where $\mathbf{w}$ is normal to the hyper plane $\frac{|b|}{|\mathbf{w}|}$ is the perpendicular distance from the hyper plane to the origin, and $\|\mathbf{w}\|$ is the Euclidean norm of $\mathbf{w}$. For the linearly separable case, the SVM algorithm simply looks for the separating hyper plane with largest margin i.e., $\mathbf{w}$ and b which can maximize the margin. Once optimal b and $\mathbf{w}$ are obtained, we simply have to determine on which side of the decision boundary a given test pattern x lies and assign the corresponding class label, i.e., we take the class of x to be $sgn(\mathbf{w} \cdot x + b)$ given as

$$x_i \cdot \mathbf{w} + b \geq 1, \text{ for } y_i = 1, \; x_i \cdot \mathbf{w} + b \leq -1, \text{ for } y_i = -1. \tag{11}$$

Equation 11 can be combined into one set of inequalities:

$$y\left(x_i \cdot \mathbf{w} + b\right) - 1 \geq 1, \text{for } i \tag{12}$$

However, in practice, it is difficult to find problems with perfectly linearly separable case. The actual SVM formulation described above for the linearly separable case is modified by introducing positive slack variables ζ_i where i=1,...,n in the constraints. Equation 11 can be be rewritten as:

$$x_i \cdot \mathbf{w} + b \geq 1 - \zeta_i, \text{ for } y_i = 1, \tag{13}$$

$$x_i \cdot \mathbf{w} + b \leq -1 - \zeta_i, \text{ for } y_i = -1 \; where \; \zeta_i \geq 0 \text{ for } \; i \tag{14}$$

NonLinear SVM

To handle cases where the decision function is not a linear function of the data, a nonlinear version of the SVM is normally used. In this case, we first map the data to some other (possibly infinite dimensional) Euclidean space H using a mapping ϕ where $\phi : \Re \rightarrow H$. SVM training algorithm would only depend on the data through dot products in H i.e., on functions of the form $\phi\left(x_i\right) \cdot \phi\left(x_j\right)$.

Now if there is a 'Kernel function' K such that, $K\left(x_i\right) \cdot K\left(x_j\right) = \phi\left(x_i\right) \cdot \phi\left(x_j\right)$. we would only need to use K in the training algorithm, and would never need to explicitly know the value(s) for ϕ. One such example is Gaussian kernel, used in the computational experiments reported here. A detailed discussion of linear and nonlinear SVMs for separable and non-separable cases, with interesting mathematical results can be found in [6].

3 Software Defect Data

The PROMISE data set includes a set of static software metrics about the *product* as a predictor of defects in the software. The data includes measurements for 145 modules (objects). There are a total of 94 attributes and one decision attribute (indicator of defect level). The defect level attribute value is TRUE if the class contains one or more defects and FALSE otherwise. The metrics at the *method level* are primarily drawn from Halstead's Software Science metrics [10] and McCabe's Complexity metrics [15]. The metrics at the *class level*, include such standard measurements as Weighted Methods per Class (WMC),

a1	a2	a3	a4	a5	a6	a7	a8	a9	a10	a11	a12	a13	a14	a15	a16	a17	a18	a19	a20	a21	a22
0	0	24	4	100	0	0	2	110	73	0	1	0	0	1	1	1	0	5.8	1.5	17	0
0	0	19	4	100	0	0	3	78	30	0	1	0	0	1	1	1	0	0	0	0	0
100	0	13	1	88	0	0	0	99	99	0	1	0	0	1	1	1	1	0	0	0	0
0	0	21	4	100	0	0	2	68	30	0	1	0	0	1	1	1	0	7.7	1.5	17	0
5	0	17	2	90	0	0	1	69	36	0	1	0	0	1	1	1	0	0	0	0	0
0	0	10	1	88	0	0	0	22	22	0	1	0	0	1	1	1	1	0	0	0	0
0	0	19	3	94	0	0	3	53	17	0	1	0	0	1	1	1	0	4.4	1.5	12	0
0	0	14	2	90	0	0	1	37	25	0	1	0	0	1	1	1	0	0	0	0	0
0	0	18	2	100	1	1	1	38	13	0	1	0	0	1	1	1	0	0	0	0	0
0	0	18	2	100	1	1	1	38	13	0	1	0	0	1	1	1	0	0	0	0	0
0	0	9	1	0	4	0	0	7	7	0	1	0	0	1	1	1	1	0	0	0	0
0	0	6	2	59	0	0	1	33	25	0	1	0	0	1	1	1	0	0	0	0	0
0	0	5	1	0	0	0	0	7	7	0	1	0	0	1	1	1	0	0	0	0	0
100	0	4	1	75	2	0	0	8	8	0	1	0	0	1	1	1	0	4.8	1	4.8	0
0	0	4	1	0	0	0	0	6	6	0	1	0	0	1	1	1	1	0	0	0	0
100	0	24	4	96	0	0	3	140	35	0	1	0	0	1	1	1	1	0	0	0	0
0	0	23	2	92	0	0	1	53	25	0	1	0	0	1	1	1	0	5.3	1	8	0
0	0	9	2	28	0	0	1	31	23	0	1	0	0	1	1	1	1	0	0	0	0
0	0	2	2	94	0	0	1	52	17	0	1	0	0	1	1	1	0	0	0	0	0
0	0	6	1	80	0	0	0	15	15	0	1	0	0	1	1	1	1	0	0	0	0
0	0	5	1	25	0	0	0	9	9	0	1	0	0	1	1	1	1	0	0	0	0
0	0	0	1	100	0	0	0	0	0	0	1	0	0	1	1	1	0	5.3	1.5	12	0
0	0	0	1	100	0	0	0	0	0	0	1	0	0	1	1	1	0	5.3	1.5	12	0
0	0	8	1	58	0	0	0	37	34	0	1	0	0	1	1	1	0	5.3	1.5	12	0
0	0	15	1	77	0	0	0	12	12	0	1	0	0	1	1	1	1	0	0	0	0
0	0	11	1	76	0	0	0	28	28	0	1	0	0	1	1	1	0	5.3	1.5	12	0
0	0	12	1	55	0	0	0	24	20	0	1	0	0	1	1	1	2	6.2	2	39	0
100	0	4	1	100	0	0	0	11	11	0	1	0	0	1	1	1	0	5.3	1.5	12	0
17	0	14	4	94	0	0	1	198	24	0	1	0	0	1	1	1	0	5.3	1.5	12	0
100	0	2	2	85	0	0	1	11	7	0	1	0	0	1	1	1	0	0	0	0	0

Fig. 2. Exemplary data set

Depth of Inheritance Tree (DIT), Number of Children (NOC), Response For a Class (RFC), Coupling Between Object Classes (CBO), and Lack of Cohesion of Methods (LCOM) [7]. A sample data set of 30 modules with 22 attributes are shown in Fig. 2.

In this chapter, for the purposes of illustration we have given a brief description of the first 22 attributes. Since the defect prediction is done at a class-level, all method level features were transformed to the class level. Transformation was achieved by obtaining min, max, sum, and avg values over all the methods in a class. Thus this data set includes four features for each method-level features.

- a1: PERCENT-PUB-DATA . The percentage of data that is public and protected data in a class.
- a2: ACCESS-TO-PUB-DATA. The amount of times that a class's public and protected data is accessed.
- a3: COUPLING-BETWEEN-OBJECTS. The number of distinct non-inheritance-related classes on which a class depends.

- a4: DEPTH. The level for a class. For instance, if a parent has one child the depth for the child is two.
- a5: LACK-OF-COHESION-OF-METHODS. For each data field in a class, the percentage of the methods in the class using that data field. This metric indicates low or high percentage of cohesion.
- a6: NUM-OF-CHILDREN. The number of classes derived from a specified class.
- a7: DEP-ON-CHILD. Whether a class is dependent on a descendant.
- a8: FAN-IN. This is a count of calls by higher modules.
- a9: RESPONSE-FOR-CLASS. A count of methods implemented within a class plus the number of methods accessible to an object class due to inheritance.
- a10: WEIGHTED-METHODS-PER-CLASS. A count of methods implemented within a class rather than all methods accessible within the class hierarchy.
- a11: minLOC-BLANK. Lines with only white space or no text content.
- a12: minBRANCH-COUNT. This metric is the number of branches for each module.
- a13: minLOC-CODE-AND-COMMENT. Lines that contain both code and comment.
- a14: minLOC-COMMENTS. Minimum lines with comments.
- a15: minDESIGN-COMPLEXITY. Design complexity is a measure of a module's decision structure as it relates to calls to other modules.
- a17: minESSENTIAL-COMPLEXITY. Essential complexity is a measure of the degree to which a module contains unstructured constructs.
- a18: minLOC-EXECUTABLE. Minimum Source lines of code that contain only code and white space.
- a19: minHALSTEAD-CONTENT. Complexity of a given algorithm independent of the language used to express the algorithm.
- a20: minHALSTEAD-DIFFICULTY. Minimum Level of difficulty in the program.
- a21: minHALSTEAD-EFFORT. Minimum estimated mental effort required to develop the program.
- a22: minHALSTEAD-ERROR-EST. Estimated number of errors in the program.

Data Preprocessing

Since all the data is real-valued with a wide variation in the values for attributes from 0.1 to 10^5, it was necessary to normalize the attribute values for experiments that were not based on fuzzy sets. What follows are some data preprocessing tasks that were performed on the defect data:

- All attributes were normalized using the WEKA[2] unsupervised instance based filter method.

[2] `http://www.cs.waikato.ac.nz/ml/weka`

- For 10-Fold CV, pairs of testing-training data sets were generated independently and used in all our experiments with different methodologies.
- All attribute values were discretized in experiments with LEM2 and J48. The version of LEM2 included in RSES[3] works only with discretized data. Discretization alogrithm implemented in WEKA was used with J48. Discretization is by simple binning in the unsupervised mode.

Experimental Setup - Discretization in RSES

Whenever the domain of a real-valued condition attribute is exceptionally large, then the number of decision rules that are generated can be unmanageably large. In such cases, discretization of attribute value sets provides a mechanism for reducing the domain space without significantly altering the quality of the rules that are derived. That is, we need to obtain approximate knowledge of a continuum by considering parts of the continum for an attribute. Discretization of a continuum entails the partition of the interval of a real-valued attribute into subintervals of reals. Let $DT = (U, A \cup d)$ be a decision table where $U = \{x_1, \ldots, x_n\}$. In addition, assume that the value set V_a for each attribute of DT is a subset of the reals. For example, consider the interval of reals $V_a = [la, ra]$ for values of an attribute $a \in A$ in a decision system $DT = (U, A \cup d)$ where $la \leq a(x) \leq ra$. Discretization of V_a entails searching for a partition P_a of V_a for any a $a \in A$ (i.e., discovering a partition of the value sets of conditional attributes into intervals). A partition of V_a is defined by a sequence of what are known as cuts $l_a = v_1 < v_2 < v_3 < \ldots << v_{n-1} < v_n = r_a$ so that $V_a = [l_a, v_2)[v_2, v_3) \ldots [v_{n-1}, r_a)$. The search for partitions of attributes into subintervals of the reals is carried out on consistent decision tables. In rough set theory, discretization leads to partitions of value sets so that if the name of the interval containing an arbitrary object is substituted for any object instead of its original value in DT, a consistent decision system is also obtained. The discretization concept defined by cuts has been generalized by using oblique hyperplanes. The boundary between each pair of decision classes is a linear plane called a hyperplane. The quality of a hyperplane has been treated by a number of measures. Measure values are viewed as energy levels of a hyperplane. During discretization of a set of numeric attributes, the search of hyperplanes is carried out using simulated annealing. A detailed description of this approach to discretization is outside the scope of this article. A complete presentation containing the details about discretization in the context of rough sets is given in [16].

Experimental Setup - Rough Set-based LEM2 and J48

LEM2 based on rough set theory learns the concept with the smallest set of rules [27]. In the results reported in [20, 21], rule-set used by the rough set classifier in RSES was quite large (average number of rules was 280). We also compared our results with a classical partial decision tree-based method (J48) method in WEKA using a variant of the well-known C4.5 revision 8 algorithm [25].

[3] `http://logic.mimuw.edu.pl/~rses`

Experimental Setup - Hybrid Methods

In all our hybrid methods involving fuzzy sets, the attribute values were not normalized. The Rough-NFDT method included i) generating reducts from rough set methods ii) using the data from the reduced set of attributes to run the NFDT algorithm. For the NFDT algorithm, after attribute fuzzification, the fuzzy ID3 algorithm with cut $\alpha = 0$ and leaf selection threshold $\beta_{th} = 0.75$ was used. The fuzzy decision trees have been tuned using the NFDT algorithm for 500 epochs with the target MSE value 0.001.

For the computational experiments reported with the FRCT, all parameters were set as described in Sect. 2.2.

Experimental Setup - SVM

The data set was classified using nonlinear SVM(SVM-NL) as well as linear SVM (SVM-L). The tuning parameters involved with nonlinear SVM are penalty parameter C and Gaussian Kernel parameter μ. These two parameters were tuned based on grid search [12]. The range for C used is 2^{-12} to 2^{12}. The range for μ used is 2^{-35} to 2^{4}. The best values of C and μ are 2^{-1} and 2^{-29}. All algorithms were implemented in MATLAB 7.3.0 (R2006b) [14] environment on a PC with Intel Core2Duo processor (2.13GHz), 1GB RAM running Ms-Windows XP operating system. The dual quadratic programming problems arising in SVM were solved using Mosek optimization toolbox [4] for MATLAB which implements fast interior point based algorithms.

4 Analysis of Classification Results

A comparison of pairs of differences in classification accuracy during one-fold of a 10Fold CV and a paired t-test is also discussed in this sect. Table 1 gives a summary of computational experiments using seven methods and Table 2 gives the average size of the rule set (and support vectors). Percentage classification accuracy has been calculated by $\frac{n_c}{n} \times 100\%$, where n is the total number of test patterns, and n_c is the number of test patterns classified correctly.

Figure 3 gives a sample LEM2 rule set for a single run of the 10fold CV. It can be observed that the most frequestly used attributes (metrics) are:

a4(DEPTH), a5(LACK-OF-COHESION-OF-METHODS), a23(minHALSTEAD-LENGTH), a35(maxLOC-COMMENTS), a53(avgLOC-BLANK), a56(avgLOC-COMMENTS) and a63(avgHALSTEAD-LENGTH).

4.1 T-Test

In this sect., we discuss whether there is any difference between the various methods in terms of classification accuracy (and the number of rules) statistically

[4] `http://www.mosek.com`

Table 1. Defect data classification I

	%Accuracy						
Run	NFDT	R-NFDT	LEM2	FRCT	J48	SVM-NL	SVM-L
1	93	71	85	86	86	76	71
2	83	93	82	86	79	76	86
3	64	71	55	71	57	57	57
4	71	71	80	93	86	86	71
5	64	57	44	71	28	64	64
6	79	79	57	79	57	93	71
7	86	71	67	86	50	64	50
8	71	79	67	71	79	64	79
9	93	100	86	93	93	93	100
10	89	89	85	89	84	89	89
Avg.Acc	80	78	71	83	73	77	74

Table 2. Defect data classification II

Average Number of Rules and Support Vectors						
NFDT	R-NFDT	LEM2	FRCT	J48	SVM-NL	SVM-L
7.2	5.6	26	17.3	12	123	49

```
Decision rules
(a35="(-Inf,1.29E-4)")&(a56="(-Inf,9.5E-6)")&(a4="(3.575E-4,Inf)")=>(decision={F[37]})
(a4="(-Inf,6.8E-5)")&(a35="(-Inf,1.29E-4)")&(a56="(-Inf,9.5E-6)")&(a53="(-Inf,1.65E-5)")=>(decision={T[23]})
(a4="(-Inf,6.8E-5)")&(a35="(-Inf,1.29E-4)")&(a23="(-Inf,4.2E-5)")&(a63="(-Inf,0.0758755)")&(a56="(-Inf,9.5E-6)")&(a53="(-Inf,1.65E-5)")=>(decision={T[14]})
(a53="(1.65E-5,Inf)")&(a35="(-Inf,1.29E-4)")&(a63="(0.0758755,Inf)")&(a4="(6.8E-5,3.575E-4)")=>(decision={F[10]})
(a4="(-Inf,6.8E-5)")&(a56="(9.5E-6,Inf)")&(a84="(0.913131,Inf)")&(a53="(1.65E-5,Inf)")&(a23="(-Inf,4.2E-5)")&(a35="(-Inf,1.29E-4)")=>(decision={T[7]})
(a4="(-Inf,6.8E-5)")&(a35="(-Inf,1.29E-4)")&(a23="(-Inf,4.2E-5)")&(a53="(1.65E-5,Inf)")&(a63="(-Inf,0.0758755)")&(a56="(9.5E-6,Inf)")=>(decision={T[6]})
(a53="(1.65E-5,Inf)")&(a56="(9.5E-6,Inf)")&(a84="(-Inf,0.913131)")&(a23="(4.2E-5,Inf)")&(a4="(6.8E-5,3.575E-4)")&(a35="(-Inf,1.29E-4)")=>(decision={F[5]})
(a56="(9.5E-6,Inf)")&(a35="(1.29E-4,Inf)")&(a53="(1.65E-5,Inf)")&(a23="(-Inf,4.2E-5)")&(a63="(-Inf,0.0758755)")=>(decision={F[5]})
(a53="(1.65E-5,Inf)")&(a4="(-Inf,6.8E-5)")&(a56="(9.5E-6,Inf)")&(a63="(0.0758755,Inf)")&(a84="(-Inf,0.913131)")&(a23="(4.2E-5,Inf)")=>(decision={T[5]})
(a53="(1.65E-5,Inf)")&(a56="(9.5E-6,Inf)")&(a23="(-Inf,4.2E-5)")&(a84="(-Inf,0.913131)")&(a4="(-Inf,6.8E-5)")&(a35="(1.29E-4,Inf)")&(a63="(-Inf,0.0758755)")=>(decision={F[3]})
(a53="(1.65E-5,Inf)")&(a63="(-Inf,0.0758755)")&(a84="(0.913131,Inf)")&(a4="(6.8E-5,3.575E-4)")&(a23="(4.2E-5,Inf)")&(a35="(-Inf,1.29E-4)")=>(decision={F[3]})
(a53="(1.65E-5,Inf)")&(a4="(-Inf,6.8E-5)")&(a23="(-Inf,4.2E-5)")&(a35="(-Inf,1.29E-4)")&(a56="(-Inf,9.5E-6)")&(a63="(0.0758755,Inf)")=>(decision={F[3]})
(a53="(1.65E-5,Inf)")&(a56="(9.5E-6,Inf)")&(a23="(4.2E-5,Inf)")&(a35="(1.29E-4,Inf)")&(a63="(0.0758755,Inf)")&(a84="(0.913131,Inf)")=>(decision={T[3]})
(a53="(1.65E-5,Inf)")&(a63="(-Inf,0.0758755)")&(a4="(-Inf,6.8E-5)")&(a56="(-Inf,9.5E-6)")&(a23="(-Inf,4.2E-5)")&(a35="(-Inf,1.29E-4)")&(a84="(-Inf,0.913131)")=>(decision={T[3]})
(a63="(-Inf,0.0758755)")&(a4="(-Inf,6.8E-5)")&(a35="(1.29E-4,Inf)")&(a84="(0.913131,Inf)")&(a23="(4.2E-5,Inf)")=>(decision={T[3]})
(a56="(9.5E-6,Inf)")&(a35="(1.29E-4,Inf)")&(a23="(4.2E-5,Inf)")&(a53="(1.65E-5,Inf)")&(a63="(-Inf,0.0758755)")&(a4="(-Inf,6.8E-5)")&(a84="(-Inf,0.913131)")=>(decision={F[2]})
(a56="(9.5E-6,Inf)")&(a35="(1.29E-4,Inf)")&(a63="(0.0758755,Inf)")&(a4="(3.575E-4,Inf)")&(a23="(4.2E-5,Inf)")&(a53="(-Inf,1.65E-5)")=>(decision={F[2]})
(a56="(9.5E-6,Inf)")&(a35="(1.29E-4,Inf)")&(a63="(0.0758755,Inf)")&(a84="(-Inf,0.913131)")&(a4="(3.575E-4,Inf)")=>(decision={F[2]})
(a53="(1.65E-5,Inf)")&(a35="(1.29E-4,Inf)")&(a56="(9.5E-6,Inf)")&(a23="(4.2E-5,Inf)")&(a63="(-Inf,0.0758755)")&(a4="(6.8E-5,3.575E-4)")&(a84="(-Inf,0.913131)")=>(decision={T[2]})
(a23="(-Inf,4.2E-5)")&(a4="(-Inf,6.8E-5)")&(a35="(1.29E-4,Inf)")&(a63="(-Inf,0.0758755)")&(a53="(-Inf,1.65E-5)")=>(decision={T[2]})
(a53="(1.65E-5,Inf)")&(a23="(-Inf,4.2E-5)")&(a35="(1.29E-4,Inf)")&(a4="(-Inf,6.8E-5)")&(a56="(-Inf,9.5E-6)")=>(decision={T[1]})
(a53="(1.65E-5,Inf)")&(a4="(6.8E-5,3.575E-4)")&(a23="(-Inf,4.2E-5)")&(a35="(-Inf,1.29E-4)")&(a56="(-Inf,9.5E-6)")=>(decision={T[1]})
```

Fig. 3. Exemplary rule set

Table 3. T-test results

	Accuracy		
Pairs	Avg. Diff.	Std. Deviation	t-stat
R-NFDT/NFDT	−1.43	9.99	−0.45
R-NFDT/LEM2	7.43	11.07	2.12
R-NFDT/J48	8.33	14.85	1.77
NFDT/LEM2	8.86	9.21	3.04
NFDT/J48	9.76	16.83	1.83
LEM2/J48	0.90	9.31	0.31
FRCT/R-NFDT	4.29	10.76	1.26
FRCT/NFDT	2.86	7.68	1.18
FRCT/LEM2	11.71	9.01	4.11
FRCT/J48	12.61	16.41	2.43
FRCT/SVM-NL	5.71	9.48	1.91
SVM-NL/NFDT	−2.86	11.33	−0.80
SVM-NL/R-NFDT	−1.43	11.65	−0.39
SVM-NL/LEM2	6.00	12.93	1.47
SVM-NL/J48	6.90	17.03	1.28
SVM-L/NFDT	−5.86	13.45	−1.38
SVM-L/R-NFDT	−4.43	8.32	−1.68
SVM-L/LEM2	3.00	12.68	0.75
SVM-L/J48	3.90	14.58	0.85

using the well-known t-test. This is done by formulating the hypothesis that the mean difference in accuracy between any two classification learning algorithms is zero. Table 3 gives the t-test results.

Let μ_d denote the mean difference in accuracy during a 10-fold classification of software defect data. Let H0 denote the hypothesis to be tested (i.e., $H0$: $\mu_d = 0$). This is our null hypothesis. The paired difference t-test is used to test this hypothesis and its alternative hypothesis ($HA : \mu_d \neq 0$). Let $\overline{d}$, S_d^2 denote the mean difference and variance in the error rates of a random sample of size n from a normal distribution N(μ_d, σ^2), where μ_d and σ^2 are both unknown. The t statistic used to test the null hypothesis is as follows:

$$t = \frac{\bar{d} - \mu_d}{S_d/\sqrt{n}} = \frac{\bar{d} - 0}{S_d/\sqrt{n}} = \frac{\bar{d}\sqrt{n}}{S_d}$$

Table 4. Null-hypothesis results for accuracy

Accept H0 ($u_d = 0$) if $\|t\ value\| < 2.262$		
Pairs	t-stat(Acc.)	Acc/Rej H0
R-NFDT/NFDT	−0.45	*Accept*
R-NFDT/LEM2	2.12	*Accept*
R-NFDT/J48	1.77	*Accept*
NFDT/LEM2	3.04	**Reject**
NFDT/J48	1.83	*Accept*
LEM2/J48	0.31	*Accept*
FRCT/R-NFDT	1.26	*Accept*
FRCT/NFDT	1.18	*Accept*
FRCT/LEM2	4.11	**Reject**
FRCT/J48	2.43	**Reject**
FRCT/SVM-NL	1.91	*Accept*
SVM-NL/NFDT	−0.80	*Accept*
SVM-NL/R-NFDT	−0.39	*Accept*
SVM-NL/LEM2	1.47	*Accept*
SVM-NL/J48	1.28	*Accept*
SVM-L/NFDT	−1.38	*Accept*
SVM-L/R-NFDT	−1.68	*Accept*
SVM-L/LEM2	0.75	*Accept*
SVM-L/J48	0.85	*Accept*

where t has a student's t-distribution with n-1 degrees of freedom [11]. In our case, $n - 1 = 9$ relative to 10 sample error rates. The significance level α of the test of the null hypothesis H0 is the probability of rejecting H0 when H0 is true (called a Type I error). Let t_{n-1}, $\alpha/2$ denote a t-value to right of which lies $\alpha/2$ of the area under the curve of the t-distribution that has n-1 degrees of freedom. Next, formulate the following decision rule with $\alpha/2 = 0.025$:

Decision Rule: Reject $H0 : \mu_d = 0$ at significance level α if, and only if $|t - value| > 2.262$
Pr-values for t_{n-1}, $\alpha/2$ can be obtained from a standard t-distribution table. It should be noted that we repeated the experiments 30 times and averages have

remained consistent. However, for the purposes of analysis, we have restricted the reporting to 10 experiments.

4.2 Analysis

In terms of the t-test for accuracy, in general the three hybrid methods (FRCT, R-NFDT and NFDT) and SVM methods are comparable in that there is no significant difference in any of the methods based on the null hypothesis. In contrast, there is a *difference* in accuracy between three pairs of methods outlined above(FRCT and LEM2, FRCT and J48 and NFDT and LEM2). This result corroborates our earlier result reported in [21]. Also of note is that SVM based methods have not had any effect in terms of classification accuracy. In terms of rules, in this chapter, we have only reported average number of rules over 10 runs (see Table 2). It is clear that the hybrid R-NFDT classifier has the smallest rule set.

The other important observation is the role that *reducts* play in defect data classification. On an average, only 6 attributes out of 95 were used by LEM2 in its rules with no significant reduction in classification accuracy. The R-NFDT and NFDT method uses an average of 4 out of 95 attributes resulting in a minimal number of rules with comparable accuracy.

The metrics at the method level that are most significant for R-NFDT and NFDT classifiers include: i) Halstead's metric of *essential complexity* which is a measure of the degree to which a module contains unstructured constructs, ii) Halstead's metric of *level* which is the level at which the program can be understood, iii) Halstead's metric of *number of unique operands* which includes variables and identifiers, constants (numeric literal or string) function names when used during calls iv) total lines of code v) *number of unique operators* is the number of branches for each module. These metrics were the most frequently occurring attributes in the rule set that contribute to the highest classification accuracy.

The metrics at the class-level that are most significant for R-NFDT, NFDT and LEM2 classifiers include: Depth of Inheritance Tree (DIT), Coupling Between Object Classes (CBO) and Lack of Cohesion of Methods (LCOM).

5 Conclusion

This chapter has presented a combination of hybrid and native methods based on rough sets, fuzzy sets, neural networks and support vector machines to classification of software defect data. The t-test shows that there is no significant difference between any of the hybrid methods in terms of accuracy at the 95% confidence level. However, in terms of rules, there is a difference between these methods. The experiments were aimed at not only comparing classification accuracy, but also collecting other useful software quality indicators such as number of rules, number of attributes (metrics) and the type of metrics (design vs. code level). In conclusion, the R-NFDT classifier seems to be the most desired classifier in terms of comparable accuracy, average number of attributes used and smallest rule set. The Rough-NFDT method consists of generating reducts (reduced set of attributes) from rough set theory and then using the data from the reduced set of attributes to run the NFDT algorithm. The desired metrics (attributes) are: COUPLING-BETWEEN-OBJECTS, DEPTH, LACK-OF-COHESION-OF-METHODS max NUM-OPERATORS, max. NUM-UNIQUE-OPERANDS, max. HALSTEAD-LEVEL and min. LOC-TOTAL.

Acknowledgments

The research of Sheela Ramanna and James F. Peters is supported by NSERC Canada grant 194376 and 185986.

References

1. Bhatt, R.B., Gopal, M.: Neuro-fuzzy decision trees. International Journal of Neural Systems 16(1), 63–78 (2006)
2. Bhatt, R.B.: Fuzzy-Rough Approach to Pattern Classification- Hybrid Algorithms and Optimization, Ph.D. Dissertation, Electrical Engineering Department, Indian Institute of Technology Delhi, India (2006)
3. Bhatt, R.B., Gopal, M.: On the Extension of Functional Dependency Degree from Crisp to Fuzzy Partitions. Pattern Recognition Letters 27(5), 487–491 (2006)
4. Bhatt, R.B., Gopal, M.: Induction of Weighted and Interpretable Fuzzy Classification Rules for Medical Informatics. International Journal of Systemics, Cybernetics, and Informatics 3(1), 20–26 (2006)
5. Bezdek, J.C.: Pattern Recognition with Fuzzy Objective Function Algorithms. Plenum, New York (1981)
6. Burges, C.J.: A Tutorial on Support Vector Machines for Pattern Recognition. Data Mining and Knowledge Discovery 2(2), 955–974 (1998)
7. Chidamber, S.R., Kemerer, F.C.: A metrics suite for object-oriented design. IEEE Trans. Soft. Eng. 20(6), 476–493 (1994)
8. Dick, S., Meeks, A., Last, M., Bunke, H., Kandel, A.: Data mining in software metrics databases. Fuzzy Sets and Systems 145, 81–110 (2004)
9. Dubois, D., Prade, H.: Rough Fuzzy Sets and Fuzzy Rough Sets. Internation Journal of General Systems 17(2-3), 191–209 (1990)
10. Halstead, M.H.: Elements of Software Science. Elsevier, New York (1977)
11. Hogg, R.V., Tanis, E.A.: Probability and Statistical Inference. Macmillan Publishing Co., Inc., New York (1977)
12. Hsu, C.W., Lin, C.J.: A Comparison on methods for multi-class support vector methods, Technical report, Department of Computer Science and Information Engineering, National Taiwan University, Taipei, Taiwan (2001)
13. Khoshgoftaar, T.M., Allen, E.B.: Neural networks for software quality prediction. In: Pedrycz, W., Peters, J.F. (eds.) Computational Intelligence in Software Engineering, pp. 33–63. World Scientific, River Edge (1998)
14. MATLAB User's Guide. The Mathworks, Inc., Natick, MA 01760 (1994-2007)
15. McCabe, T.: A complexity measure. IEEE Trans. on Software Engineering SE 2(4), 308–320 (1976)

16. Nguyen, H.S., Nguyen, S.H.: Discretization methods in data mining. In: Polkowski, L., Skowron, A. (eds.) Rough Sets in Knowledge Discovery, vol. 1, pp. 451–482. Physica-Verlag, Berlin (1998a)
17. Pawlak, Z.: Rough sets. International J. Comp. Information Science 11(3), 341–356 (1982)
18. Peters, J.F., Ramanna, S.: Towards a software change classification system: A rough set approach. Software Quality Journal 11, 121–147 (2003)
19. Peters, J.F., Pedrycz, W.: Software Engineering: An Engineering Approach. John Wiley and Sons, New York (2000)
20. Ramanna, S., Bhatt, R., Biernot, P.: A rough-hybrid approach to software defect classification. In: An, A., Stefanowski, J., Ramanna, S., Butz, C.J., Pedrycz, W., Wang, G. (eds.) RSFDGrC 2007. LNCS, vol. 4482, pp. 79–86. Springer, Heidelberg (2007)
21. Ramanna, S., Bhatt, R.: Software defect classification: A comparative study with rough hybrid approaches. In: Kryszkiewicz, M., Peters, J.F., Rybinski, H., Skowron, A. (eds.) RSEISP 2007. LNCS, vol. 4585, pp. 630–638. Springer, Heidelberg (2007)
22. Safavian, S.R., Landgrebe, D.: A survey of decision tree classifier methodology. IEEE Transactions on Systems, Man and Cybernetics 21(3), 660–674 (1991)
23. Quinlan, J.R.: Induction of decision trees. Machine Learning 1(1), 81–106 (1986)
24. Quinlan, J.R.: Decision trees and decision making. IEEE Transactions on Systems, Man and Cybernetics 20(2), 339–346 (1990)
25. Quinlan, J.R.: C.4.5 Programs for machine learning. Morgan Kauffmann, San Mateo (1993)
26. Tsang, E.C.C., Yeung, D.S., Lee, J.W.T., Huang, D.M., Wang, X.Z.: Refinement of generated fuzzy production rules by using a fuzzy neural network. IEEE Trans. on Transactions on Systems, Man and Cybernetics 34(1), 409–418 (2004)
27. Gryzmala-Busse, J.W.: A New Version of the Rule Induction System LERS. Fundamenta Informaticae 31(1), 27–39 (1997)
28. Wang, X.Z., Yeung, D.S., Tsang, E.C.C.: A comparative study on heuristic algorithms for generating fuzzy decision trees. IEEE Trans. on Transactions on Systems, Man and Cybernetics 31, 215–226 (2001)
29. Yuan, X., Khoshgoftaar, T.M., Allen, E.B., Ganesan, K.: An application of fuzzy clustering to software quality prediction. In: Proc. 3rd IEEE Symp. on Application-Specific Software Engineering Technology, pp. 85–90 (2000)
30. Yuan, Y., Shaw, M.J.: Induction of fuzzy decision trees. Fuzzy Sets and Systems 69, 125–139 (1995)

Part III

Rough Hybrid Models to Classification and Attribute Reduction

Rough Sets and Evolutionary Computation to Solve the Feature Selection Problem

Rafael Bello[1], Yudel Gómez[1], Yailé Caballero[2], Ann Nowe[3], and Rafael Falcón[1]

[1] Computer Science Department, Central University of Las Villas (UCLV)
Carretera Camajuaní km 5.5, Santa Clara, Cuba
{rbellop,ygomezd,rfalcon}@uclv.edu.cu

[2] Informatics Department, University of Camagüey, Cuba
yaile@inf.reduc.edu.cu

[3] CoMo Lab, Computer Science Department, Vrije Universiteit Brussel, Belgium
asnowe@info.vub.ac.be

Summary. The feature selection problem has been usually addressed through heuristic approaches given its significant computational complexity. In this context, evolutionary techniques have drawn the researchers' attention owing to their appealing optimization capabilities. In this chapter, promising results achieved by the authors in solving the feature selection problem through a joint effort between rough set theory and evolutionary computation techniques are reviewed. In particular, two new heuristic search algorithms are introduced, i.e. Dynamic Mesh Optimization and another approach which splits the search process carried out by swarm intelligence methods.

Keywords: meta-heuristic, evolutionary computation, feature selection.

1 Introduction

The solution of a great deal of problems can be formulated as an optimization problem. The quest for the problem's solution is often stated as finding the optimum of an objective function $f : D \rightarrow \Re$; i.e., finding a point $x_0 \in D$ such that $f(x_0) \leq f(x)\ \forall x \in D$, for the minimization case. The Feature Selection Problem (FSP) can well illustrate this point.

The relevance of the feature selection methods has been widely acknowledged [15, 28]. These methods search throughout the space of feature subsets aiming to find the best subset out of the $2^N - 1$ possible feature subsets (N stands for the number of attributes characterizing the problem). The search is guided by an evaluation measure. Every state denotes a subset of features in the search space.

All feature selection techniques share two crucial components: an evaluation function (used for numerically assessing the quality of a candidate feature subset) and a search engine (an algorithm responsible for the generation of the feature subsets).

The evaluation function attempts to estimate the capability of an attribute or a subset of attributes to discriminate between the collection of existing classes. A

A. Abraham, R. Falcón, and R. Bello (Eds.): Rough Set Theory, SCI 174, pp. 235–260.
springerlink.com

subset is said to be optimal with regards to a given evaluation function. Several categories of evaluation functions stand nowadays, like distance measures, information measures (e.g., entropy), dependency measures, consistency measures and classification error measures [8]. More recently, the "quality of the classification" measure borrowed from rough set theory (RST) has been employed as a numerical estimator of the quality of reducts [11, 1, 2, 24, 25]. A reduct is a minimal subset of attributes which preserves the partition over a universe, as stated in [14] wherein also the role played by reducts in feature selection and reduction is explained.

The second component of a feature selection algorithm is the search engine, which acts as a procedure for the generation of the feature subsets. The search strategies are important because this type of problem can be extremely time-consuming and an exhaustive search of a rather "optimal" subset can be proved infeasible, even for moderate values of N. Algorithms to feature selection are usually designed by using heuristics or random search strategies in order to reduce complexity. Heuristic search is very fast because it is not necessary to wait until the search ends but it doesn't guarantee to find the best solution although a better one is known when it is found in the process. An illustrative example of search strategies is given by the evolutionary methodologies.

Evolutionary algorithms perform on the basis of a subset of prospective solutions to the problem, called "population", and they locate the optimal solution through cooperative and competitive activities among the potential solutions. Genetic Algorithms (GA) [10], Ant Colony Optimization (ACO) [9] and Particle Swarm Optimization (PSO) [12] are genuine exemplars of this sort of powerful approaches. They have also been termed as "bioinspired computational models" owing to the natural processes and behaviors they have been built upon.

Diverse studies have been carried out concerning the performance of the above meta-heuristics in the feature selection problem. Some of them have exhibited good results, mainly attained by using ACO- or PSO-based approaches, such as [11, 1, 2, 24, 25, 26]. In [3] and [4], a new approach to feature selection based on the ACO and PSO methodologies is presented. The chief thought is the split of the search process accomplished by the agents (ants or particles) into two stages, such that an agent is commanded in the first stage to find a partial solution to the problem, which in turn is afterwards used as an initial state during the upcoming phase. The application of the two-step approach to the feature selection problem provokes that, after finishing the first stage, agents hold feature subsets which are prospective reducts of the system. They are taken as initial states for the agents during the remaining phase.

The new meta-heuristic named Dynamic Mesh Optimization (DMO) also falls under the umbrella of the evolutionary computation techniques. A set of nodes characterizing potential solutions of an optimization problem make up a mesh which dynamically expands itself and moves across the search space. To achieve this, intermediate nodes are generated at each cycle (iteration) between the mesh nodes and those nodes regarded as local optima, as well as between the mesh nodes and the global optimum. Moreover, new nodes are also generated out of the most external mesh nodes, thus allowing for a broader covering of the search space. The fittest nodes of the ensuing mesh are promoted to make up the mesh at the next cycle. The performance achieved by the DMO procedure in the context of feature selection is studied.

In this chapter, we study the integration between rough set theory and the aforementioned evolutionary algorithms for working out the feature selection problem. No attempt has been made to cover all approaches currently existing in literature but our intention has been to highlight the role played by rough set theory across several evolutionary algorithms in the quest for really meaningful attributes.

The study is structured as follows: after enunciating the fundamentals of rough set theory in section 2, the application of a greedy algorithm to feature selection is presented. Next we elaborate on the way this challenge is tackled by Genetic Algorithms and Swarm Intelligence meta-heuristic approaches, like ACO and PSO. Section 6 discusses the novel DMO evolutionary optimization method whereas section 7 is devoted to unfold a comparative study between the different algorithmic models under consideration. Finally, some conclusions are derived.

2 Rough Set Theory: Basic Concepts

Rough set theory (RST) was proposed by Z. Pawlak [19]. The rough set philosophy is anchored on the assumption that some information is associated with every object of the universe of discourse [21]. Rough set data analysis is one of the main techniques arising from RST; it provides a manner for gaining insight into the underlying data properties [29]. The rough set model has several appealing advantages for data analysis. It is based on the original data only and does not rely on any external information, i.e. no assumptions about data are made. It is suitable for analyzing both quantitative and qualitative features leading to highly interpretable results [23].

In RST a training set can be represented by a table where each row represents an object and each column represents an attribute. This table is called an "information system"; more formally, it is a pair $S = (U, A)$, where U is a non-empty finite set of objects called the universe and A is a non-empty finite set of attributes. A decision system is a pair $DS = (U, A \cup \{d\})$, where $d \in A$ is the decision feature or class attribute. The basic concepts of RST are the lower and upper approximations of a subset $X \subseteq U$. These were originally introduced with reference to an indiscernibility relation $IND(B)$, where objects x and y belong to IND(B) if and only if x and y are indiscernible from each other by features in B.

Let $B \subseteq A$ and $X \subseteq U$. It can be proved that B induces an equivalence relation. The set X can be approximated using only the information contained in B by constructing the B-lower and B-upper approximations of X, denoted by $\underline{B}X$ and $\overline{B}X$ respectively, where $\underline{B}X = \{x \in U : [x]_B \in X\}$ and $\overline{B}X = \{x \in U : [x]_B \cap X \neq \emptyset\}$ and $[x]_B$ denotes the equivalence class of x according to the B-indiscernible relation. The objects in $\underline{B}X$ are guaranteed to be members of X

while those in $\overline{B}X$ are possible members of X. The boundary region $BNX = \overline{B}X - \underline{B}X$ determines the roughness of a concept X, namely X is said to be rough if its boundary region is not empty, otherwise it is said to be a crisp (precise) concept.

RST offers several measures for gauging the quality of a decision system. Among them one can find the "quality of classification", displayed in expression (1). It quantifies the percentage of objects which are correctly classified into the given decision classes $Y = \{Y_1, \ldots, Y_n\}$ employing only the knowledge induced by the set of features in B.

$$\gamma_B(Y) = \frac{\sum_{i=1}^{n} |\underline{B}Y_i|}{|U|} \tag{1}$$

An important issue concerning RST is attribute reduction based on the reduct concept. A reduct is a minimal set of features that preserves the partitioning of the universe and hence the ability to perform classifications. The subset B is a reduct if $IND(A) = IND(B)$; that is, $\gamma_A(Y) = \gamma_B(Y)$. The notion of reduct is one of the most important concepts within rough set theory.

However, their practical use is limited because of the heavy workload involved in computing the reducts. The problem of finding a globally minimal reduct for a given information system is NP-hard. For that reason, methods for calculating reducts have been developed on the basis of heuristic-driven approaches [22].

3 A Greedy Algorithm to Feature Selection

The incorporation of rough sets to a greedy approach for finding reducts has been studied in [7]. The method begins with an empty set of attributes and constructs good reducts in an acceptable time. The heuristic search performed by the algorithm adds the fittest attributes to the solution according to some predefined criterion.

The criterion for assessing the quality of an attribute is borrowed from the ID3 classifier with respect to the normalized entropy and the gain of the attributes as well as the degree of dependency between attributes, this latter indicator coming from RST. In this algorithm we use the terms $R(A)$ and $H(A)$ proposed in [20]. $R(A)$ lies within [0,1] and stands for the relative importance of attribute A while $H(A)$ represents heuristic information about a subset of candidate features.

$R(A)$ can be computed by the following expression:

$$R(A) = \sum_{i=1}^{k} \frac{|S_i|}{|S|} \cdot e^{1-C_i} \tag{2}$$

where k is the number of different values of attribute A whereas C_i represents the number of different classes present in the objects having the i-th value for the feature A. Moreover, $|S_i|$ indicates the amount of objects with the value i in the feature A and $|S|$ the total number of objects.

On the other hand, the term $H(A)$ is obtained by the procedure below:

1. Calculate $R(A)$ for each attribute in the problem and make up a vector with the best n attributes (n selected by the user) according to the $R(A)$ indicator. As a result of it, the vector $\mathbf{BR} = (R(A_i), R(A_j), \ldots)$ with $n = |\mathbf{BR}|$ is obtained.
2. Create another vector holding the combinations of n in p (this value also inputted by the user) of the attributes in **BR**. The combination vector looks like $\mathbf{Comb} = (\{A_i, A_j, A_k\}, \ldots, \{A_i, A_t, A_p\})$
3. Compute the degree of dependency of the decision classes with respect to every combination lying in **Comb**. Let us denote by $\mathbf{DEP}(d)$ the vector containing the degree of dependency of decision class d with respect to every subset of attributes in **Comb**, that is $\mathbf{DEP}(d) = (k(Comb_1, d), \ldots, k(Comb_{|Comb|,d}))$
4. Compute $H(A) = \sum_{\forall i: A \in \mathbf{Comb_i}} k(Comb_i, d)$

where $k = \dfrac{|POS_B(d)|}{|U|}$ and $POS_B(d) = \bigcup_{X \in U/B} \underline{B}X$

If $k = 1$ then d totally depends on B else it partially depends on it.

Another alternative measure that has been used successfully is the gain ratio [17] which is defined in terms of the following measure:

$$\text{SplitInformation}(S, A) = -\sum_{i=1}^{c} \frac{|S_i|}{|S|} \cdot log_2 \frac{|S_i|}{|S|} \tag{3}$$

where c is the cardinality of the domain of values of attribute A. This measure is the entropy of S with respect to A.

The gain ratio $G(A)$ quantifies how much information gain attribute A produces or how important it is to the data set. The formal expression is shown below:

$$G(A) = \frac{G(S, A)}{\text{SplitInformation}(S, A)} \tag{4}$$

$$G(S, A) = \text{Entropy}(S) - \sum_{v \in V_A} \frac{|S_v|}{|S|} \cdot \text{Entropy}(S_v) \tag{5}$$

where V_A is the set of values of attribute A and S_v is the subset of S for which attribute A has the value v, namely $S_v = \{s \in S \mid A(s) = v\}$

$$\text{Entropy}(S) = \sum_{i=1}^{c} - P_i \cdot \log_2 P_i \tag{6}$$

where P_i is the ratio of objects in S belonging to the i-th decision class.

The cost of an attribute can be defined using expressions (7) or (8):

$$C(A) = \frac{G^2(S, A)}{Cost(A)} \tag{7}$$

where $Cost(A)$ is the cost of attribute A (say, for instance, the cost of running a medical exam). This value ranges between 0 and 1 and must be specified by the user.

$$C(A) = \frac{2^{G(S,A)} - 1}{(Cost(A) + 1)^w} \tag{8}$$

where $Cost(A)$ is just like in (7) and w is a constant value also in [0,1] that determines the relative importance of the cost versus information gain.

Bearing the measures $R(A)$, $H(A)$, $G(A)$ and $C(A)$ in mind, the RSReduct algorithm was devised and implemented as shown in Algorithm 1.

Algorithm 1. RSReduct

---

procedure RSREDUCT( )

STEP 1 Form the distinction table with a binary matrix B $(m^2 - m)/2 \times (N + 1)$. Each row corresponds to a pair of different objects. Each column of this matrix corresponds to an attribute; the last column corresponds to the decision value (treated as an attribute).

For each attribute, let $b((k, n), i) \in B$ corresponding to the pair (O_k, O_n) and attribute i be defined as

$$b((k,n),i) = \begin{cases} 1, \text{ if } a_i(O_k) \neg R\, a_i(O_n) \;\; \forall i = \{1, \ldots, N\} \\ 0, \text{ otherwise} \end{cases}$$

$$b((k,n),N+1) = \begin{cases} 0, \text{ if } d_i(O_k) \neq d_i(O_n) \\ 1, \text{ otherwise} \end{cases}$$

where R is a similarity relation depending on the type of attribute a_i

STEP 2 For each attribute A calculate the value of $RG(A)$ for any of the following three heuristics and then form an ordered list of attributes starting from the most relevant attribute (that which maximizes $RG(A)$):

- Heuristic 1: $RG(A) = R(A) + H(A)$
- Heuristic 2: $RG(A) = H(A) + G(A)$
- Heuristic 3: $RG(A) = H(A) + C(A)$

STEP 3 With $i = 1$, $R = \emptyset$ and $(A_1, A_2, \ldots, A_n)$ an ordered list of attributes according to step 2, consider if $i \leq n$ then $R = R \cup A_i$, $i = i + 1$.

STEP 4 If R satisfies condition I (see below) then Reduct=minimal subset $R' \subseteq R$ does meet condition I, so stop otherwise go to step 3.

Condition I uses the following relation between objects x and q for attribute a: $q_a R x_a \Leftrightarrow sim(x_a, q_a) \geq \varepsilon$ where $0 \leq \epsilon \leq 1$

end procedure

The RSReduct approach has been tested with several data sets from the UCI machine learning repository [6] that are available at the ftp site of the University of California. Some of the databases belong to real-world data such as Vote, Iris, Breast Cancer, Heart and Credit while the other ones represent results obtained in labs such as Balloons-a, Hayes-Roth, LED, M-of-N, Lung Cancer and Mushroom. The results portrayed in Table 1 were obtained after using RSReduct with the three heuristic functions defined in step 2 of Algorithm 1. Furthermore, the execution time of the algorithm has been recorded in each case.

Table 1. Average length of reducts and computational time required by the RSReduct approach

Data set name	Heuristic 1		Heuristic 2		Heuristic 3	
(cases#, attr#)	Time(s)	Avg. len	Time(s)	Avg. len	Time(s)	Avg. len
Ballons-a (20,4)	5.31	2	3.12	2	16.34	2
Iris (150,4)	40.15	3	30.79	3	34.73	3
Hayes-Roth(133,4)	36.00	3	32.30	3	39.00	3
Bupa(345,6)	74.20	6	89.00	6	89.00	6
E-Coli(336,7)	57.00	5	41.15	5	46.60	5
Heart(270,13)	30.89	9	16.75	9	54.78	10
Pima(768,8)	110.00	8	110.00	8	110.00	8
Breast-Cancer(683,9)	39.62	4	31.15	4	32.56	5
Yeast(1484,8)	82.00	6	78.00	6	85.70	6
Dermatology(358,34)	148.70	8	125.9	8	190.00	9
Lung-Cancer(27,56)	25.46	7	18.59	7	31.5	8
LED(226,25)	78.10	9	185.00	8	185.00	9
M-of-N(1000,14)	230.26	6	162.50	6	79.4	6
Exactly(780,13)	230.00	11	215.00	11	230.00	11
Mushroom(3954,22)	86.20	8	64.10	8	67.2	8
Credit(876,20)	91.20	14	86.01	14	90.2	15
Vote(435,16)	37.93	12	21.25	11	26.9	12

In the experiments displayed in Table 1, $C(A)$ has been computed as in (8), $Cost(A)$ takes random values and $w = 0.1$.

4 Feature Selection by Using a Genetic Approach

Arising as a true exemplar of evolutionary computation techniques, Genetic Algorithms (GAs) have been widely utilized for attribute reduction. GAs are stochastic search methods based on populations. First, a population of random individuals is generated and the best individuals (in accord with some predefined criterion) are selected. Then, the new individuals making up the population will be generated using the mutation, crossover and (possibly) inversion operators. In [27], three methods for finding short reducts are presented. They use genetic algorithms together with a greedy approach and have defined the adaptability functions f_1, f_2 and f_3.

An adaptation of the GA plan is the Estimation of Distribution Algorithms (EDA) [18] but most of them don't use crossover or mutation because the new population is generated from the distribution of the probability estimated from the selected set. The principal problem of the EDA is the estimation of $ps(x, t)$

Table 2. Results obtained with the proposed Estimation Distribution Algorithms (EDA)

Data set name (cases#, attr#)	Algorithms with Wróblewski's functions								
	f_1			f_2			f_3		
	AT	ARL	ANR	AT	ARL	ANR	AT	ARL	ANR
Ballons-a(20,4)	0.167	2	1	1.860	2	1	0.260	2	1
Iris(150,4)	82.390	3	4	3.540	3	4	17.250	3	4
Hayes-Roth(133,4)	40.830	4	1	30.100	4	1	22.450	4	1
Bupa(345,6)	436	3	6.85	995.300	3	8	466	3	8
E-Coli(336,7)	64.150	3	6.85	1514	3	7	169.200	3	7
Heart(270,13)	337	3	8	2782	3	18	1109	3	17
Pima(768,8)	2686	3	17	6460	3	18.4	4387	3	18.6
Breast-Cancer(683,9)	1568	4	6.55	8250	4	7.83	2586	4	8
Yeast(1484,8)	1772	4	2	12964	4	2	2709	4	2
Dermatology(358,34)	1017	6.05	10.15	15553	6	14.90	30658	6	47
Lung-Cancer(27,56)	7.780	4.2	9.55	0.0956	4	15.95	264.200	4	38.6

and the generation of new points according to this distribution in a way that yields reasonable computational efforts. For this reason, different manners to determine $ps(x,t)$ have been crafted.

One of the members of this family is the Univariate Marginal Distribution Algorithm (UMDA) for discrete domains [18], which takes into account univariate probabilities alone. This algorithm is capable of optimizing non-linear functions as long as the additive (linear) variance of the problem has an acceptable weight in the total variance. The UMDA version for continuous domains [16] was introduced in 2000. In every generation and for each variable, UMDA carries out statistic tests to find the density function that best fits to the variable. The continuous variant of UMDA is an algorithm of structure identification in the sense that the density components are identified through hypotheses tests.

We have defined a method [7] for calculating reducts starting from the integration of the adaptability functions (f_1, f_2, f_3) of the methods reported by Wróblewski in [27] and the UMDA approach, thus leading to encouraging results which are shown in Table 2. The values of the parameters used were: $N = 100$; $g = 3000$; $e = 50$; $T = 0.5$ where N is the number of individuals, g is the maximum number of evaluations that will be executed, e is the number of elite (best fitting) individuals which pass directly to the next generation and T is the percentage of the best individuals that were selected to do all the calculations.

In Table 2, AT means the average time required to calculate the reducts (measured in seconds), ARL stands for the average length of the reducts found and ANR their average number.

The use of the three functions reported in [27] in the Estimation of Distribution Algorithms turned out successful. EDA performed the calculation of small reducts in little time when the set of examples was not very large (say, less than 600 cases) even though the number of attributes characterizing the data set was huge. The best combination was accomplished with f_1 when it comes to the execution time; however f_3 found a larger number of reducts in a reasonable time frame.

5 Swarm Intelligence in Feature Selection

This section will unfold the potential of major swarm intelligence approaches to be applied in the feature selection (attribute reduction) problem.

5.1 Particle Swarm Optimization

Particle swarm optimization (PSO) is a heuristic method which uses a population of particles and is strongly inspired by the natural behavior of bird flocks and fish schools. Each particle symbolizes a potential solution to the optimization problem. The system begins with an initial population (most of the times, random individuals) and searches for optima according to some fitness function by updating particles over generations; that is, particles "fly" through the N-dimensional problem search space by following the current best-performing particles.

Each particle records its own best position X_{pbest} (that is, its fittest function value ever achieved) as well as the global best position X_{gbest} ever reached by the swarm. As shown in expression (9), the particles are drawn to some degree by X_{pbest} and X_{gbest}. At each iteration the velocity vector $\mathbf{V}$ associated with every particle is updated according to (9). Acceleration constants c_1 and c_2 are empirically determined and used to set up a tradeoff between the exploration and convergence capabilities of the algorithm. The particle's new position is calculated by means of (10).

$$\mathbf{V_i'} = w \cdot \mathbf{V_i} + c_1 \cdot r_1 \cdot (\mathbf{X_{pbest}} - \mathbf{X_i}) + c_2 \cdot r_2 \cdot (\mathbf{X_{gbest}} - \mathbf{X_i}) \tag{9}$$

$$\mathbf{X_i'} = \mathbf{X_i} + \mathbf{V_i} \tag{10}$$

where $\mathbf{V_i}$, $\mathbf{X_i}$, $\mathbf{X_{pbest}}$ and $\mathbf{X_{gbest}}$ are N-dimensional vectors and w is the inertia weight. A suitable selection of w provides a balance between global and local exploration. Random numbers r_1 and r_2 usually follow a normal distribution within [0,1] and outfit the algorithm with the stochastic component.

In feature selection we have a N-dimensional search space, where N is the number of features characterizing the problem. The optimal position along the search space is the shortest subset of features with the highest quality of classification. Being this so, the configuration of the PSO meta-heuristic is as follows: each particle encodes a N-dimensional binary vector with the i-th bit set to one if the corresponding feature is part of the subset and zero otherwise.

The algorithm seeks for minimal reducts R, that is, minimal subsets of features whose quality of the classification $\gamma_R(Y)$ is equal to that yielded by the whole set of features $\gamma_A(Y)$ (Y being the set of decision classes). In this case, the fitness function is the same used in [26], see expression (11), which takes into account the quality of classification and length of the reducts for deeming the worth of a prospective solution. The ensuing optimization activities attempt to maximize the fitness function value.

$$\text{fitness} = \alpha \cdot \gamma_R(Y) + \beta \cdot \frac{N - |R|}{N} \tag{11}$$

In light of the particle encoding scheme used in this proposal, it is necessary to redefine expression (10). The movement of the particle is realized by the flip of the bit value and the velocity is no longer a change ratio of its position but a change probability of it. We propose expression (12) in [4] to calculate the j-dimension of the i-th particle. This is based on the position and velocity update equations of the particle as shown in [13] and [30].

$$X'_{ij} = \begin{cases} 1, \text{ if } \mathrm{rand}() \leq \dfrac{1}{1 + e^{1.5 \cdot N \cdot V_{ij}}} \\ 0, \text{ otherwise} \end{cases} \tag{12}$$

The value of the inertia weight w is defined by a positive linear function changing according to the current iteration, as shown below:

$$w = w_{max} - \frac{w_{max} - w_{min}}{NC} \times k \tag{13}$$

where w_{max} is the initial value of the inertia weight, w_{min} its final value, NC the maximal number of cycles (iterations) allowed and k denotes the current iteration number.

The PSO-driven approach is outlined in Algorithm 2.

Algorithm 2. PSO-RST-FS

```
procedure PSO-RST-FS( )
    Generate initial population by setting the Xi and Vi vectors
    repeat
        Compute Xpbest for each particle
        Compute Xgbest for the swarm
        Reducts = Reducts ∪ {R} such that γR(Y) = γA(Y)
        Update velocity and position of every particle by (9) and (12)
    until k = NC
    Output the set Reducts
end procedure
```

A new approach concerning PSO is introduced in [4] . The Two-Step Particle Swarm Optimization (TS-PSO) algorithm is rooted on the idea of splitting the search process carried out by the particles into two stages so that, in the first stage, preliminary results are reached which could be subsequently used to make up the initial swarm for the second stage. In the case of FSP, this means that subsets of features which are potential reducts of the information system are generated along the first stage. These subsets are used to modify the swarm resulting from the last cycle in the first stage; the modified swarm is used as the initial population of the second stage.

Determining the state by which the search process should commence has long been an interesting problem in heuristic search. It is well known that setting up the initial state has an important bearing over the global search process. The purpose is to be able to approach the initial state to the goal state. Of course, it is necessary to consider an adequate balance between the computational cost of obtaining that initial state and the total cost; in other words, the sum of the cost of approaching the initial state towards the goal state plus the cost of finding the solution from that "improved" location should not be greater than the cost of looking for the solution from a random initial position.

More formally, the aim is the following. Let E_i be the initial state which has been either randomly generated or produced after the execution of any other method without a significant computational cost, E_i^* the initial state generated by some method M that approaches it to the goal state. By $CM(E_i, E_i^*)$ we denote the cost of getting E_i^* from state E_i by means of M and $CCHSA(x)$ is the computational cost involved in finding a solution from state x using a Heuristic Search Algorithm (HSA). Then, the goal is that $CM(E_i, E_i^*) + CCHSA(E_i^*) < CCHSA(E_i)$.

In the two-step approach proposed here, the procedures to generate E_i^* and the HSA are both the PSO algorithm, so the objective is $CPSO(E_i, E_i^*) + CCPSO(E_i^*) < CCPSO(E_i)$. Since PSO is used in both phases, some parameters of the model are meant to distinguish between them. A ratio r is introduced in order to establish the relative setting of the values of the algorithm parameters in both stages; the ratio indicates the proportion of the overall search that will be carried out during the first stage. For example, if $r = 0.3$ for the NC parameter, it means that the first part of the search process will involve 30% of the total number of iterations whereas the subsequent step will be responsible for realizing the remaining 70%.

The parameters that establish the differences between stages are the following: $ratio_Q$ and $ratio_C$. The first one is related to the definition of a quality threshold according to expression (14) and is involved in the selection of the candidate feature subsets. In the first stage, each candidate reduct R whose quality of classification exceeds the quality threshold is selected as a potential reduct. The $ratio_C$ parameter is used to compute the number of cycles in each stage according to (15) and (16).

$$\phi = ratio_Q \cdot \gamma_A(Y) \tag{14}$$

$$nc_1 = \text{round}(ratio_C \cdot NC) \tag{15}$$

$$nc_2 = NC - nc_1 \tag{16}$$

where round(x) denotes the closest integer to x.

Algorithm 3. TS-PSO-RST-FS

```
procedure TS-PSO-RST-FS( )
    *****STAGE 1*****
    Generate initial population by setting the Xi and Vi vectors
    repeat
        Compute Xpbest for each particle
        Compute Xgbest for the swarm
        PR = PR ∪ {R} such that γR(Y) ≥ φ
        Update velocity and position of every particle by (9) and (12)
    until k = nc1
    *****POST PROCESSING STAGE*****
    Compute UsedFeatures and NotUsedFeatures using PR
    currentSwarm ← last swarm of stage 1
    for each particle Xi in currentSwarm do
        if rand() ≤ 0.5 then
            Xi ← UsedFeatures
        else
            Modify Xi by resetting all features in NotUsedFeatures
        end if
    end for
    *****STAGE 2*****
    repeat
        Compute Xpbest for each particle
        Compute Xgbest for the swarm
        Reducts = Reducts ∪ {R} such that γR(Y) = γA(Y)
        Update velocity and position of every particle by (9) and (12)
    until k = nc2
    Output the set Reducts
end procedure
```

The TS-PSO-RST-FS algorithm introduces a step between the first and second phases (called "postprocessing step") in which the set of potential reducts PR is used to build the **UsedFeatures** and **NotUsedFeatures** N-dimensional binary vectors; the features highlighted in these vectors have value 1 in their corresponding vector component. The **UsedFeatures** vector sets to one its i-th component provided that the i-th feature in the data set belongs to a number of candidate reducts greater than a given percentage threshold, called PerUsed, of the total number of potential reducts found; for instance, if PerUsed=75%, this means that only features which belong to at least the 75% of the potential reducts will receive a signaling in their associated bit within **UsedFeatures**. On the other hand, the **NotUsedFeatures** vector highlights all features belonging to at most PerNotUsed of potential reducts; for instance, if PerNotUsed=30% this means that only features which are included in at most the 30% of the potential reducts become signalized in **NotUsedFeatures**.

By means of the **UsedFeatures** and **NotUsedFeatures** vectors, the optimal swarm of the first stage is modified to give rise to the startup swarm of

Table 3. Results obtained with the proposed PSO-based approaches. Columns 4 and 6 display the average length of the reducts while columns 5 and 7 show the number of times the algorithm found minimal reducts.

Data set name	Features	Instances	PSO	PSO	TS-PSO	TS-PSO
(1)	(2)	(3)	(4)	(5)	(6)	(7)
Breast cancer	9	699	4.95	6	4.6	6
Heart	13	294	7.97	4	6.8	6
Exactly	13	1000	6	6	6	6
Credit	20	1000	12.4	4	10.3	5
Dermatology	34	358	15.3	3	12.6	5
Lung	56	32	15.6	3	12.8	5

the second stage in the following way. Each particle $\mathbf{X_i}$ in the optimal swarm is replaced by the vector **UsedFeatures** or is modified by using the vector **NotUsedFeatures** in a random way: if $rand() \leq 0.5$ then replace else modify. Modify means that all features included in **NotUsedFeatures** are reset in the particle encoding.

Greater values of the inertia weight during both the first and processing steps help to find good seeds to build the initial swarm for the second stage. The entire description of the two-step approach can be found in Algorithm 3.

The algorithms PSO-RST-FS and TS-PSO-RST-FS were executed by using the following parameters: NC$= 120$, $c_1 = c_2 = 2$, population size $= 21$ and $\alpha = 0.54$. In the case of the TS-PSO-RST-FS algorithm, $ratio_Q = 0.75$, $ratio_C = 0.3$, PerUsed=66% and PerNotUsed=30%.

In the two-step approach, the values of the ratios have important bearing over the desired outcome. A low value of $ratio_Q$ yields many low-quality potential reducts, consequently the **UsedFeatures** and **NotUsedFeatures** vectors include useless information about the features; therefore, the effect of using **UsedFeatures** and **NotUsedFeatures** to modify the swarm is poor. On the other side, a value near to one produces subsets close to the definition of reducts in the first stage. As to $ratio_C$, a low value allows to perform a greater quantity of cycles in the second stage from the modified swarm.

The two algorithms were tested and compared using six data sets from UCI Repository. Each algorithm was executed six times on every data set and the average results are offered in Table 3. The performances obtained were compared in terms of the average length of the resulting reduct set (columns 4 and 6) and the number of times in which the algorithm found the minimal reducts (columns 5 and 7). The length of a reduct is defined by the number of features in it.

These results are very interesting because they shed light on the fact that the two-step PSO approach ends up with shorter reducts than the PSO-RST-FS algorithm. So, the certainty of finding minimal length reducts increases by using the TS-PSO-RST-FS method.

5.2 Ant Colony Optimization

Another very popular swarm intelligence technique is Ant Colony Optimization (ACO). ACO is a generic strategy (meta-heuristic) [9] used to guide other heuristics in order to obtain superior solutions than those generated by local optimization methods. In the early ACO model, a colony of artificial ants cooperates to look for good solutions to discrete optimization problems. Artificial ants are simple agents that incrementally build a solution by adding components to a partial solution under construction.

Ant System (AS) [9] is considered as the first ACO algorithm and was introduced using the Travel Salesman Problem (TSP). In TSP, we have a set of N fully connected cities $c_1, \ldots, c_n$ by edges (i, j). Edges have associated pheromone trails τ_{ij} which denote the desirability of visiting city j directly from city i. Also, the function $\eta_{ij} = 1/d_{ij}$ indicates the heuristic desirability of going from i to j, where d_{ij} is the distance between cities i and j. Initially, ants are randomly associated to cities. In the successive steps, ant k applies a random proportional rule to decide which city to visit next according to (17):

$$p_{ij}^k = \frac{(\tau_{ij})^\alpha \cdot (\eta_{ij})^\beta}{\sum\limits_{l \in N_i^k} (\tau_{il})^\alpha \cdot (\eta_{il})^\beta} \qquad \text{if } j \in N_i^k \tag{17}$$

where N_i^k is the neighborhood of the k-th ant while α and β are two parameters that point out the relative importance of the pheromone trail and the heuristic information, respectively. After all ants have built their tours, the values τ_{ij} are updated in two different ways. First, τ_{ij} values decrease because of the evaporation ($\tau_{ij} = (1 - \rho) \cdot \tau_{ij}$). The ρ parameter is meant to prevent unlimited pheromone accumulation along the edges. Second, all ants reinforce the value of τ_{ij} on the edges they have passed on in their tours ($\tau_{ij} = \tau_{ij} + Inc_{ij}$), where Inc_{ij} is the amount of pheromone deposited by all ants which included edge (i, j) in their tour. Usually, the amount of pheromone deposited by the k-th ant is equal to $1/C_k$, where C_k is the length of the tour of ant k.

Some direct successor algorithms of Ant Systems are: Elitist AS, Rank-based AS and MAX-MIN AS [9]. A more different ACO approach is Ant Colony System (ACS) which employs the following pseudo-random proportional rule to select the next city j from city i:

$$j = \begin{cases} \arg\max\limits_{l \in N_i^k} \{\tau_{ij} \cdot (\eta_{il})^\beta\}, & \text{if } q < q_0 \\ \\ \text{random selection as in (17)}, & \text{otherwise} \end{cases} \tag{18}$$

where q is a random variable uniformly distributed in [0,1] and $0 \leq q_0 \leq 1$, controls the amount of exploration. In ACS, ants have a local pheromone trail update which is defined as ($\tau_{ij} = (1-\rho)\cdot\tau_{ij}+\rho\cdot\tau_{ij}(0)$) and is applied after crossing an edge (i,j), where $\tau_{ij}(0)$ represents the initial pheromone value. Furthermore, a global pheromone trail update ($\tau_{ij} = (1-\rho)\cdot\tau_{ij} + \rho \cdot Inc_{ij}$) is executed only by the best-so-far ant.

The feature selection problem is an example of a tough discrete optimization problem which can be represented as a graph problem; this is why the ACO model is well suited to solve it.

For this study we used the ACS-RST-FS according to results showed in [1] and [2]. Let $A = \{a_1, a_2, \ldots, a_{nf}\}$ be a set of features. One can think of this set as an undirected graph in which nodes represent features and all nodes are connected by bidirectional links. Pheromone values τ_i are associated to nodes a_i. The amount of pheromone is a function of the dependency of the feature associated to that node to all other features. The pheromone stands for the absolute contribution of that feature to a reduct.

The solution consists of reducts which have to be gradually constructed by the system agents (ants). Initially the ants are distributed over the nodes of the graph and each one stores an empty subset which has to become a candidate reduct. The behavior of a single ant can be described as follows. In the first step, the ant is assigned to one of the nodes, from which it will move to some other node in the network. By doing so, the ant performs a forward selection in which it expands its subset step-by-step by adding new features. To select the next node to visit, the ant looks for all features which are not yet included in the subset and selects the next one according to the ACS rule. On the one hand, it is drawn by the pheromone the other ants have already put down in the graph and, on the other hand, by the heuristic function. We have confined ourselves to choose the standard quality of classification (see expression 1) as the heuristic function for our problem. It is used too for determining whether the candidate subset is a reduct or not. Over time, the quality of the subsets constructed by the ants will improve, which is supported by the monotonicity property of classical RST; these converge to nearly optimal reducts.

The initial deployment of the ants during each cycle (iteration) is governed by the following rules. Recall that m is the population size (number of ants) whereas nf is the number of features present in the data set.

1. If $m <$ nf then perform a random initial distribution of ants.
2. If $m =$ nf then one ant is assigned to each feature.
3. If $m >$ nf then assign the first m ants according to (2) and the remaining ones as in (1).

The process of finding the candidate reduct sets B happens in a sequence of cycles $NC = 1, 2, \ldots$ In each cycle, all ants build their own set B_k. The process stop criterion is met (PSC = true) once the maximal number of cycles has been exceeded $NC > NC_{max}$. Each ant k keeps adding one feature at a time to its current partial set B_k until $\gamma_{B_k}(Y) = \gamma_A(Y)$. This is known as the ant stopping criterion (ASC_k=true). The population size is envisioned as a function of the number of features $m = f(\text{nf})$ where round(x) denotes the closest integer to x.

Algorithm 4. ACS-RST-FS

```
procedure ACS-RST-FS( )
    PSC ← false, NC ← 1
    Calculate τ_i(0), i = 1,...,nf (random initial values for trail intensity)
    repeat
        Each ant k is assigned to an attribute a_i, ∀k ∈ {1,...,m} and B^k ← {a_i}
        ASC_k ← false ∀k ∈ {1,...,m}
        repeat
            for k ← 1 to m do
                if ASC_k = false then
                    Select new feature a_i^* according to (1)
                    B_k = B_k ∪ {a_i^*}
                    τ_i ← (1 − ξ) · τ_i + ξ · τ_i(0)          ▷ i is the index of a_i^*
                    Update ASC_k                  ▷ Did ant k complete a reduct B_k?
                end if
            end for
        until ASC_k = true ∀k ∈ {1,...,m}
        B_k^* ← best B_k     ▷ Now that all ants have finished, select the best reduct
        for each a_i ∈ B_k^* do
            τ_i ← (1 − ρ) · τ_i + ρ · γ_B^k(Y)          ▷ update global pheromone trail
        end for
        For each feature i do τ_i = τ_i / Σ_{j=1}^{n} τ_j
        NC ← NC + 1
        Update PSC
    until PSC = true
end procedure
```

R₁: If nf $<$ 19 then m = nf
R₂: If 20 $\leq$ nf $\leq$ 49 then if 2/3 nf $\leq$ 24 then m=24 else m=round(2/3 nf)
R₃: If nf $>$ 50 then if nf/2 $\leq$ 33 then m = 33 else m = round(nf/2)

The above rules are the direct outcome of a thorough experimental analysis conducted with the purpose in mind of setting the population size on the basis of the number of features describing the data set.

Now we are ready to present the ACS-RST-FS approach in a more formal way. Let us look at Algorithm 4.

A new approach in Ant Colony Optimization for solving the feature selection problem was introduced in [3] . The two-step ACO algorithm is also based on the idea of splitting the process of finding reducts into two stages. The algorithm dynamically constructs candidate feature subsets during the first stage which shall be afterwards used as starting points for each ant's own candidate feature subset in the second stage. The number of cycles, the number of ants and the desired quality of the subsets are degrees of freedom of the model related to each stage. We use the same ratio r that affects the three aforementioned parameters.

Algorithm 5. TS-ACS-RST-FS

procedure TS-ACS-RST-FS( )
  Compute the population size (m) on the basis of the number of features (nf)
  Compute the quality of classification using (1) and $B = A$
  STAGE 1
  Calculate parameter values in the first stage as follows:
  $NC_{max1} = r \cdot NC_{max},\ m_1 = r \cdot m,\ \gamma_{B1}(Y) = r \cdot \gamma_B(Y)$
  Run the ACS-RST-FS approach
  CS ← output of ACS-RST-FS ▷ CS holds the candidate reducts
  STAGE 2
  Calculate parameter values in the second stage as follows:
  $NC_{max2} = NC_{max} - NC_{max1},\ m_2 = m - m_1,\ \gamma_{B2}(Y) = \gamma_B(Y)$
  Run the ACS-RST-FS approach but assign in each cycle a random subset from CS as initial subset for each ant
end procedure

For instance, suppose we are interested in carrying out 100 cycles as the overall algorithm's execution and we will use 30 ants for generating B_k subsets with the maximum possible quality of classification ($NC_{max} = 100, m = 30, \gamma_B(Y) = 1$). Setting $r = 0.3$ means that the first stage will last only 30 iterations, involving 9 ants and will settle for reducts whose quality would be 0.3 or above. Being this so, the values of these parameters during the second phase will be $NC_{max} = 70$, $m = 21$ and the algorithm will look for subsets with the maximum possible quality of classification. The workflow of activities of the TS-ACS-RST-FS proposed approach is depicted in Algorithm 5.

Of course, any other alternative ACO-based implementation can be used rather than the ACS-RST-FS algorithm. An important issue in this approach is to study which is the most suitable value for ratio r. High values of r (near to 1) cause the two-step algorithm to obtain candidate subsets close to the definition of reducts in the first stage, therefore ants in the second step swiftly find reducts but using very limited iterations and a scarce number of search agents (ants). On the contrary, if the ratio value is low, the quality of the candidate feature subsets computed during the first stage is poor yet there are more ants to work for a larger number of cycles in the second stage. We have developed an experimental study which is concerned with this tradeoff.

The following values for the ratio parameter have been proposed $r \in \{0.2, 0.3, 0.36, 0.5, 0.6, 0.8\}$ and the impact of each of these values over the number of reducts obtained, their length as well as the computational time needed to produce the output has been observed. Table 4 reports the average results achieved after 20 iterations. A synthetic repository comprised of 20 objects which are described by 16 features provides the data for conducting the experiments. The maximum number of cycles is 21.

We can see that $r = 0.3$ bears the best results. This setting implies a number of reducts similar to ACS-RST-FS but only in the 23% of the time. Similar results can be witnessed across other data sets. For instance, in Table 5 we display

Table 4. Results obtained with the proposed ACO-based approaches. The last two columns portray the average number of reducts, the average length of the reducts and the computational time (in seconds), these three indicators separated by backslash.

Algorithm	NC_{max1}	NC_{max2}	m_1	m_2	$\beta = 5, q_0 = 0.9$	$\beta = 1, q_0 = 0.3$
ACS	–	–	–	–	46.7/3.95/228	123/4.19/274
TS-ACS $r = 0.2$	4	17	3	13	32.7/4.2/82	76.3/4.2/89.9
TS-ACS $r = 0.3$	6	15	5	11	43.3/4.1/53	71.3/4.2/64
TS-ACS $r = 0.36$	8	13	6	10	38.7/3.9/39	67.3/4.1/47
TS-ACS $r = 0.5$	10	11	8	8	29.7/3.8/32	43.3/4.1/44
TS-ACS $r = 0.6$	13	8	10	6	20.33/3.8/41	37/4.2/49
TS-ACS $r = 0.8$	17	4	13	3	9/3.8/82	10.67/4.2/97

Table 5. A comparison between ACS and several configurations of the two-step ACO approach using $r = 0.3$, $NC_{max1} = 6$ and $NC_{max2} = 15$

Algorithm	m_1	m_2	$\beta = 5, q_0 = 0.9$	$\beta = 1, q_0 = 0.3$
ACS ($m = 16$)	–	–	46.7/228	123/274
TS-ACS ($m = 16$)	5	11	92%/23%	58%/23%
TS-ACS ($m' = 1.33m = 21$)	6	15	96%/31%	81%/37%
TS-ACS ($m' = 1.5m = 24$)	7	17	109%/38%	83%/42%
TS-ACS ($m' = 1.8m = 29$)	9	20	120%/52%	89%/55%
TS-ACS ($m' = 2.1m = 34$)	10	24	126%/66%	99%/69%

the results using the Breast Cancer database from UCI Repository. The result here is not surprising, since the value $r = 0.3$ provides a good balance between both stages; a higher number of ants and cycles in the second stage allows the algorithms to perform a larger exploration of the search space departing from initial subsets with an acceptable quality.

Another point worthwhile stressing is that the time complexity of TS-ACS-RST-FS is very low. In light of this, we propose a second idea: to increase the number of ants in order to bring about a greater exploration of the search space. In Table 6 the same data set than in Table 4 was used but now the population size is increased by the factors 1.33, 1.5, 1.8 and 2.1, respectively. In columns 4 and 5 the relationship between each alternative and the ACS-RST-FS benchmark algorithm is reported in terms of the amount of reducts achieved and the computational time needed. For instance, when the number of ants is $1.8m$, the TS-ACS-RST-FS approach gets 120% of reducts with respect to the number of reducts computed via ACS-RST-FS only in 52% of the time required by the

Table 6. A comparison between ACS and several configurations of the two-step ACO approach using $NC_{max1} = 6$ and $NC_{max2} = 15$

Algorithm	m_1	m_2	$\beta = 5, q_0 = 0.9$	$\beta = 1, q_0 = 0.3$
ACS ($m = 9$)	–	–	46.7/228	123/274
TS-ACS ($r = 0.2$)	2	7	60%/34%	70%/49%
TS-ACS ($r = 0.3$)	3	6	109%/31%	73%/37%
TS-ACS ($r = 0.36$)	3	6	105%/25%	77%/31%
TS-ACS ($r = 0.5$)	4	9	100%/22%	73%/26%
TS-ACS ($r = 0.6$)	5	4	65%/13%	50%/20%
TS-ACS ($r = 0.8$)	7	2	33%/27%	31%/26%
TS-ACS ($r = 0.3, m' = 1.8m = 16$)	5	11	102%/58%	98%/74%
TS-ACS ($r = 0.3, m' = 2.1m = 19$)	6	13	124%/67%	103%/83%

latter one (for $\beta = 5$ and $q_0 = 0.9$). Here we have set $r = 0.3$ because this value accomplishes the most encouraging results throughout several experimental studies. In the case of $\beta = 1$ and $q_0 = 3$, the TS-ACS-RST-FS method reached the same number of reducts (99%) but only using 69% of the CPU time than its counterpart, the ACS-RST-FS model.

Table 6 sketches a similar study using the Breast Cancer database. These results are very interesting because the two-step ACO approach enables us to obtain the same or an even greater number of reducts in less time than ACS-RST-FS, hence the feasibility of splitting the search process is empirically confirmed once again.

6 Dynamic Mesh Optimization in Feature Selection

We want to elaborate now on a novel optimization technique called "Dynamic Mesh Optimization" (DMO) [5] which follows some patterns already present in earlier evolutionary approaches but provides a unique framework for managing both discrete and continuous optimization problems.

The essentials behind the DMO method is the creation of a mesh of points in the multi-dimensional space wherein the optimization of the objective function is being carried out. The mesh endures an expansion process toward the most promising regions of the search space but, at the same time, becomes finer in those areas where there exist points that constitute local ends of the function. The dynamic nature of the mesh is given by the fact that its size (number of nodes) and configuration both change over time. When it comes to the feature selection problem, nodes can be visualized as binary vectors $\mathbf{n} = (n_1, n_2, \ldots, n_N)$ of N components, one per attribute, with the component $n_i = 1$ if the i-th attribute is being considered as part of the solution or zero otherwise. This is the same representation adopted in the previously discussed evolutionary approaches.

At every cycle, the mesh is created with an initial number of nodes. Subsequently, new nodes are generated until an upper boundary in the number of nodes is reached. The mesh at the next cycle is comprised of the fittest nodes of the mesh in the current iteration. Along the search process, the node carrying the best value of the objective (evaluation) function so far is recorded, so n_g denotes the global end attained up to now by the search algorithm.

In the case of the feature selection problem, the evaluation (fitness) function for the DMO meta-heuristic is expression (11), which attempts to achieve a tradeoff between the classificatory ability of a reduct and its length.

The dynamic nature of our proposal manifests in the generation of (i) the initial mesh; (ii) intermediate nodes oriented toward the local optima; (iii) intermediate nodes in the direction of the global optimum and (iv) nodes aiming at expanding the dimensions of the current mesh.

The model gives rise to the following parameters: (i) $N_i \rightarrow$ size of the initial mesh, (ii) $N \rightarrow$ maximum size of the mesh across each cycle ($N_i < N$) and (iii) $M \rightarrow$ number of cycles.

The DMO method is defined in the following manner:

STEP 1. Generate the initial mesh for each cycle: At the beginning of the algorithm's execution, the initial mesh will be made up of N_i randomly generated nodes while in the remaining iterations, the initial mesh is built upon the selection of the best (in terms of evaluation measure) N_i nodes of the mesh in the preceding cycle.

STEP 2. Node generation toward local optima: The aim of this step is to come up with new nodes laid in the direction of the local optima found by the algorithm.

For each node **n**, its K-nearest neighbor nodes are computed (the Hamming distance is a suitable option for the FSP). If none of the neighbors surpasses **n** in fitness function value, then **n** is said to be a local optimum and no nodes are begotten out of it in this step. Conversely, suppose that node **ne** is "better" than **n** and the rest of its neighbors. In this case, a new node arises somewhere between **n** and **ne**.

The proximity of the newly generated node $\mathbf{n}^*$ to the current node **n** or to the local optimum **ne** is contingent upon a factor r which is calculated based on the fitness function values at both nodes **n** and **ne**. Each component of $\mathbf{n}^*$ takes either the value of n_i or ne_i according to a rule involving a stochastic component. The threshold r determining how every component n_i^* must be fixed is calculated as in (19).

$$r = 1 - 0.5\,\frac{\text{Eval}(n)}{\text{Eval}(ne)} \tag{19}$$

$f(\mathbf{n}, \mathbf{ne}, r)$: For each component n_i: If random() $< r$ then $n_i^* = ne_i$ otherwise $n_i^* = n_i$

Algorithm 6. The DMO meta-heuristic

```
procedure D(M)O
    Randomly generate N_i nodes to build the initial mesh
    Evaluate all the mesh nodes
    repeat
        for each node n in the mesh do
            Find its K-nearest neighbors
            n_best ← the best of its neighbors
            if n_best is better than n then
                Generate a new node by using function f
            end if
        end for
        for each initial node in the current mesh do
            Generate a new node by using function g
        end for
        repeat
            Select the most outward node of the mesh
            Generate a new node by using function h
        until MeshSize = N
        Select the best N_i nodes of the current mesh and set up the next mesh
    until CurrentIteration = M
end procedure
```

Notice from (19) that the lower the ratio between Eval(n) and Eval(ne), the more likely it is that n_i^* takes the value of the i-th component of the local optimum.

STEP 3. Node generation toward global optimum: Here the idea is the same as in the previous step but now r is computed differently and a new function g is introduced. Needless to say that **ng** represents the global optimum found thus far by the algorithm.

$$r = 1 - 0.5\,\frac{\text{Eval}(n)}{\text{Eval}(ng)} \tag{20}$$

$g(\mathbf{n}, \mathbf{ng}, r)$: For each component n_i: If random() $< r$ then $n_i^* = ng_i$ otherwise $n_i^* = n_i$

STEP 4. Mesh expansion: In this step, the mesh is stretched from its outer nodes using function h, i.e. using nodes located at the boundary of the initial mesh in each cycle. The weight w depicted in (13) assures that the expansion declines along the search process (i.e., a bigger expansion is achieved at the early cycles and it fades out as the algorithm progresses). To determine which nodes lie in the outskirts of the mesh, we turn to the norm of a vector. Those nodes exhibiting the lowest and greatest norm values are picked. Remark that, in this step, as many outer nodes as needed are selected so as to fill out the maximum mesh size N. The rules regulating this sort of node generation can be found next:

For each node **nl** in the lower boundary (those with the lowest norm):
$h(\mathbf{nl}, w)$: For each component n_i: If random() $< w$ then $n_i^* = 0$ otherwise $n_i^* = nl_i$

For each node **nu** in the upper boundary (those with the greatest norm):
$h(\mathbf{nu}, w)$: For each component n_i: If random() $< w$ then $n_i^* = 1$ otherwise $n_i^* = nu_i$

In the context of feature selection, the norm of a node (vector) is the number of components set to one. Algorithm 6 outlines the workflow of the DMO approach. It is also worth remarking that no direct search algorithm guarantees to find the global optimum no matter how refined the heuristic search might be.

7 A Comparative Study

The conducted experimentation embraces a comparison between DMO and existing ACO- and PSO-based approaches. The chosen criteria were the number and length of the reducts found as well as the computational time required by every method.

Concerning ACO, the Ant Colony System (ACS) model was picked for benchmarking following the advice in [1] and [2], for it reported the most encouraging outcomes. As to the parameter setting, we stuck to the guidelines provided in the aforesaid studies, i.e. $\beta = 5, q_0 = 0.9, NC_{max} = 21$ and the population size (number of ants) depending on the number of features as in the previously enunciated rules.

Regarding the TS-ACS-RST-FS approach, the value of the ratio r used for determining the number of ants, number of cycles and threshold of the quality of the classification in each stage was set to 0.3 whereas the number of ants m is increased 2.1 times, i.e. $m' = 2.1m$

Moving on to the PSO-driven approaches' configuration, each individual was shaped as a binary vector whose length matches the number of attributes in the system. The parameters associated with the PSO-RST-FS were fixed as $c_1 = c_2 = 2, maxCycles = 120$ and swarmSize $= 21$. The inertia weight w keeps its dynamic character as reflected in (13). As to the TS-PSO-RST-FS method, the factor used to calculate the quality of the classification in the first stage ($ratio_Q$) takes 0.75 while the parameter involved in the computation of the number of cycles ($ratio_C$) for each phase was set to 0.3.

The configuration of the DMO-RST-FS (DMO + RST to feature selection) has been defined as follows: a mesh with 30 nodes is used, 9 of them regarded as initial nodes (which means that it is necessary to generate 21 nodes per cycle, just the same number of particles than in the PSO-based models) and the computations lasted for 90 iterations.

Table 7 reports the experimental results obtained after applying the above methods over the Breast Cancer, Heart and Dermatology data sets coming from the UCI Repository. Each table entry holds the average number of reducts found, the average length (number of attributes) of the reducts in addition to the length

Table 7. Quality of the reducts found by different evolutionary algorithms. First datum is the average number of reducts found, followed by their average length, the length of the shortest reduct and, finally, the percentage of times a reduct of the same length was found throughout the different executions of the algorithm.

Method	BreastCancer	Heart	Dermatology
DMO-RST-FS	18.3/5.1/4,100%	14.8/8.29/6, 83%	179.5/20.9/9,50%
TS-PSO-RST-FS	11/4.6/4,100%	3/6.8/6,100%	39.3/12.6/9, 50%
PSO-RST-FS	14/4.95/4,100%	6/7.97/6,67%	78.2/15.3/9, 50%
TS-ACS-RST-FS	12.7/4.74/4,100%	7/7/6,67%	249/13/9,33%
ACS-RST-FS	11.75/4.94/4,100%	14.3/7.53/6,100%	300/14.17/10,66%

Table 8. Average number of evaluations of the fitness function in each algorithm

Algorithm	Avg. number of times
DMO-RST-FS	2530
TS-PSO-RST-FS	2542
PSO-RST-FS	2968
TS-ACS-RST-FS	17222
ACS-RST-FS	13487

of the shortest reduct and the number of times it was found with regards to the number of runs performed by the algorithm. Every algorithm was executed six times per data set. From the information in Table 7 we notice, for instance, that the DMO-RST-FS algorithm discovered 18.3 reducts on average for the Breast Cancer data set, the reducts having average length of 5.1 and the shortest reduct found is composed of four attributes, having a reduct of such length always (100%) been found throughout the different runs of the algorithm.

From the outlook of the computational cost, one may notice that the DMO-RST-FS, TS-PSO-RST-FS and PSO-RST-FS algorithms have a very similar performance. This is clearly understood if we keep in mind that the greater the number of times expression (1) is computed, the more time-consuming the algorithm turns into. While PSO-based and DMO approaches compute this indicator roughly $P \times Q$ times (P being the number of cycles and Q the number of agents engaged in the search process, viz particles in PSO and nodes in DMO), the ACO-based models evaluate this function a far greater number of times, i.e. roughly $P \times Q \times k$ (Q being the number of ants and k the average length of the reducts found, since every time an ant adds a node to the solution, it must evaluate all possible alternatives at hand, namely, all attributes still not considered so far).

Regarding the average amount of times the fitness function was calculated by all approaches under discussion, Table 8 presents the corresponding magnitudes.

8 Conclusions

An study on the performance of several evolutionary techniques for tackling the feature selection problem has been outlined. The common denominator has been the pivotal role played by rough set theory in assessing the quality of a feature subset as a prospective reduct of the system under consideration. Therefore this criterion has been successfully incorporated to the fitness function of all the studied algorithms and the preliminary results allow to confirm the feasibility and efficiency of this sort of techniques for attribute reduction.

Moreover, the introduction of the two-step search paradigm for the swarm intelligence methods translated into a substantial reduction of the computational time needed to find the reducts of the information system.

Under empirical evidence we can also conclude that the Dynamic Mesh Optimization approach explores the search space in a similar way to the algorithms based on the Ant Colony Optimization model but with a computational cost very close to that of Particle Swarm Optimization.

Acknowledgment

The authors would like to thanks VLIR (Vlaamse InterUniversitaire Raad, Flemish Interuniversity Council, Belgium) for supporting this work under the IUC Program VLIR-UCLV.

References

1. Bello, R., Nowe, A.: Using ACO and rough set theory to feature selection. In: WSEAS Trans. on Information Science and Applications, pp. 512–517 (2005)
2. Bello, R., Nowe, A.: A model based on ant colony system and rough set theory to feature selection. In: Proc. of Genetic and Evolutionary Computation Conference (GECCO 2005), pp. 275–276 (2005)
3. Bello, R., Puris, A., Nowe, A., Martínez, Y., García, M.: Two step ant colony system to solve the feature selection problem. In: Martínez-Trinidad, J.F., Carrasco Ochoa, J.A., Kittler, J. (eds.) CIARP 2006. LNCS, vol. 4225, pp. 588–596. Springer, Heidelberg (2006)
4. Bello, R., Gómez, Y., Nowé, A., García, M.: Two-step particle swarm optimization to solve the feature selection problem. In: Proc. of 7th Int'l Conf. on Intelligent Systems Design and Applications, pp. 691–696. IEEE Computer Society, Washington (2007)
5. Bello, R., Puris, A., Falcón, R.: Feature selection through dynamic mesh optimization. In: Ruiz-Shulcloper, J., Kropatsch, W.G. (eds.) CIARP 2008. LNCS, vol. 5197. Springer, Heidelberg (2008)
6. Blake, C.L., Merz, C.J.: UCI repository of machine learning databases (1998), http://www.ics.uci.edu/~mlearn/MLRepository.html

7. Caballero, Y., Bello, R.: Two new feature selection algorithms with rough sets theory. In: Bramer, M. (ed.) Artificial Intelligence in Theory and Practice. Springer, Boston (2006)
8. Dash, M., Liu, H.: Consistency-based search in feature selection. Artificial Intelligence 151, 155–176 (2003)
9. Dorigo, M., Stützle, T.: Ant colony optimization. Cambridge University Press, New York (2005)
10. Goldberg, D.E.: Genetic algorithms in search, optimization and machine learning. Addison-Wesley Longman Publishing Co. (1989)
11. Jensen, R., Shen, Q.: Finding rough set reducts with ant colony optimization. In: Proceedings of 6th Annual UK Workshop on Computational Intelligence, pp. 15–22 (2003)
12. Kennedy, J., Eberhart, R.C.: Particle swarm optimization. In: Proc. of IEEE Int'l. Conf. on Neural Networks, pp. 1942–1948. IEEE Service Center, Piscataway (1995)
13. Kennedy, J., Eberhart, R.C.: A discrete binary version of the particle swarm optimization algorithm. In: IEEE International Conference on Neural Networks, pp. 4104–4108 (1997)
14. Komorowski, J., Polkowski, L., Skowron, A.: Rough sets: a tutorial. In: Pal, S., Skowron, A. (eds.) Rough-Fuzzy Hybridization: A New Trend in Decision-Making. Springer, New York (1999)
15. Kudo, M., Sklansky, J.: Comparison of algorithms that select features for pattern classifiers. Pattern Recognition 33, 25–41 (2000)
16. Larrañaga, P., Etxeberria, R., Lozano, J.A., Peña, J.M.: Optimization in continuous domains by learning and simulation of Gaussian networks. In: Wu, A.S. (ed.) Proceedings of the 2000 Genetic and Evolutionary Computation Conference, pp. 201–204 (2000)
17. Mitchell, T.M.: Machine learning. McGraw Hill, New York (1997)
18. Mühlenbein, H., Mahnig, T., Ochoa, A.: Schemata, distributions and graphical models on evolutionary optimization. Journal of Heuristics 5(2), 215–247 (1999)
19. Pawlak, Z.: Rough sets. International Journal of Information & Computer Sciences 11, 341–356 (1982)
20. Piñero, P., Arco, L., García, M., Caballero, Y.: Two new metrics for feature selection in pattern recognition. In: Sanfeliu, A., Ruiz-Shulcloper, J. (eds.) CIARP 2003. LNCS, vol. 2905, pp. 488–497. Springer, Heidelberg (2003)
21. Polkowski, L.: Rough sets: mathematical foundations. Physica-Verlag, Berlin (2002)
22. Stefanowski, J.: An experimental evaluation of improving rule-based classifiers with two approaches that change representations of learning examples. Engineering Applications of Artificial Intelligence 17, 439–445 (2004)
23. Tay, F.E., Shen, L.: Economic and financial prediction using rough set model. European Journal of Operational Research 141, 641–659 (2002)
24. Wang, X., Yang, Y., Peng, N., Teng, X.: Finding minimal rough set reducts with particle swarm optimization. In: Ślęzak, D., Wang, G., Szczuka, M.S., Düntsch, I., Yao, Y. (eds.) RSFDGrC 2005. LNCS, vol. 3641, pp. 451–460. Springer, Heidelberg (2005)
25. Wang, X., Yang, J., Jensen, R., Liu, X.: Rough set feature selection and rule induction for prediction of malignancy degree in brain glioma. Computer Methods and Programs in Biomedicine 83, 147–156 (2006)
26. Wang, X., Yang, J., Teng, X., Xia, W., Jensen, R.: Feature selection based on rough sets and particle swarm optimization. Pattern Recognition Letters 28, 459–471 (2007)

27. Wroblewski, J.: Finding minimal reducts using genetic algorithms. In: Wang, P.P. (ed.) Proceedings of the 2nd Annual Joint Conference on Information Sciences, pp. 186–189 (1995)
28. Xing, H., Xu, L.: Feature space theory -a mathematical foundation for data mining. Knowledge-based systems 14, 253–257 (2001)
29. Xu, Z.B., Liang, J., Dang, C., Chin, K.S.: Inclusion degree: a perspective on measures for rough set data analysis. Information Sciences 141, 227–236 (2002)
30. Yuan, S., Chu, F.L.: Fault diagnostics based on particle swarm optimization and support vector machines. Mechanical Systems and Signal Processing 21, 1787–1798 (2007)

Nature Inspired Population-Based Heuristics for Rough Set Reduction

Hongbo Liu[1], Ajith Abraham[2], and Yanheng Li[1]

[1] School of Computer Science and Engineering, Dalian Maritime University, 116026 Dalian, China
lhb@dlut.edu.cn
http://hongboliu.torrent.googlepages.com

[2] Centre for Quantifiable Quality of Service in Communication Systems, Norwegian University of Science and Technology, Trondheim, Norway
ajith.abraham@ieee.org
http://www.softcomputing.net

Summary. Finding reducts is one of the key problems in the increasing applications of rough set theory, which is also one of the bottlenecks of the rough set methodology. The population-based reduction approaches are attractive to find multiple reducts in the decision systems. In this chapter, we introduce two nature inspired population-based computational optimization techniques, Particle Swarm Optimization (PSO) and Genetic Algorithm (GA) for rough set reduction. Particle Swarm Optimization (PSO) is particularly attractive for the challenging problem as a new heuristic algorithm. The approach discover the best feature combinations in an efficient way to observe the change of positive region as the particles proceed throughout the search space. We evaluated the performance of the two algorithms using some benchmark datasets and the corresponding computational experiments are discussed. Empirical results indicate that both methods are ideal for all the considered problems and particle swarm optimization technique outperformed the genetic algorithm approach by obtaining more number of reducts for the datasets. We also illustrate a real world application in fMRI data analysis, which is helpful for cognition research.

1 Introduction

Rough set theory [1, 2, 3] provides a mathematical tool that can be used for both feature selection and knowledge discovery. It helps us to find out the minimal attribute sets called '*reducts*' to classify objects without deterioration of classification quality and induce minimal length decision rules inherent in a given information system. The idea of reducts has encouraged many researchers in studying the effectiveness of rough set theory in a number of real world domains, including medicine, pharmacology, control systems, fault-diagnosis, text categorization, social sciences, switching circuits, economic/financial prediction, image processing, and so on [4, 5, 6, 7, 8, 9, 10].

Usually real world objects are the corresponding tuple in some decision tables. They store a huge quantity of data, which is hard to manage from a computational point of view. Finding reducts in a large information system is still an

A. Abraham, R. Falcón, and R. Bello (Eds.): Rough Set Theory, SCI 174, pp. 261–278.
springerlink.com

NP-hard problem [11]. The high complexity of this problem has motivated investigators to apply various approximation techniques to find near-optimal solutions. Many approaches have been proposed for finding reducts, e.g., discernibility matrices, dynamic reducts, and others [12, 13]. The heuristic algorithm is a better choice. Hu et al. [14] proposed a heuristic algorithm using discernibility matrix. The approach provided a weighting mechanism to rank attributes. Zhong and Dong [15] presented a wrapper approach using rough sets theory with greedy heuristics for feature subset selection. The aim of feature subset selection is to find out a minimum set of relevant attributes that describe the dataset as well as the original all attributes do. So finding reduct is similar to feature selection. Zhong's algorithm employed the number of consistent instances as heuristics. Banerjee et al. [16] presented various attempts of using Genetic Algorithms in order to obtain reducts. Although several variants of reduct algorithms are reported in the literature, at the moment, there is no accredited best heuristic reduct algorithm. So far, it is still an open research area in rough sets theory.

Conventional approaches for knowledge discovery always try to find a good reduct or to select a set of features [17]. In the knowledge discovery applications, only the good reduct can be applied to represent knowledge, which is called a single body of knowledge. In fact, many information systems in the real world have multiple reducts, and each reduct can be applied to generate a single body of knowledge. Therefore, multi-knowledge based on multiple reducts has the potential to improve knowledge representation and decision accuracy [18]. However, it would be exceedingly time-consuming to find multiple reducts in an instance information system with larger numbers of attributes and instances. In most of strategies, different reducts are obtained by changing the order of condition attributes and calculating the significance of different condition attribute combinations against decision attribute(s). It is a complex multi-restart processing about condition attribute increasing or decreasing in quantity. Population-based search approaches are of great benefits in the multiple reduction problems, because different individual trends to be encoded to different reduct. So it is attractive to find multiple reducts in the decision systems.

Particle swarm algorithm is inspired by social behavior patterns of organisms that live and interact within large groups. In particular, it incorporates swarming behaviors observed in flocks of birds, schools of fish, or swarms of bees, and even human social behavior, from which the Swarm Intelligence (SI) paradigm has emerged [19]. The swarm intelligent model helps to find optimal regions of complex search spaces through interaction of individuals in a population of particles [20, 21, 22]. As an algorithm, its main strength is its fast convergence, which compares favorably with many other global optimization algorithms [23, 24]. It has exhibited good performance across a wide range of applications [25, 26, 27, 28, 29]. The particle swarm algorithm is particularly attractive for feature selection as there seems to be no heuristic that can guide search to the optimal minimal feature subset. Additionally, it can be the case that particles discover the best feature combinations as they proceed throughout the search space.

The main focus of this chapter is to introduce how particle swarm optimization algorithm may be applied for the difficult problem of finding multiple reducts. The rest of the chapter is organized as follows. Some related terms and theorems on rough set theory are explained briefly in Sect. 2. The proposed approach based on particle swarm algorithm is presented in Sect. 3. In Sect. 4, experiment results and discussions are provided in detail. In Sect. 5, we illustrate an application in fMRI data analysis. Finally conclusions are made in Sect. 6.

2 Rough Set Reduction

The basic concepts of rough set theory and its philosophy are presented and illustrated with examples in [1, 2, 3, 15, 30, 31, 17]. Here, we illustrate only the relevant basic ideas of rough sets that are relevant to the present work.

In rough set theory, an information system is denoted in 4-tuple by $S = (U, A, V, f)$, where U is the universe of discourse, a non-empty finite set of N objects $\{x_1, x_2, \cdots, x_N\}$. A is a non-empty finite set of attributes such that $a : U \to V_a$ for every $a \in A$ (V_a is the value set of the attribute a).

$$V = \bigcup_{a \in A} V_a$$

$f : U \times A \to V$ is the total decision function (also called the information function) such that $f(x, a) \in V_a$ for every $a \in A$, $x \in U$. The information system can also be defined as a decision table by $S = (U, C, D, V, f)$. For the decision table, C and D are two subsets of attributes. $A = \{C \cup D\}$, $C \cap D = \emptyset$, where C is the set of input features and D is the set of class indices. They are also called condition and decision attributes, respectively.

Let $a \in C \cup D$, $P \subseteq C \cup D$. A binary relation $IND(P)$, called an equivalence (indiscernibility) relation, is defined as follows:

$$IND(P) = \{(x, y) \in U \times U \quad | \quad \forall a \in P, f(x, a) = f(y, a)\} \tag{1}$$

The equivalence relation $IND(P)$ partitions the set U into disjoint subsets. Let $U/IND(P)$ denote the family of all equivalence classes of the relation $IND(P)$. For simplicity of notation, U/P will be written instead of $U/IND(P)$. Such a partition of the universe is denoted by $U/P = \{P_1, P_2, \cdots, P_i, \cdots\}$, where P_i is an equivalence class of P, which is denoted $[x_i]_P$. Equivalence classes U/C and U/D will be called condition and decision classes, respectively.

Lower Approximation: Given a decision table $T = (U, C, D, V, f)$. Let $R \subseteq C \cup D$, $X \subseteq U$ and $U/R = \{R_1, R_2, \cdots, R_i, \cdots\}$. The R-lower approximation set of X is the set of all elements of U which can be with certainty classified as elements of X, assuming knowledge R. It can be presented formally as

$$APR_R^-(X) = \bigcup\{R_i \quad | \quad R_i \in U/R, R_i \subseteq X\} \tag{2}$$

Positive Region: Given a decision table $T = (U, C, D, V, f)$. Let $B \subseteq C$, $U/D = \{D_1, D_2, \cdots, D_i, \cdots\}$ and $U/B = \{B_1, B_2, \cdots, B_i, \cdots\}$. The B-positive region of D is the set of all objects from the universe U which can be classified with certainty to classes of U/D employing features from B, i.e.,

$$POS_B(D) = \bigcup_{D_i \in U/D} APR_{\overline{B}}(D_i) \tag{3}$$

Positive Region: Given a decision table $T = (U, C, D, V, f)$. Let $B \subseteq C$, $U/D = \{D_1, D_2, \cdots, D_i, \cdots\}$ and $U/B = \{B_1, B_2, \cdots, B_i, \cdots\}$. The B-positive region of D is the set of all objects from the universe U which can be classified with certainty to classes of U/D employing features from B, i.e.,

$$POS_B(D) = \bigcup_{D_i \in U/D} B_{-}(D_i) \tag{4}$$

Reduct: Given a decision table $T = (U, C, D, V, f)$. The attribute $a \in B \subseteq C$ is $D - dispensable$ in B, if $POS_B(D) = POS_{(B-\{a\})}(D)$; otherwise the attribute a is $D - indispensable$ in B. If all attributes $a \in B$ are $D - indispensable$ in B, then B will be called $D - independent$. A subset of attributes $B \subseteq C$ is a $D - reduct$ of C, iff $POS_B(D) = POS_C(D)$ and B is $D - independent$. It means that a reduct is the minimal subset of attributes that enables the same classification of elements of the universe as the whole set of attributes. In other words, attributes that do not belong to a reduct are superfluous with regard to classification of elements of the universe. Usually, there are many reducts in an instance information system. Let 2^A represent all possible attribute subsets $\{\{a_1\}, \cdots, \{a_{|A|}\}, \{a_1, a_2\}, \cdots, \{a_1, \cdots, a_{|A|}\}\}$. Let RED represent the set of reducts, i.e.,

$$RED = \{B \mid POS_B(D) = POS_C(D), POS_{(B-\{a\})}(D) < POS_B(D)\} \tag{5}$$

Multi-knowledge: Given a decision table $T = (U, C, D, V, f)$. Let RED represent the set of reducts, Let φ is a mapping from the condition space to the decision space. Then multi-knowledge can be defined as follows:

$$\Psi = \{\varphi_B \mid B \in RED\} \tag{6}$$

Reduced Positive Universe and *Reduced Positive Region*: Given a decision table $T = (U, C, D, V, f)$. Let $U/C = \{[u_1^{'}]_C, [u_2^{'}]_C, \cdots, [u_m^{'}]_C\}$, Reduced Positive Universe $U^{'}$ can be written as:

$$U^{'} = \{u_1^{'}, u_2^{'}, \cdots, u_m^{'}\}. \tag{7}$$

and

$$POS_C(D) = [u_{i_1}^{'}]_C \cup [u_{i_2}^{'}]_C \cup \cdots \cup [u_{i_t}^{'}]_C. \tag{8}$$

Where $\forall u_{i_s}^{'} \in U^{'}$ and $|[u_{i_s}^{'}]_C/D| = 1 (s = 1, 2, \cdots, t)$. Reduced positive universe can be written as:

$$U_{pos}^{'} = \{u_{i_1}^{'}, u_{i_2}^{'}, \cdots, u_{i_t}^{'}\}. \tag{9}$$

and $\forall B \subseteq C$, reduced positive region

$$POS'_B(D) = \bigcup_{X \in U'/B \wedge X \subseteq U'_{pos} \wedge |X/D|=1} X \tag{10}$$

where $|X/D|$ represents the cardinality of the set X/D. $\forall B \subseteq C$, $POS_B(D) = POS_C(D)$ if $POS'_B = U'_{pos}$ [31]. It is to be noted that U' is the reduced universe, which usually would reduce significantly the scale of datasets. It provides a more efficient method to observe the change of positive region when we search the reducts. We didn't have to calculate U/C, U/D, U/B, $POS_C(D)$, $POS_B(D)$ and then compare $POS_B(D)$ with $POS_C(D)$ to determine whether they are equal to each other or not. We only calculate U/C, U', U'_{pos}, POS'_B and then compare POS'_B with U'_{pos}.

3 Nature Inspired Heuristics for Reduction

Combinatorial optimization problems are important in many real life applications and recently, the area has attracted much research with the advances in nature inspired heuristics and multi-agent systems.

3.1 Particle Swarm Optimization for Reduction

Given a decision table $T = (U, C, D, V, f)$, the set of condition attributes, C, consist of m attributes. We set up a search space of m dimension for the reduction problem. Accordingly, each particle's position is represented as a binary bit string of length m. Each dimension of the particle's position maps one condition attribute. The domain for each dimension is limited to 0 or 1. The value '1' means the corresponding attribute is selected while '0' not selected. Each position can be "decoded" to a potential reduction solution, an subset of C. The particle's position is a series of priority levels of the attributes. The sequence of the attribute will not be changed during the iteration. But after updating the velocity and position of the particles, the particle's position may appear real values such as 0.4, etc. It is meaningless for the reduction. Therefore, we introduce a discrete particle swarm optimization for this combinatorial problem.

During the search procedure, each individual is evaluated using the fitness. According to the definition of rough set reduct, the reduction solution must ensure the decision ability is the same as the primary decision table and the number of attributes in the feasible solution is kept as low as possible. In our algorithm, we first evaluate whether the potential reduction solution satisfies $POS'_E = U'_{pos}$ or not (E is the subset of attributes represented by the potential reduction solution). If it is a feasible solution, we calculate the number of '1' in it. The solution with the lowest number of '1' would be selected. For the particle swarm, the lower number of '1' in its position, the better the fitness of the individual is. $POS'_E = U'_{pos}$ is used as the criterion of the solution validity.

As a summary, the particle swarm model consists of a swarm of particles, which are initialized with a population of random candidate solutions. They move iteratively through the d-dimension problem space to search the new solutions, where the fitness f can be measured by calculating the number of condition attributes in the potential reduction solution. Each particle has a position represented by a position-vector $\mathbf{p}_i$ (i is the index of the particle), and a velocity represented by a velocity-vector $\mathbf{v}_i$. Each particle remembers its own best position so far in a vector $\mathbf{p}_i^{\#}$, and its j-th dimensional value is $p_{ij}^{\#}$. The best position-vector among the swarm so far is then stored in a vector $\mathbf{p}^*$, and its j-th dimensional value is p_j^*. When the particle moves in a state space restricted to zero and one on each dimension, the change of probability with time steps is defined as follows:

$$P(p_{ij}(t)=1) = f(p_{ij}(t-1), v_{ij}(t-1), p_{ij}^{\#}(t-1), p_j^*(t-1)). \tag{11}$$

where the probability function is

$$sig(v_{ij}(t)) = \frac{1}{1+e^{-v_{ij}(t)}}. \tag{12}$$

At each time step, each particle updates its velocity and moves to a new position according to Eqs.(13) and (14):

$$v_{ij}(t) = wv_{ij}(t-1)+\phi_1 r_1(p_{ij}^{\#}(t-1)-p_{ij}(t-1))+\phi_2 r_2(p_j^*(t-1)-p_{ij}(t-1)). \tag{13}$$

$$p_{ij}(t) = \begin{cases} 1 & \text{if } \rho < sig(v_{ij}(t)); \\ 0 & \text{otherwise.} \end{cases} \tag{14}$$

Algorithm 1. A Rough Set Reduct Algorithm Based on Particle Swarm Optimization

Calculate U', U'_{pos} using Eqs.(7) and (9)
Initialize the size of the particle swarm n, and other parameters
Initialize the positions and the velocities for all the particles randomly
while the stop criterion is not met **do**
  $t \leftarrow t+1$
  Calculate the fitness value of each particle
  if $POS'_E \neq U'_{pos}$ **then**
    the fitness is punished as the total number of the condition attributes
  else
    the fitness is the number of '1' in the position
  end if
  $\mathbf{p}^* = argmin_{i=1}^{n}(f(\mathbf{p}^*(t-1)), f(\mathbf{p}_1(t)), f(\mathbf{p}_2(t)), \cdots, f(\mathbf{p}_i(t)), \cdots, f(\mathbf{p}_n(t)))$
  for $i = 1$ **to** n **do**
    $\mathbf{p}_i^{\#}(t) = argmin_{i=1}^{n}(f(\mathbf{p}_i^{\#}(t-1)), f(\mathbf{p}_i(t))$
    for $j = 1$ **to** d **do**
      Update the j-th dimension value of $\mathbf{p}_i$ and $\mathbf{v}_i$
      according to Eqs.(13) and (14)
    end for
  end for
end while

Where ϕ_1 is a positive constant, called as coefficient of the self-recognition component, ϕ_2 is a positive constant, called as coefficient of the social component. r_1 and r_2 are the random numbers in the interval [0,1]. The variable w is called as the inertia factor, which value is typically setup to vary linearly from 1 to near 0 during the iterated processing. ρ is random number in the closed interval [0, 1]. From Eq.(13), a particle decides where to move next, considering its current state, its own experience, which is the memory of its best past position, and the experience of its most successful particle in the swarm. The pseudo-code for the particle swarm search method is illustrated in Algorithm 1.

3.2 Genetic Algorithms for Reduction

In nature, evolution is mostly determined by natural selection, where individuals that are better are more likely to survive and propagate their genetic material. The encoding of genetic information (genome) is done in a way that admits asexual reproduction, which results in offspring's that are genetically identical to the parent. Sexual reproduction allows some exchange and re-ordering of chromosomes, producing offspring that contain a combination of information from each parent. This is the recombination operation, which is often referred to as crossover because of the way strands of chromosomes crossover during the exchange. Diversity in the population is achieved by mutation. A typical evolutionary (genetic) algorithm procedure takes the following steps: A population of candidate solutions (for the optimization task to be solved) is initialized. New solutions are created by applying genetic operators (mutation and/or crossover).

Algorithm 2. A Rough Set Reduct Algorithm Based on Genetic Algorithm

Calculate U', U'_{pos} using Eqs.(7) and (9)
Initialize the population randomly, and other parameters
while the stop criterion is not met **do**
  Evaluate the fitness of each individual in the population
  if $POS'_E \neq U'_{pos}$ **then**
    the fitness is punished as the total number of the condition attributes
  else
    the fitness is the number of '1' in the position
  end if
  Select best-ranking individuals to reproduce
  Breed new generation through crossover operator and give birth to offspring
  Breed new generation through mutation operator and give birth to offspring
  Evaluate the individual fitnesses of the offspring
  if $POS'_E \neq U'_{pos}$ **then**
    the fitness is punished as the total number of the condition attributes
  else
    the fitness is the number of '1' in the position
  end if
  Replace worst ranked part of population with offspring
end while

The fitness (how good the solutions are) of the resulting solutions are evaluated and suitable selection strategy is then applied to determine which solutions will be maintained into the next generation. The procedure is then iterated [38]. The pseudo-code for the genetic algorithm search method is illustrated in Algorithm 2.

4 Experiments Using Some Benchmark Problems

For all experiments, Genetic algorithm (GA) was used to compare the performance with PSO. The two algorithms share many similarities [33, 34]. Both methods are valid and efficient methods in numeric programming and have been employed in various fields due to their strong convergence properties. Specific parameter settings for the algorithms are described in Table 1, where D is the dimension of the position, i.e., the number of condition attributes. Besides the first small scale rough set reduction problem shown in Table 2, the maximum number of iterations is $(int)(0.1 * recnum + 10 * (nfields - 1))$ in each trial, where $recnum$ is the number of records/rows and $nfields - 1$ is the number of condition attributes. Each experiment (for each algorithm) was repeated 3 times with different random seeds. If the standard deviation is larger than 20%, the times of trials were set to larger, 10 or 20.

To analyze the effectiveness and performance of the considered algorithms, first we tested a small scale rough set reduction problem shown in Table 2. In the experiment, the maximum number of iterations was fixed as 10. Each experiment was repeated 3 times with different random seeds. The results (the

Table 1. Parameter settings for the algorithms

Algorithm	*ParameterName*	*Value*
GA		
	size of the population	$(int)(10 + 2 * sqrt(D))$
	Probability of crossover	0.8
	Probability of mutation	0.08
PSO		
	Swarm size	$(int)(10 + 2 * sqrt(D))$
	Self coefficient ϕ_1	1.49
	Social coefficient ϕ_2	1.49
	Inertia weight w	$0.9 \rightarrow 0.1$
	Clamping Coefficient ρ	0.5

Table 2. A decision table

Instance	c_1	c_2	c_3	c_4	d
x_1	1	1	1	1	0
x_2	2	2	2	1	1
x_3	1	1	1	1	0
x_4	2	3	2	3	0
x_5	2	2	2	1	1
x_6	3	1	2	1	0
x_7	1	2	3	2	2
x_8	2	3	1	2	3
x_9	3	1	2	1	1
x_{10}	1	2	3	2	2
x_{11}	3	1	2	1	1
x_{12}	2	3	1	2	3
x_{13}	4	3	4	2	1
x_{14}	1	2	3	2	3
x_{15}	4	3	4	2	2

number of reduced attributes) for 3 GA runs were all 2. The results of 3 PSO runs were also all 2. The optimal result is supposed to be 2. But the reduction result for 3 GA runs is $\{2, 3\}$ while the reduction results for 3 PSO runs are $\{1, 4\}$ and $\{2, 3\}$. Table 3 depicts the reducts for Table 2. Figure 1 shows the performance of the algorithms for Table 2. For the small scale rough set reduction problem, GA has faster convergence than PSO. There seems like a conflict between the instances 13 and 15. It depends on conflict analysis and how to explain the obtained knowledge, which is beyond the scope of this chapter.

Further we consider the datasets in Table 4 from AFS[1], AiLab[2] and UCI[3]. Figures 2, 3, 4 and 5 illustrate the performance of the algorithms for lung-cancer, lymphography and mofn-3-7-10 datasets, respectively. For lung-cancer dataset, the results (the number of reduced attributes) for 3 GA runs were 10: {1, 3, 9, 12, 33, 41, 44, 47, 54, 56} (The number before the colon is the number of condition attributes, the numbers in brackets are attribute index, which represents a reduction solution). The results of 3 PSO runs were 9: { 3, 8, 9,

[1] http://sra.itc.it/research/afs/
[2] http://www.ailab.si/orange/datasets.asp
[3] http://www.datalab.uci.edu/data/mldb-sgi/data/

Table 3. A reduction of the data in Table 2

Reduct	*Instance*	c_1	c_2	c_3	c_4	d
$\{1,4\}$						
	x_1	1			1	0
	x_2	2			1	1
	x_4	2			3	0
	x_6	3			1	0
	x_7	1			2	2
	x_8	2			2	3
	x_9	3			1	1
	x_{13}	4			2	1
	x_{14}	1			2	3
	x_{15}	4			2	2
$\{2,3\}$						
	x_1		1	1		0
	x_2		2	2		1
	x_4		3	2		0
	x_6		1	2		0
	x_7		2	3		2
	x_8		3	1		3
	x_9		1	2		1
	x_{13}		3	4		1
	x_{14}		2	3		3
	x_{15}		3	4		2

12, 15, 35, 47, 54, 55}, 10: {2, 3, 12, 19, 25, 27, 30, 32, 40, 56}, 8: {11, 14, 24, 30, 42, 44, 45, 50}. For zoo dataset, the results of 3 GA runs all were 5: {3, 4, 6, 9, 13}, the results of 3 PSO runs were 5: {3, 6, 8, 13, 16, }, 5: {4, 6, 8, 12, 13}, 5: {3, 4, 6, 8, 13}. For lymphography dataset, the results of 3 GA runs all were 7: {2, 6, 10, 13, 14, 17, 18}, the results of 3 PSO runs were 6: {2, 13, 14, 15, 16, 18}, 7: {1, 2, 13, 14, 15, 17, 18}, 7: {2, 10, 12, 13, 14, 15, 18}. For mofn-3-7-10 dataset, the results of 3 GA runs all were 7: {3, 4, 5, 6, 7, 8, 9} and

Table 4. Data sets used in the experiments

Dataset	*Size*	*ConditionAttributes*	*Class*	*GA*	*PSO*
lung-cancer	32	56	3	10	8
zoo	101	16	7	5	5
corral	128	6	2	4	4
lymphography	148	18	4	7	6
hayes-roth	160	4	3	3	3
shuttle-landing-control	253	6	2	6	6
monks	432	6	2	3	3
xd6-test	512	9	2	9	9
balance-scale	625	4	3	4	4
breast-cancer-wisconsin	683	9	2	4	4
mofn-3-7-10	1024	10	2	7	7
parity5+5	1024	10	2	5	5

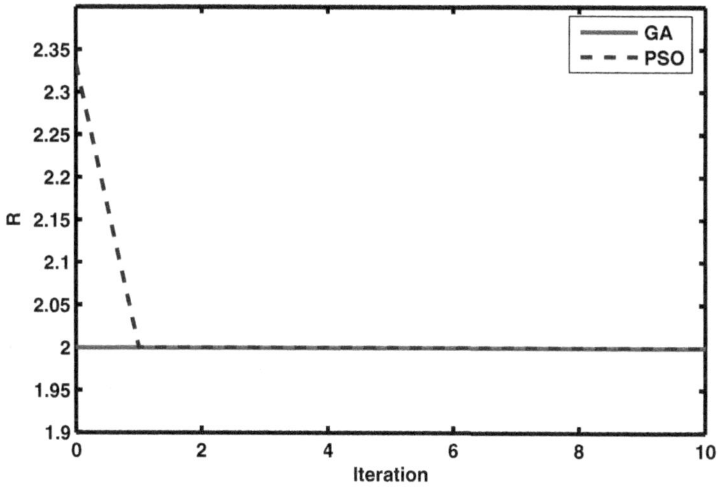

Fig. 1. Performance of rough set reduction for the data in Table 2

the results of 3 PSO runs were 7: {3, 4, 5, 6, 7, 8, 9}. Other results are shown in Table 4, in which only the best objective results are listed. PSO usually obtained better results than GA, specially for the large scale problems. Although GA and

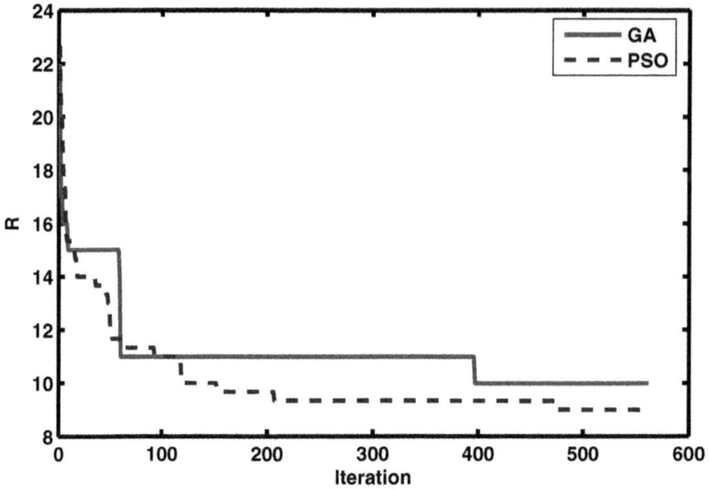

Fig. 2. Performance of rough set reduction for lung-cancer dataset

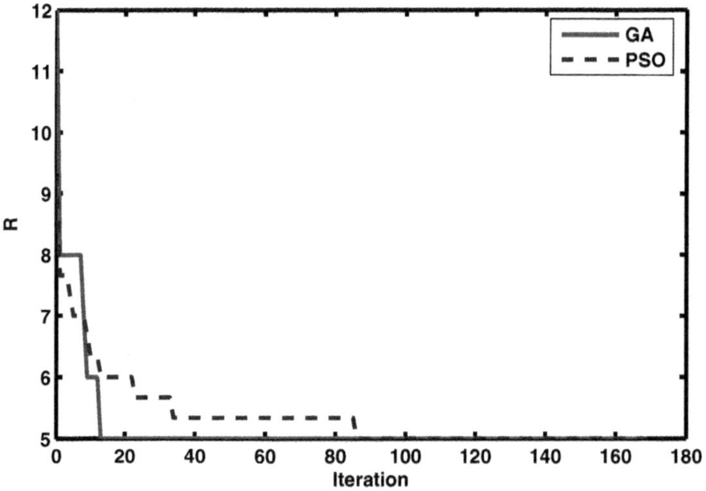

Fig. 3. Performance of rough set reduction for zoo dataset

PSO achieved the same results, PSO usually requires only very few iterations, as illustrated in Fig. 4. It indicates that PSO have a better convergence than GA for the larger scale rough set reduction problem, although PSO is worst for some small scale rough set reduction problems. It is to be noted that PSO usually can obtain more candidate solutions for the reduction problems.

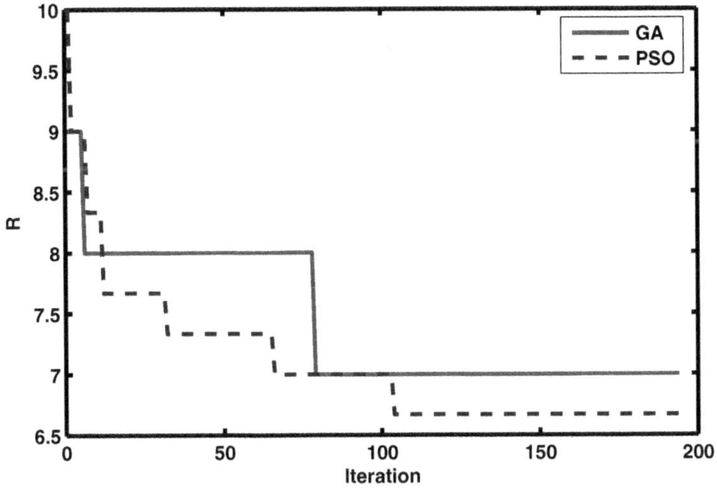

Fig. 4. Performance of rough set reduction for lymphography dataset

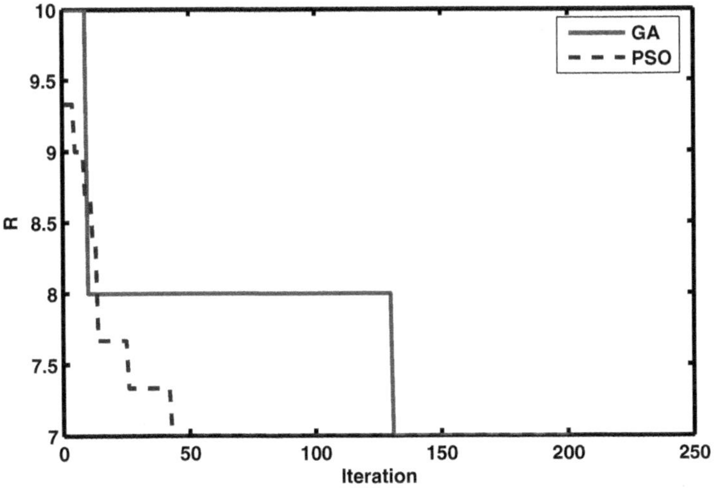

Fig. 5. Performance of rough set reduction for mofn-3-7-10 dataset

5 Application in fMRI Data Analysis

Functional Magnetic Resonance Imaging (fMRI) is one of the most important tools for Neuroinformatics, which combines neuroscience and informatics science and computational science to develop approaches needed to understand human brain [35]. The study of human brain function has received a tremendous boost in recent years due to the advent of the new brain imaging technique.

With the development of the new technology, a mass of fMRI data is collected ceaselessly. These datasets implicate very important information, which need to be extracted and translated to intelligible knowledge. Recently most of the research are focused on the activation features on the Region of Interest (ROI) through statistical analysis for single experiment or using only a few data. Neuroscientists or psychologists provide explanation for the experimental results, which depends strongly on their accumulative experience and subjective tendency. What is more, it is difficult to deal with slightly large datasets. So it is exigent to develop some computational intelligence methods to analyze them effectively and objectively. Rough set theory provides a novel approach to reduct the fMRI data and extract meaningful knowledge. There are usually many reducts in the information system, which can be applied to generate multi-knowledge. The rough set approach consists of several steps leading towards the final goal of generating rules [36].

The main steps of the rough set approach are: (1)mapping of the information from the original database into the decision system format; (2) completion of data; (3) discretization of data; (4) computation of reducts from data; (5) derivation of rules from reducts; (6) filtering of rules. One of most important task is the data reduction process.

Algorithm 3. Feature selection & extraction algorithm for fMRI data

Step 1. Find out the most active voxels in several regions of brain under the t-test of basic models in SPM99 and save their coordinates
Step 2. Scan fMRI image and search the voxels according to the coordinates saved
Step 3. Respectively average all voxels in the spherical region whose center is corresponding saved voxel and whose radius is a predefined constant. These results of a single image are formed one feature vector
Step 4. If the image isn't the last one, go to Step 2, otherwise, end

A typical normalized image contains more than 500,000 voxels, so it is impossible that feature vector can contain so immense voxels. We transform datasets from MNI template to Talairach coordinate system. Then we can use the region information in Talairach as features to reduce the dimensionality of the images. We used a SPM99 software package[4] and in-house programs for image processing, including corrections for head motion, normalization and global fMRI signal shift [37]. A simplified workflow is illustrated in Fig. 6. Feature selection & extraction algorithm for fMRI data is described in Algorithm 3. The location for feature selection & extraction is shown in Fig. 7.

We analyzed the fMRI data from three cognition experiments: Tongue movement experiment, Associating Chinese verb experiment, and Looking at or silent reading Chinese word experiment. They are involved in 9 tasks: 0 - Control task; 1 - Tongue movement; 2 - Associating verb from single noun; 3 - Associating verb

[4] http://www.fil.ion.ucl.ac.uk/spm/

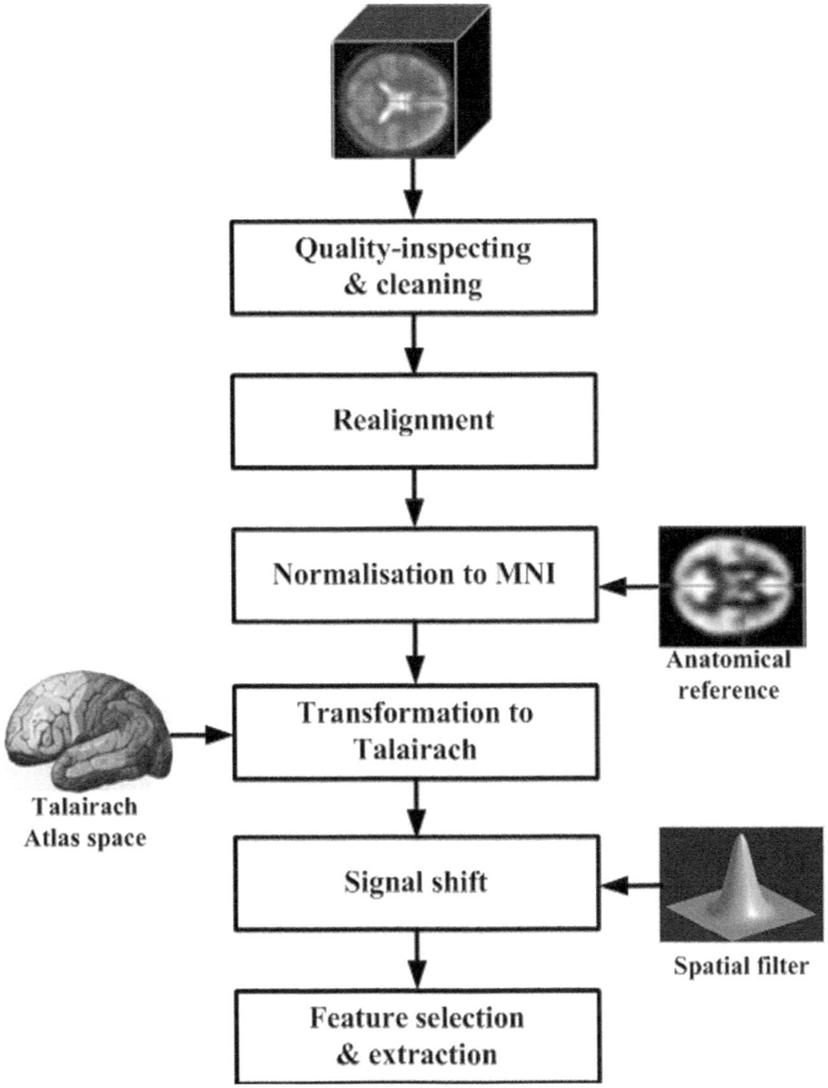

Fig. 6. Pre-processing workflow for fMRI data

from single non-noun; 4 - Making verb before single word; 5 - Looking at number; 6 - Silent reading Number; 7 - Looking at Chinese word; 8 - Silent reading Chinese word. Some of rules are described as follows:

Rule1: if M1=2, SMA=2, Broca=2 then Task=1;
Rule2: if BAs { 7,19,20,40,44,45 } =3, BSC=2 then Task=2;
Rule3: if BAs { 10,11,13,44,45 } =3, BSC=1 then Task=3;

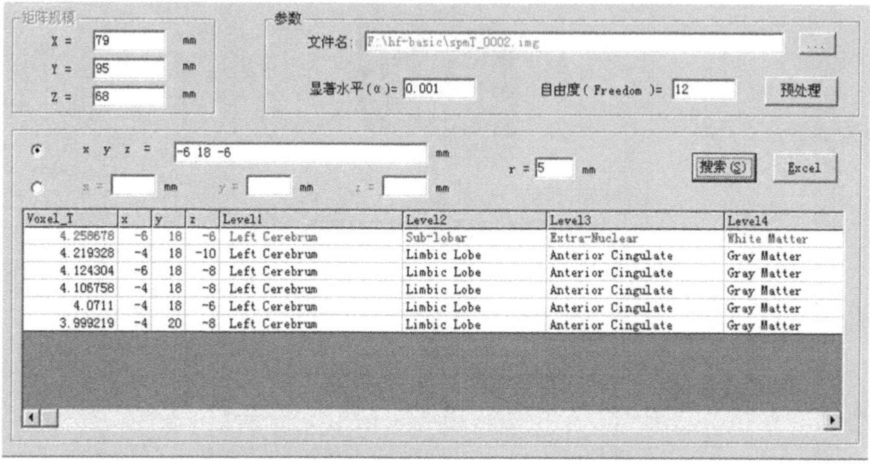

Fig. 7. Developed software interface for feature selection and extraction

Rule4: if BAs { 7,19,40 } =3, BSC=3 then Task=4;
Rule5: if SMA=2, Broca=3 then Task=6;
Rule6: if SMA=2, Broca=2, Wernike=3 then Task=8.

6 Conclusions

In this Chapter, we introduced the problem of finding optimal reducts using particle swarm optimization and genetic algorithm approaches. The considered approaches discovered the good feature combinations in an efficient way to observe the change of positive region as the particles proceed throughout the search space. Population-based search approaches are of great benefits in the multiple reduction problems, because different individual trends to be encoded to different reduct. Empirical results indicate that PSO usually required shorter time to obtain better results than GA, specially for large scale problems, although its stability need to be improved in further research. PSO have a better convergence than GA for the larger scale rough set reduction problem, although PSO is worst for some small scale rough set reduction problems. PSO also can obtain more candidate solutions for the reduction problems. The population-based algorithms could be ideal approaches for solving the reduction problem. We also illustrated an application in fMRI data analysis. Although the correctness of the rules need neuroscientists to analyze and verify further, the approach is helpful for cognition research.

Acknowledgements

This work was partly supported by NSFC (60373095) and DLMU (DLMU-ZL-200709).

References

1. Pawlak, Z.: Rough Sets. International Journal of Computer and Information Sciences 11, 341–356 (1982)
2. Pawlak, Z.: Rough Sets: Present State and The Future. Foundations of Computing and Decision Sciences 18, 157–166 (1993)
3. Pawlak, Z.: Rough Sets and Intelligent Data Analysis. Information Sciences 147, 1–12 (2002)
4. Kusiak, A.: Rough Set Theory: A Data Mining Tool for Semiconductor Manufacturing. IEEE Transactions on Electronics Packaging Manufacturing 24, 44–50 (2001)
5. Shang, C., Shen, Q.: Rough Feature Selection for Neural Network Based Image Classification. International Journal of Image and Graphics 2, 541–555 (2002)
6. Francis, E.H., Tay, S.L.: Economic And Financial Prediction Using Rough Sets Model. European Journal of Operational Research 141, 641–659 (2002)
7. Świniarski, R.W., Skowron, A.: Rough Set Methods in Feature Selection and Recognition. Pattern Recognition Letters 24, 833–849 (2003)
8. Beaubouef, T., Ladner, R., Petry, F.: Rough Set Spatial Data Modeling for Data Mining. International Journal of Intelligent Systems 19, 567–584 (2004)
9. Shen, L., Francis, E.H.: Tay Applying Rough Sets to Market Timing Decisions. Decision Support Systems 37, 583–597 (2004)
10. Gupta, K.M., Moore, P.G., Aha, D.W., Pal, S.K.: Rough set feature selection methods for case-based categorization of text documents. In: Pal, S.K., Bandyopadhyay, S., Biswas, S. (eds.) PReMI 2005. LNCS, vol. 3776, pp. 792–798. Springer, Heidelberg (2005)
11. Boussouf, M.: A Hybrid Approach to Feature Selection. In: Żytkow, J.M. (ed.) PKDD 1998. LNCS, vol. 1510. Springer, Heidelberg (1998)
12. Skowron, A., Rauszer, C.: The Discernibility Matrices and Functions in Information Systems. In: Świniarski, R.W. (ed.) Handbook of Applications and Advances of the Rough Set Theory, pp. 331–362. Kluwer Academic Publishers, Dordrecht (1992)
13. Zhang, J., Wang, J., Li, D., He, H., Sun, J.: A new heuristic reduct algorithm base on rough sets theory. In: Dong, G., Tang, C., Wang, W. (eds.) WAIM 2003. LNCS, vol. 2762, pp. 247–253. Springer, Heidelberg (2003)
14. Hu, K., Diao, L., Shi, C.: A heuristic optimal reduct algorithm. In: Leung, K.-S., Chan, L., Meng, H. (eds.) IDEAL 2000. LNCS, vol. 1983, pp. 139–144. Springer, Heidelberg (2000)
15. Zhong, N., Dong, J.: Using Rough Sets with Heuristics for Feature Selection. Journal of Intelligent Information Systems 16, 199–214 (2001)
16. Banerjee, M., Mitra, S., Anand, A.: Feature Selection Using Rough Sets. Studies in Computational Intelligence 16, 3–20 (2006)
17. Wu, Q., Bell, D.: Multi-knowledge Extraction and Application. In: Wang, G., Liu, Q., Yao, Y., Skowron, A. (eds.) RSFDGrC 2003. LNCS, vol. 2639, pp. 574–575. Springer, Heidelberg (2003)
18. Wu, Q.: Multiknowledge and Computing Network Model for Decision Making and Localisation of Robots. Thesis, University of Ulster (2005)
19. Kennedy, J., Eberhart, R.: Swarm Intelligence. Morgan Kaufmann Publishers, San Francisco (2001)
20. Clerc, M., Kennedy, J.: The Particle Swarm-Explosion, Stability, and Convergence in a Multidimensional Complex Space. IEEE Transactions on Evolutionary Computation 6, 58–73 (2002)

21. Clerc, M.: Particle Swarm Optimization. ISTE Publishing Company, London (2006)
22. Liu, H., Abraham, A., Clerc, M.: Chaotic Dynamic Characteristics in Swarm Intelligence. Applied Soft Computing Journal 7, 1019–1026 (2007)
23. Parsopoulos, K.E., Vrahatis, M.N.: Recent approaches to global optimization problems through particle swarm optimization. Natural Computing 1, 235–306 (2002)
24. Abraham, A., Guo, H., Liu, H.: Swarm Intelligence: Foundations, Perspectives and Applications. In: Nedjah, N., Mourelle, L. (eds.) Swarm Intelligent Systems. Studies in Computational Intelligence, pp. 3–25. Springer, Heidelberg (2006)
25. Salman, A., Ahmad, I., Al-Madani, S.: Particle Swarm Optimization for Task Assignment Problem. Microprocessors and Microsystems 26, 363–371 (2002)
26. Sousa, T., Silva, A., Neves, A.: Particle Swarm Based Data Mining Algorithms for Classification Tasks. Parallel Computing 30, 767–783 (2004)
27. Liu, B., Wang, L., Jin, Y., Tang, F., Huang, D.: Improved Particle Swarm Optimization Combined With Chaos. Chaos, Solitons and Fractals 25, 1261–1271 (2005)
28. Schute, J.F., Groenwold, A.A.: A Study of Global Optimization Using Particle Swarms. Journal of Global Optimization 3, 108–193 (2005)
29. Liu, H., Abraham, A., Clerc, M.: An Hybrid Fuzzy Variable Neighborhood Particle Swarm Optimization Algorithm for Solving Quadratic Assignment Problems. Journal of Universal Computer Science 13(7), 1032–1054 (2007)
30. Wang, G.: Rough Reduction in Algebra View and Information View. International Journal of Intelligent Systems 18, 679–688 (2003)
31. Xu, Z., Liu, Z., Yang, B., Song, W.: A Quick Attibute Reduction Algorithm with Complexity of Max$(O(|C||U|), O(|C|^2|U/C|))$. Chinese Journal of Computers 29, 391–399 (2006)
32. Wu, Q.X., Bell, D.A.: Multi-Knowledge Extraction and Application. In: Wang, G., Liu, Q., Yao, Y., Skowron, A. (eds.) RSFDGrC 2003. LNCS, vol. 2639, pp. 274–279. Springer, Heidelberg (2003)
33. Cantú-Paz, E.: Efficient and Accurate Parallel Genetic Algorithms. Kluwer Academic publishers, Netherland (2000)
34. Abraham, A.: Evolutionary Computation. In: Sydenham, P., Thorn, R. (eds.) Handbook for Measurement Systems Design, pp. 920–931. John Wiley and Sons Ltd., London (2005)
35. Arbib, M.A., Grethe, J.S. (eds.): Computing the Brain: A Guide to Neuroinformatics. Academic Press, London (2001)
36. Røed, G.: Knowledge Extraction from Process Data: A rough Set Approach to Data Mining on Time Series. Knowledge Extraction from Process data. Thesis, Norwegian University of Science and Technology (1999)
37. Ji, Y., Liu, H., Wang, X., Tang, Y.: Congitive States Classification from fMRI Data Using Support Vector Machines. In: Proceedings of The Third International Conference on Machine Learning and Cybernetics, pp. 2920–2923. IEEE Computer Society Press, Los Alamitos (2004)
38. Goldberg, D.E.: Genetic Algorithms in search, optimization, and machine learning. Addison-Wesley Publishing Corporation, Inc., Reading (1989)

Developing a Knowledge-Based System Using Rough Set Theory and Genetic Algorithms for Substation Fault Diagnosis

Ching Lai Hor[1], Peter Crossley[2], Simon Watson[3], and Dean Millar[1]

[1] University of Exeter in Cornwall, CSM Renewable Energy Group, School of Geography, Archaeology and Earth Resources, Tremough Campus, Penryn, TR10 9EZ, United Kingdom
c.l.hor@exeter.ac.uk, dlmillar@csm.ex.ac.uk

[2] University of Manchester, Electrical Energy and Power Systems, B7 Ferranti Building, School of Electrical and Electronic Engineering, P.O. Box 88, Manchester, M60 1QD, United Kingdom
p.crossley@manchester.ac.uk

[3] University of Loughborough, CREST, AMREL, Department of Electronic and Electrical Engineering, Main Building, Garendon Wing, Holywell Park, Leicestershire, LE11 3TU, United Kingdom
s.j.watson@lboro.ac.uk

"Intelligent Electronic Devices (IEDs) can provide a flood of data. What is lacking are the tools to convert that data into useful information and knowledge".
~ William J. Ackerman, ABB Automation USA (2002)

Summary. Supervisory Control and Data Acquisition (SCADA) systems are fundamental tools for quick fault diagnosis and efficient restoration of power systems. When multiple faults, or malfunctions of protection devices occur in the system, the SCADA system issues many alarm signals rapidly and relays these to the control center. The original cause and location of the fault can be difficult to determine for operators under stress without assistance from a computer aided decision support system. In cases of power system disturbances, network operators in the control center must use their judgement and experience to determine the possible faulty elements as the first step in the restoration procedures. If a breaker or its associated relays fail to operate, the fault is removed by backup protection. In such cases, the outage area can be large and it is then difficult for the network operators to estimate the fault location. Multiple faults, events and actions may eventually take place with many breakers being tripped within a short time. In these circumstances, many alarms need to be analysed by the operators to ensure that the most appropriate actions are taken [1]. Therefore, it is essential to develop software tools to assist in these situations.

This chapter proposes a novel and hybrid approach using Rough Set Theory and a Genetic Algorithm (RS-GA) indexrough hybrid to extract knowledge from a set of events captured by (microprocessor based) protection, control and monitoring devices (referred to as Intelligent Electronic Devices (IED)). The approach involves formulating a set of rules that identify the most probable faulty section in a network. The idea of this work is

A. Abraham, R. Falcón, and R. Bello (Eds.): Rough Set Theory, SCI 174, pp. 279–320.
springerlink.com

to enhance the capability of substation informatics and to assist real time decision support so that the network operators can diagnose the type and cause of the events in a time frame ranging from a few minutes to an hour. Building knowledge for a fault diagnostic system can be a lengthy and costly process. The quality of knowledge base is sometimes hampered by extra and superfluous rules that lead to large knowledge based systems and serious inconveniences to rule maintenance. The proposed technique not only can induce the decision rules efficiently but also reduce the size of the knowledge base without causing loss of useful information. Numerous case studies have been performed on a simulated distribution network [2] that includes relay models [3]. The network, modelled using a commercial power system simulator; PSCAD (**P**ower **S**ystems **C**omputer **A**ided **D**esign/EMTDC (**E**lectro**M**agnetic **T**ransients including **DC**), was used to investigate the effect of faults and switching actions on the protection and control equipment. The results have revealed the usefulness of the proposed technique for fault diagnosis and have also demonstrated that the extracted rules are capable of identifying and isolating the faulty section and hence improves the outage response time. These rules can be used by an expert system in supervisory automation and to support operators during emergency situations, for example, diagnosis of the type and cause of a fault event leads to network restoration and post-emergency repair.

1 Introduction

With the advent of Artificial Intelligence (AI), rule-based expert systems offer capability of powerful inference and explanation for the knowledge intensive problem of fault diagnosis. However, they suffer from some bottlenecks. The size of a conventional knowledge base for a substation can be very large depending on its duty. The process of knowledge acquisition, knowledge base construction and maintenance for a great number of rules can be quite tedious and time consuming. Often the process involves interviews with experts and engineers and thus significant effort is required to establish a rule-based system with good performance. As such, the cost of developing and/or maintaining a knowledge based system is generally high. Owing to the nature of conventional knowledge representation and inference mechanisms, the on line response time required by an expert system is often unsatisfactory in a real time context. Once a rule-based expert system has been built, it is difficult for it to improve its performance by learning from new experiences. To compound matters, some rule-based systems may be well suited for one substation but not for others and the situation can get a lot more complicated with distributed generation integrated into the network as it no longer behaves as a passive circuit. This will not only have an impact on the protection system design but also on the conventional rule-based systems designed to operate upon a traditional passive network.

In [4, 5, 6], the use of supervised and unsupervised rough classification, was proposed for handling large numbers of messages received during an emergency, in order to reduce the quantity of data, while maintaining useful and concise information. This is crucial to alarm processing as the network operator could be overwhelmed by the large quantity of alarm messages during an emergency. Without methods for reducing these messages, operators will not respond effectively to the emergency within required time limits. In this chapter, we are concerned about knowledge base construction for on-line intelligent switching, fault identification and service restoration in which detailed rules covering every possible scenario that may occur in a substation are needed. A classical information system may not cope well with high volumes of data and the existing system may rely on particular data relating to the state of the power system. If the latter are not available or missing, such a system may not perform accurately. Consequently, power system state estimation is required to calculate missing voltage and power data across the entire network to enable accurate diagnosis of faults.This chapter serves as an improvement of our published work in [7]; a classification method based on rough sets for knowledge base reduction and rule induction.

With many IEDs being deployed on the network, we now have significant volumes of redundant information. If one particular IED is missing/out of service, there is still information available in other IEDs within the network that can help diagnose the problem. In this chapter, a new classification method using rough sets and a genetic algorithm is proposed to make use of the redundant information from all IED sources to improve the operation of the power system. The model is capable of extracting good quality inference rules from these sets of event data. The idea is to build a scheme for autonomous rule induction followed by subsequent classification to verify the extracted rules without the presence of an expert. It is believed that ultimately experts will still be required to perform a final verification check before these rules are applied. The proposed technique can reduce the time needed to install such a knowledge based system whenever a new substation configuration arises and hence reduces the set up and maintenance costs. For an effective rule synthesis and induction, the simulated event data should be as complete as possible, so that the rule induction system can recognise a large variety of events for classification.

Almost all of the IEDs can provide some level of automation in collecting and storing the recorded data from a substation. However, most of the data analysis systems today still require the operators and engineers to perform the manual operation associated with selecting and viewing the files. The future trend of substation analysis is for these to operate automatically with minimum interaction from utility staff. This requires conventional substations to be refurbished in order to provide automated data collection, storage, processing of the recorded data and distribution of the analysis reports. Figure 1 depicts a modern digital control system (DCS) integrated with an information management unit (IMU) to deliver useful information to the appropriate manpower groups in a utility. Each group uses the monitored data for a different purpose and consequently has varied information requirements. All the requirements need to be satisfied; the right information must be delivered to the right people, at the right place, at the right time.

The future trend of data concentration and processing is to decentralise the intelligence. The concept of distributed intelligence related to protection, automation and control moves centralised data processing closer to the data collection point. Devices at higher levels are allowed to access and use a subset of the data

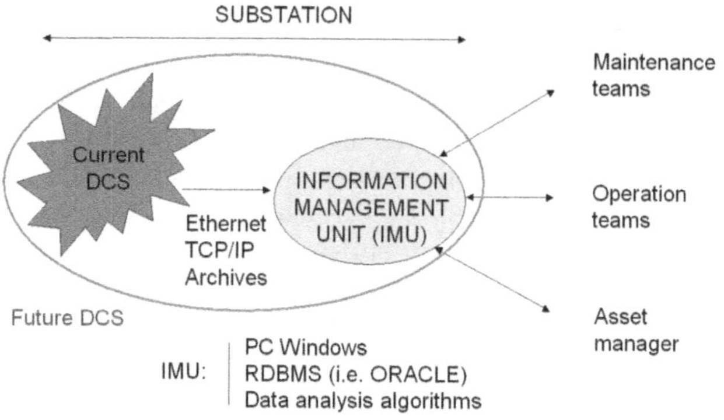

Fig. 1. Integration of future DCS with IMU

from the lower level devices, which reduces data traffic over the substation network. The processed information is then transferred to the right end users. This can improve local response and reduce the amount of data transmitted to the control centre. Also, the information collected by protection and control IEDs can be processed locally to estimate for example, the remaining lifetime until maintenance is required. Maintenance can then be performed when needed, instead of at regular intervals. This can lead to reduced maintenance effort and restrict the number of unexpected failures [8]. Some recent developments of substation automation can be found in reference papers [9][10][11][12][13].

This chapter consists of four following sections. Section 2 introduces, defines and explains the theoretical basis of proposed Rough Sets and Genetic Algorithms. Sections 3 and 4 present two case studies and the results obtained from the data collected using the power system simulation. The former illustrates how the knowledge is extracted from a simple bus bar substation while the latter presents and explains the results based on a more complex double bus bar distribution substation. Two different sets of data have been considered to demonstrate and evaluate the proposed algorithm in extracting knowledge from a substation; one is based on the fault data in time series and the other is based on various fault scenarios on defined protection zones. Section 5 consolidates the ideas from the previous parts and suggests future directions for the research as well as conclude the achievements of this application in power engineering domain.

2 Rule Induction System

The proposed rule induction system consists of a hybrid model of rough sets, a genetic algorithm and a standard voting classifier. Detailed descriptions of rough sets and the genetic algorithm are not included in this chapter. They can be found in the reference papers [14][15][16][17][18].

2.1 Rough Set Theory - An Overview

Rough Set Theory was first introduced by Prof. Zdzislaw Pawlak and his co-workers in the early 1980s. This relatively new theory has become one of the crucial and fundamental approaches in the area of machine learning, knowledge discovery from databases and decision support systems [19]. It offers an opportunity to explore data, which is neither accommodated by classical statistics nor by conventional analytical models[20]. The results simplify the search for dominating and minimal attributes leading to specific rules[21].

The concept of rough sets applies the idea of using approximations of a set to deal with uncertainty (or inconsistencies). The approximation is based on the assumption that every object (event) of universe is associated with some information (data, knowledge). Events that can be characterized by the same information are said to be indiscernible in view of the available information about them. This can also be interpreted as the fact that our knowledge about elements of the universe is limited and therefore we are unable to discern them [22]. The approximations of a set consist of the positive (lower approximation), negative (complement of the upper approximation) and boundary (difference between the upper and lower approximation) regions. A lower approximation defines the collection of events in which the equivalence classes are fully contained in the set of events which is to be reduced to its essential attributes. The upper approximation, however, defines the collection of events in which the equivalence classes are at least partially contained in the set of events to be reduced. These two approximations provide *crisp* and *rough* descriptions of a data set respectively. If a universe can be formed as a union of some elementary sets, it is called **crisp**, otherwise it is **rough**.

The theory of rough sets is made up by two main parts; dispensable attribute reduction and rule extraction. 'Discernibility' is the main theme of rough set analysis, defined as the ability to discern events from each other. It requires understanding of how the characteristics of one event differ from another before the two events can be classified. To achieve this, a discernibility matrix is used. A discernibility matrix is a symmetric $n \times n$ matrix where n denotes the number of elementary sets [23]. All the events in the rows and the columns are listed in the same order. In each entry of the matrix, the differences between the event corresponding to the row and the event corresponding to the column are compared and recorded. Naturally, the matrix will be symmetric due to the fact that the attribute, which differs in value for events a and b differs the other way around in value for events b and a [23]. However, it is important to remark that the symmetry of the discernibility matrix is only valid when considering symmetric indiscernibility relations, which is the assumption made in this study. Two examples will be introduced in this chapter to demonstrate how rough sets and a discernibility matrix are used to compute **reducts**. A reduct is a reduced set of attributes, e.g voltages and currents, that conveys the same amount of information about a data set as a complete set of attributes. A relative discernibility matrix is then applied to this minimal attribute set to look for the **core** before any rules are extracted. A core is the set of relations occurring in every reduct, i.e. the set of all indispensable relations that characterise the equivalence relation.

Decision rules are generated on the basis of the computed reducts via the discernibility matrix for classification of objects. They constitute one of the most important results of rough set data analysis. Given a decision system $\mathcal{D} = (U, A \cup \{d\})$, a *descriptor* is defined as a simple expression $a = v$ where $a \in A$ and $v \in V_a$. A decision rule is denoted $\alpha \rightarrow \beta$, where α is a conjunction of descriptors, formed by overlaying a reduct $B \in RED(A)$ over an object $x \in U$ and reading off the value of x for every $a \in B$, and β is the corresponding descriptors $d = d(x)$. α is called the rule's *antecedent* and β is called the rule's *consequent* [24]. Each row of the decision table specifies a decision rule, which determines the decision (action) that should be performed if specific conditions are satisfied. A decision rule can be expressed as a logical statement: IF some conditions are met, THEN some decisions rule can be commended. The condition part is a conjunction of elementary conditions, and the decision part is a disjunction of recommended decisions. A rule is said to cover an object if all conditions in the condition part are matched by the attribute values of an object.

The advantages of using Rough Set approach is that it is able to discover the dependencies that exist within the data, remove superfluous and redundant data not required for the analysis and finally generate decision rules. It does this by classifying the events of interest into similar classes indistinguishable to each other. It generally constitutes a smaller size of the substation database allowing faster knowledge acquisition and significantly reduces the time and costs for rules based maintenance. Unlike other techniques e.g. probability in statistics and grade of membership in fuzzy sets, the rough set approach does not require additional information about the data to justify the outcome of interest [19].

2.2 Genetic Algorithms

Genetic algorithms (GAs) are the stochastic global search algorithm that reflects in a primitive way some of the processes of natural biological evolution [25]. They are generally effective for rapid search of large, nonlinear and poorly understood spaces. Unlike classical feature, selection strategies where one solution is optimised, a population can be modified at the same time. This can result in several optimal feature subsets as output [26]. The motivation for using GAs in this work is because they provide us flexibility to generate a number of rules according to the parameter settings. They have been used successfully to generate shorter reducts in knowledge processing and has good adaptive capability[27]. A GA starts by generating a large set of possible solutions to a given problem. It then evaluates each of those solutions, and decides on a fitness level for each solution set. These solutions then breed new solutions. The parent solutions that have better fitness level are more likely to reproduce, while those that have less fitness level are more unlikely to do so. GAs evolve the search space scope over time or generation to a point where the solution can be found. The reduct computation process using GAs begins with initialising the population of chromosomes (or individuals) from the discernibility functions *f(B)* via rough sets. This is represented in the following steps: -

1. **Generate an initial population.** An initial population is created from a random selection of solutions. Bit-vectors are used to represent the sets and their hitting sets. For a collection of subsets S of a universe U, a hitting set is a non empty subset H of U that intersects/hits every set in S. A minimal hitting set is the element that can not be removed from S without violating the hitting set property. These bit-vectors are called *chromosomes*, and each bit is called *gene*, and all of the chromosomes are called *population.* Each of the chromosomes is assigned a weight to record the number of times it appears in f(B) before the fitness of all the chromosomes is evaluated. A straightforward choice of population is a set P of elements from the power set of A, written P(A) or 2^A encoded as bit-vectors where each bit indicates the presence of a particular element in the set. Given a set A, 2^A is the set of all subsets of A. As 2 can be defined as $\{0,1\}$, 2^A is the set of all functions from A to $\{0,1\}$. Assume that we have 10 IED relays in our network $\{R1, R2, \ldots, R10\}$ and a candidate reduct A $= \{R1, R4, R9\}$. Then, 2^A is $\{\}, \{R1\}, \{R4\}, \{R9\}, \{R1, R4\}$, $\{R1, R9\}, \{R4, R9\}, \{R1, R4, R9\}$. The candidate reduct can be represented as 1001000010.
2. **Evaluate fitness.** Fitness is a value assigned to each solution (chromosome) depending on how close it actually is to solving the problem. These solutions are possible characteristics that the system would employ to reach the answer.

   The fitness function $f\left(B\right)$ evaluates an individual candidate solution based on how well it performs at solving the problem. The fitness function depends on the number of attributes (should be kept as low as possible) and decision ability (should be kept as high as possible). It drives the search in GAs by optimising the fitness that determines how good a solution is. The better a solution performs, the higher rated fitness value it receives. The fitness function in use by the algorithm is defined in Eq. (1), where B$\subseteq$ A and S is the set of sets corresponding to the discernibility function. The cost function specifies the cost of an attribute subset. If no cost information is used, then a default unit cost is effectively defined as $cost\left(B\right) = |B|$ [28].

$$f\left(B\right) = (1-\rho) \times \frac{\cos t\left(A\right) - \cos t\left(B\right)}{\cos t\left(A\right)} + \rho \times hf\left(B, S\right) \qquad (1)$$

$$\text{where :} \quad hf\left(B, S\right) = \min\left\{\varepsilon, \frac{|[S_i \text{ in } S | S_i \cap B \neq \emptyset]|}{|S|}\right\}$$

   The subsets B of A are found through the evolutionary search driven by the fitness function. The parameter ρ defines a weighting between subset cost and hitfraction (*hf*). The first term of Eq. (1) rewards the shorter elements and the second term ensures that hitting sets are rewarded to guarantee the decision ability.

3. **Reproduction, selection, mutation, inversion and crossover.** In the reproduction, crossover, mutation and inversion are performed to create offspring. The mutation operator chooses a random location in the bit string and changes that particular bit. The inversion operator inverts the order of the elements between two randomly chosen points on a single chromosome. It rearranges the bits in a string allowing linked bits to migrate and move close together by keeping track of the original positions of the genes. The crossover operator recombines two individuals by exchanging parts of their strings to form two new individuals (offspring).
4. **Control next generation.** If the new generation contains a solution that produces an output that is close enough or equal to the desired answer then the problem has been solved. Otherwise, the process is repeated until the predetermined number of generations has been exceeded or an acceptable performance level has been reached.

2.3 Rule Accuracy and Assessment

The rules generated must be assessed to validate their accuracy. The metrics described below are used to evaluate the quality of a given decision rule [28]:

1. **Support:** the number of events that possess both property α then β. The pattern α is called the rule's antecedent while the pattern β is called the rule's consequent, which can be represented in metric as follows: -

$$Support\,(\alpha) = |\,||\alpha||\,| \tag{2}$$

   where $||\alpha||$ is the norm of the element α of a normal vector space.
2. **Accuracy:** A decision rule $\alpha \rightarrow \beta$, read as "if α then β", may only reveal partially the overall picture of the derived decision system. Given pattern α, the probability of the conclusion β can be assessed by measuring how trustworthy the rule is in drawing the conclusion β on the basis of evidence α using Eq. (3).

$$Accuracy\,(\alpha \rightarrow \beta) = \frac{support\,(\alpha \cdot \beta)}{support\,(\alpha)} \tag{3}$$

3. **Coverage:** The strength of the rule relies upon the large support basis that describes the number of events, which support each rule. The quantity coverage $(\alpha \rightarrow \beta)$ is required in order to measure how well the evidence α describes the decision class. It can be defined by Eq. (4): -

$$Coverage\,(\alpha \rightarrow \beta) = \frac{support\,(\alpha \cdot \beta)}{support\,(\beta)} \tag{4}$$

2.4 Standard Voting Algorithm

A voting algorithm is used to resolve the rule conflicts and rank the predicted outcomes. This works reasonably well for rule-based classification. The concept of the voting algorithm can be divided into three parts [28]:

1. Let RUL denote an unordered set of decision rules. The set of rules RUL searches for applicable rules RUL(x) that match the attributes of event x (i.e. rules that *fire*) in which RUL(x) $\subseteq$ RUL.
2. If no rule is found, i.e. RUL(x) $= \emptyset$, no classification will be made. The most frequently occurring decision is chosen. If more than one rule fires, this means that more than one possible outcome exists.
3. The voting process is a way of employing RUL to assign a certainty factor to each decision class for each event. It is performed in three stages: -

- **Casting the votes:** Let a rule $r \in$ RUL(x) cast as many votes, *votes(r)* in favour of its outcomes associated with the support counts as given by Eq. (5):-

$$votes(r) = |\ ||\alpha \cap \beta||\ | = |\ ||\alpha|| \cap ||\beta||\ | \tag{5}$$

- **Compute a normalisation vector norm(x).** The normalization factor is computed as the total number of votes cast and only serves as a scaling factor in Eq. (6): -

$$norm(x) = \sum_{r \in RUL(x)} votes\,(r_i) \tag{6}$$

- **Certainty Coefficient:** The votes from all the decision rules β are accumulated before they are divided by the normalisation factor norm(x) to yield a numerical certainty coefficient. Certainty(x, β) for each decision class is given in Eq. (7): -

$$Certainty\,(x, \beta) = \left(\frac{votes\,(\beta)}{norm\,(x)}\right) \tag{7}$$

in which the $votes\,(\beta) = \sum \{votes\,(r)\}$ and $r \in RUL(x) \wedge r \equiv (\alpha \rightarrow \beta)$. The certainty coefficient for every decision class decides which label (decision class) the unknown event is assigned to.

2.5 Classifier's Performance

The set of rules derived from the reducts must be evaluated on its classification performance, readability and usefulness before it can be effectively used for on line fault diagnosis. Usually, a domain expert shall be the one who evaluates the usefulness of the rules because of his/her knowledge about the power system and experience from operating and monitoring the process. When classifying a new and unseen event using a set of rules, we generally expect to see a return value for each event from the classification algorithm. If rules are matched and able to classify all events, that decision is definitely chosen. However, in those cases where the rules are not able to classify all events, particularly when more than one classification is possible, the algorithm will then have to make an educated guess. A good guess may indeed correctly classify some of the undefined events. However, a wrong guess could result in incorrect classification and action. In power systems, event classification is crucial. Wrong classification may lead to a dangerous situation in the worst case. If the rules are not able to classify all events, the operators should be informed. This is because the operators would stand a better chance and are more qualified to make an educated guess than a classification algorithm.

For assessing the classifier performance, the data set is divided into a training set and a test set. The training set is a set of examples used for learning that is used to fit the parameters, whereas the test set is a set of examples used only to assess the performance of a classifier. Rules are mined from a selection of events in each training set using a discernibility matrix based rough set approach with genetic algorithms to calculate the minimal reducts. These rules are then used to classify the events in the test set. If the rules cannot classify the events in the test set satisfactorily, the rules must be notified to the user and refined to suit the real application. This method can be carried out for two purposes. First, the rule set can be viewed as a classifier, used for the purpose of classifying only. Second, the computed reducts and the generated rules can be used by domain experts to learn more about the data. The last approach often requires a small set of rules for the human expert to examine, thus rule filtering can be carried out [24].

Real time data is the main asset for a substation control system but often it is very difficult to obtain for research purposes. Because of the complexity of a power network, it is almost impossible to anticipate or provide an infinite case of problems to investigate every scenario in a substation. To partially solve the problems, the primary and secondary systems of 132/11kV substations given in Figs. 2 and 3 have been simulated using PSCAD/EMTDC [2].

3 Example I

The objective of this example is to demonstrate how the knowledge (in the form of rules) is extracted from substation data and thus, the rule assessment and classification will not be considered in this demonstration. The simulation data were collected from a simple 132/11kV substation model with a single bus arrangement as given in Fig. 2.

A selection of fault scenarios was applied to the network model protected by several types of relay models, each of which includes one or more of the protection function listed in Table 1. The relay R5 includes time graded earth fault elements to protect the 11kV busbar and provides backup for the 11kV load feeders, supervised by relays R1 and R2. To simplify the studies, we ignore the unit protection relays that are used to protect the transformer and the auto re-closers at the two load feeders L1 and L2 are also disabled.

Table 2 lists the information of the auxiliary contacts, 52A and 52B, which are used to determine the contact condition of a breaker and relay. '01' indicates that the contact of the breaker/relay is closed. '10' indicates that the contact of breaker/relay is open/tripped, '00' indicates failure of the breaker/relay and '11' indicates an undefined breaker/relay state. These information are essential for assessing protection system failures.

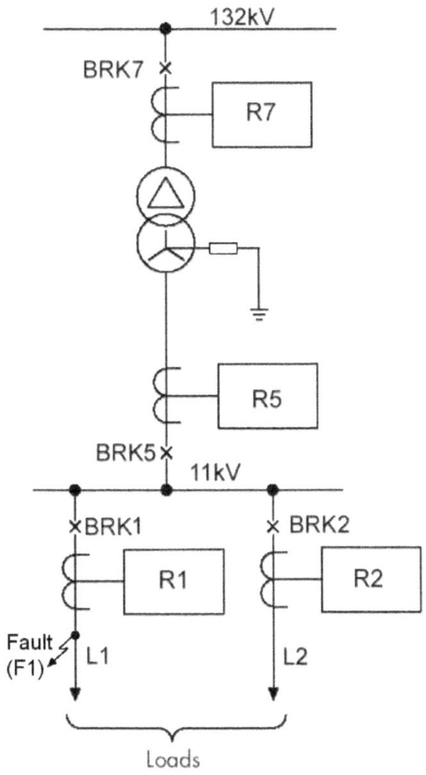

Fig. 2. 132/11kV substation model with a single bus arrangement

Table 1. List of IED relay models in the 132/11kV single bus substation

IED Plate numbers	**1,2**	**5**	**7**
Instantaneous Overcurrent, 50	√	√	√
Instantaneous Earth Fault, 50N	√	√	√
Time delayed Overcurrent, 51	√	√	√
Time delayed Earth Fault, 51N	√	√	√
Balanced Earth Fault, 51G	×	×	√
Auto-recloser, 79 (disabled)	×	×	×

√: available, ×: not available.

3.1 Fault Scenarios on Defined Protection Zones

The substation data consists of a list of recorded voltage and current patterns and the switching actions caused by the protection systems subject to various

Table 2. Auxiliary contacts of circuit breakers

Case	**Auxiliary contacts**		**Breaker**
	52a	52b	**Status**
1	Closed (0)	Closed (0)	Failure
2	Closed (0)	Open (1)	Closed
3	Open (1)	Closed (0)	Open
4	Open (1)	Open (1)	Unknown

Table 3. List of voltage and current patterns with defined fault zones in Fig. 2

R1		R2		R5		R7		ZONE
V_1	I_1	V_2	I_2	V_5	I_5	V_7	I_7	
L	H	L	L	L	H	N	H	Z11
L	L	L	H	L	H	N	H	Z12
L	L	L	L	L	H	N	H	Z25
L	L	L	L	L	L	L	H	Z27

Table 4. List of switching actions with defined fault zones in Fig. 2

R1	R2	R5	R7	B1	B2	B5	B7	ZONE
10	01	01	01	10	01	01	01	Z11
01	10	01	01	01	10	01	01	Z12
01	01	10	01	01	01	10	01	Z25
01	01	01	10	01	01	01	10	Z27

faults at different locations in the substation. As some situations in the event data set may occur in more than one decision class can specify, we thus consider the two different sets of data separately. For instance, a new event might occur at the same time as an existing event and consequently the decision classes overlap [29]. To overcome such problem and prevent the loss of information, the data set is split into two different tables; measurement data in Table 3 and protection data in Table 4. The rules extracted from the measurement data table are merged with the protection data to produce robust rules for fault diagnosis. This is done by incorporating the change of protection trip signals with respect to each decision attribute.

Table 5. Discernibility matrix for Table 3

Index	1	2	3	4
1	∅			
2	R1,R2	∅		
3	R1	R2	∅	
4	R1,R5,R7	R2,R5,R7	R5,R7	∅

The voltage and current are used because they are the fundamental components in the power system. The normal operating voltage range (N) is typically from 0.90 to 1.10 p.u. of the nominal value. Lower than 0.90p.u., the voltage is considered as Low (L) and above 1.10p.u., it is high (H). As the current varies significantly more than the voltage, a wider range of threshold is used. The nominal current range (N) is considered to be between 0.50 and 1.50p.u, meaning that if the current is lower than 0.50pu, it is low (L) and if it is higher than 1.50p.u., it is high (H).

Vx and Iy in which x = $\{1,\ldots,8\}$ and y = $\{1,\cdots,12\}$ represent the three phase voltage and current. To assist our fault section estimation, we defined two protection zones in Fig. 2. Breakers BRK5 and BRK7 are regarded one breaker zone labelled as "BZ2". Zone 1 represents the protection zones of R1 and R2. Zone 2 includes the protection zones of R5 and R7. Zone 25 indicates the regional Zone 2 supervised by the relay R5.

Before defining the discernibility function, Table 3 must be converted into a discernibility matrix as shown in Table 5. Then, a discernibility function f(B) is used to express how an event (or a set of events) can be discerned from a certain subset of the full universe of events. Assume the attribute B $\subseteq$ A and the decision table is represented as $\mathcal{D} = (\mathrm{U}, \mathrm{B} \cup \{d\})$. The discernibility matrix of a decision system, $M^d(B)$ can be defined as: -

$$M^d(B) = \left\{m^d_B(x_i, x_j)\right\}_{n\times n}, \text{ for } i,\ j = 1,\ldots,n \quad \text{and} \quad n = |U/IND(B)|$$

$$m^d_B(x_i,x_j) = \begin{cases} \emptyset & \text{if } \partial^d_B(x_i) = \partial^d_B(x_j) \\ m_B(x_i,x_j) - \{d\} = \{\, b \in B : b(x_i) \neq b(x_j)\} & \text{if } \partial^d_B(x_i) \neq \partial^d_B(x_j) \end{cases}$$

where the notion $\partial^d_B(x)$ denotes the set of possible decisions for a given class $x \in \mathrm{U}/IND(B)$. The entry $m^d_B(x_i, x_j)$ in the discernibility matrix is the set of all (condition) attributes from B that classify events x_i and x_j into different classes in U/IND(B) if $\partial^d_B(x_i) \neq \partial^d_B(x_j)$. Empty set ∅ denotes that this case does not need to be considered. All disjuncts of minimal disjunctive form of this function define the reducts of B.

A discernibility function $f(B)$ is a boolean function that expresses how an object (or a set of events) can be discerned from a certain subset of the full universe of events [28]. A boolean expression normally consists of Boolean variables and constants, linked by disjunction ($\bigvee$) and conjunction ($\bigwedge$). The discernibility function derived from a decision system is generally less complex and constraining than a discernibility function of an information system. This is because the decision functions become equal when the decision attribute d assigns unique decision values to all events and there will be no product terms to eliminate. Given a decision system $\mathcal{D} = (\mathrm{U}, \mathrm{B} \cup \{d\})$, the discernibility function of $\mathcal{D}$ is: -

$$f_B^d(x_i) = \bigwedge \left\{ \bigvee \bar{m}_B^d(x_i, x_j) \; : \; 1 \leq j \leq i \leq n, \; m_B^d(x_i, x_j) \neq \emptyset \right\}$$

where $n = |\mathrm{U}/\mathrm{IND}(\mathrm{B})|$, and $\bigvee \bar{m}_B^d(x_i, x_j)$ is the disjunction taken over the set of Boolean variables $\bar{m}_B^d(x_i, x_j)$ corresponding to the discernibility matrix $m_B^d(x_i, x_j)$ [14].

To find an approximation of the decision d, the decision relative discernibility function of B can be constructed to discern an object from events belonging to another class. The decision relative discernibility function for an object class $x_k = (1 \leq k \leq n)$ over attributes B is a Boolean function of m variables:

$$f(x_k, B) = \bigwedge \left\{ \bigvee \bar{m}_B^d(x_k, x_j) \; : \; 1 \leq j \leq n, \; m_B^d(x_k, x_j) \neq \emptyset \right\}$$

This function computes the minimal sets of attributes B in the decision system that are necessary to distinguish x_k from other object classes defined by B [14].

For better interpretation, the Boolean function attains the form '+' for the operator of disjunction ($\bigvee$) and '·' for the operator of conjunction ($\bigwedge$). The final discernibility function obtained is: -

$$\begin{aligned} f(B) &= f(1) \cdot f(2) \cdot f(3) \cdot f(4) \\ &= \mathrm{R1} \cdot \mathrm{R2} \cdot (\mathrm{R5} + \mathrm{R7}) \end{aligned}$$

The final reducts can be either $\{\mathrm{R1} \cdot \mathrm{R2} \cdot \mathrm{R5}\}$ or $\{\mathrm{R1} \cdot \mathrm{R2} \cdot \mathrm{R7}\}$, which are represented respectively in Tables 6 and 7.

Table 6. Reduct (R1 · R2 · R5)

R1		R2		R5		ZONE
V_1	I_1	V_2	I_2	V_5	I_5	
L	H	L	L	L	H	Z11
L	L	L	H	L	H	Z12
L	L	L	L	L	H	Z25
L	L	L	L	L	L	Z27

Table 7. Reduct (R1 · R2 · R7)

R1		R2		R7		ZONE
V_1	I_1	V_2	I_2	V_7	I_7	
L	H	L	L	N	H	Z11
L	L	L	H	N	H	Z12
L	L	L	L	N	H	Z25
L	L	L	L	L	H	Z27

Table 8. RDM for Table 6

Index	Z11	Z12	Z25	Z27
Z11	∅	R1,R2	R1	R1,R5
Z12	R1,R2	∅	R2	R2,R5
Z25	R1	R2	∅	R5
Z27	R1,R5	R2,R5	R5	∅

$f(\text{Z11}) = \mathbf{R1}$; $f(\text{Z12}) = \mathbf{R2}$;
$f(\text{Z25}) = \mathbf{R1{\cdot}R2{\cdot}R5}$;$f(\text{Z27}) = \mathbf{R5}$.

Table 9. RDM for Table 7

Index	Z11	Z12	Z25	Z27
Z11	∅	R1,R2	R1	R1,R7
Z12	R1,R2	∅	R2	R2,R7
Z25	R1	R2	∅	R7
Z27	R1,R7	R2,R7	R7	∅

$f(\text{Z11}) = \mathbf{R1}$; $f(\text{Z12}) = \mathbf{R2}$;
$f(\text{Z25}) = \mathbf{R1{\cdot}R2{\cdot}R7}$; $f(\text{Z27}) = \mathbf{R7}$.

Table 10. Rules for Reduct:(R1 · R2 · R5)

R1		R2		R5		ZONE	Yes
V_1	I_1	V_2	I_2	V_5	I_5		
L	H	•	•	•	•	Z11	√
•	•	L	H	•	•	Z12	√
L	L	L	L	L	H	Z25	√
•	•	•	•	L	L	Z27	√

Table 11. Rules for Reduct (R1 · R2 · R7)

R1		R2		R7		ZONE	Yes
V_1	I_1	V_2	I_2	V_7	I_7		
L	H	•	•	•	•	Z11	R
•	•	L	H	•	•	Z12	R
L	L	L	L	N	H	Z25	√
•	•	•	•	L	H	Z27	√

•: do not care. √: good rules. ×: bad rules. 'R': redundant rules.

The discernibility function obtained may still include some unnecessary values of the condition attributes in the decision table. To solve this, we calculate a relative reduct and core of attribute values using the relative discernibility function. For simplicity, the demonstration is based on the relative discernibility matrix (RDM) constructed for the subspace $\{\text{R1} \cdot \text{R2} \cdot \text{R5}\}$ and $\{\text{R1} \cdot \text{R2} \cdot \text{R7}\}$ as shown in Tables 8 and 9. For instance in Table 8, $f(\text{Z11}) = \{\text{R1} + \text{R2}\} \cdot \text{R1} \cdot \{\text{R1} + \text{R5}\} = \text{R1}$. Note: individual voltage and current measurements can be considered to yield more concise rules.

Four concise decision rules can be formed by refining the list of the rules extracted from Tables 10 and 11 and by combining the switching action data in Table 4. The rules can be read as: -

Rule 1: IF LOW voltage and HIGH current at LOAD BUS L1 **AND** breaker BRK1 OPEN, THEN FAULT in ZONE 11

Rule 2: IF LOW voltage and HIGH current at LOAD BUS L2 **AND** breaker BRK2 OPEN, THEN FAULT in ZONE 12

Rule 3: IF LOW voltage and LOW current at LOAD BUS L1 and L2 **AND** HIGH CURRENT at BUSES L5 and L7 **AND** breaker BRK5 OPEN, THEN FAULT in ZONE 25

Table 12. GA Reducts for Case 3.1

No	Reducts	Support	Length
1	$\{V_1, I_1\} = \{R1\}$	100	1
2	$\{V_2, I_2\} = \{R2\}$	100	1
3	$\{V_5, I_5\} = \{R5\}$	100	1
4	$\{V_7, I_7\} = \{R7\}$	100	1
5	$\{V_1, I_1, V_2, I_2, V_5, I_5\} = \{R1, R2, R5\}$	100	3
6	$\{V_1, I_1, V_2, I_2, V_7, I_7\} = \{R1, R2, R7\}$	100	3

Rule 4: IF LOW voltage and HIGH current at BUS L7 **AND** LOW VOLTAGE at BUS L5 **AND** breaker BRK7 OPEN, THEN FAULT in ZONE 27

To optimise the search of reducts, we employed a GA, in which the candidates are converted to bit vectors; that is 1110 or 1101. Note: $\{R1, R2, R5, R7\} = 1111$, $\{R1, R2, R5\} = 1110$ and $\{R1, R2, R7\} = 1101$. The performance of an individual solution is evaluated using the fitness function given in Eq. (1). Tables 12 and 13 show the reducts and the rules generated using the GA, which is similar to the combined decision rules of Tables 10 and 11.

Each reduct in Table 12 has an associated support count that measures the *strength* of the reduct equivalent to the reduct's hitting fraction multiplied by 100. This is the percentage of sets in the collection of sets that the attribute of interest has a non-empty intersection with [28]. *Length* indicates the number of attributes in the reducts.

The quality of rules from Table 13 can be assessed based on the metrics: RHS and LHS support, accuracy coverage and length. The **LHS** ("*left hand side*") support signifies the number of events in the data set. The **RHS** ("*right hand side*") support signifies the number of events in the data set that matches the 'if' part of the decision rule and has the decision value of the 'then' part (consequent). For an inconsistent rule, the 'then' part shall consist of several decisions. *Accuracy* and *coverage* are computed from the support counts using Eq. (3) and Eq. (4). Since there is no inconsistency in the decision system, the accuracy of rules is equal to 1.0. *Length* (not shown due to the space constraints in the tables) indicates the number of attributes in the LHS or RHS. For instance, Rule 2 in Table 13 has LHS length $= 1$ $\{(V_2, I_2)\}$ and RHS length $= 1$ $\{(Z11)\}$, whereas Rule 3 has LHS length $= 3$ $\{(V_1, I_1), (V_2, I_2), (V_5, I_5)\}$ and RHS length $= 1$ $\{(Z25)\}$.

Some rules generated are redundant due to similar patterns or carrying the same information. Thus, they are denoted as 'R'. Depending on data availability, these redundant reducts can be used to classify the events if there are some IED relays not available or out of service. For example, if R5 is not available, we can use the data from R7 to classify the fault zone in Zone 25 and Zone 27.

Table 13. GA Rules generated from Table 12

No	List of rules	LSup	RSup	Acc	LCov	RCov	Zone	Yes
1	$\{V_1(L), I_1(H)\}$	1.0	1.0	1.0	0.25	1.000	Z11	√
2	$\{V_2(L), I_2(H)\}$	1.0	1.0	1.0	0.25	1.000	Z12	√
3	$\{V_1(L), I_1(L)\} \wedge \{V_2(L), I_2(L)\} \wedge \{V_5(L), I_5(H)\}$	1.0	1.0	1.0	0.25	3.000	Z25	√
4	$\{V_1(L), I_1(L)\} \wedge \{V_2(L), I_2(L)\} \wedge \{V_7(N), I_7(H)\}$	1.0	1.0	1.0	0.25	3.000	Z25	R
5	$\{V_7(L), I_7(H)\}$	1.0	1.0	1.0	0.25	1.000	Z27	√
6	$\{V_5(L), I_5(L)\}$	1.0	1.0	1.0	0.25	1.000	Z27	R

'LSup': LHS Support. 'RSup': RHS Support. 'Acc': Accuracy. 'LCov': LHS Coverage. 'RCov': RHS Coverage.

3.2 Fault Data in Time Series

Unlike the data given in Tables 3 and 4, the real time operational data captured in a substation is usually in a time series format. Thus, it is necessary to extract knowledge from this type of data set. This example considers a single phase to ground fault (F1) applied to the bus L1 in Fig. 2 using the data collected from the simulation. The fault initiated at 1.000s results in the operation of the relay R1, the tripping of BRK1 and the isolation of bus L1.

Table 14 shows that the IED1 relay has operated whilst the other relays remain stable. 0.139s is the time required for the simulated system to reach the steady state condition from its initial zero value. The IED1 relay picked up the fault at 1.004s, tripped at 1.937s and reset at 2.010s after the current has dropped below the threshold value. The breaker BRK1 opened at 2.007s.

Table 15 presents a simple modified time series data set, which is composed of a set of discrete voltages and currents over a time period of 0.139s to 2.012s. **U** is the universe of events and **B** is the set of condition attributes $\{R1, R2, R5, R7\}$. A set of decision values is pre-classified in the decision attribute $\boldsymbol{d}=d_1 d_2$. d_1 consists of four state classification i.e. Normal (**N**), Alert (**A**), Emergency (**E**) and Safe (**S**). Normal (**N**) indicates that all the constraints and loads are satisfied, i.e. the voltages and currents are nominal. Alert (**A**) indicates at least one current is high and the voltages are nominal, or the currents are nominal but at least one voltage is abnormal. Emergency (**E**) indicates at least two physical operating limits are violated (e.g. under voltages and over currents). Safe (**S**) is when those parts of the power system that remain are operating normally, but one or more loads are not satisfied after a breaker has opened [30]. d_2 is used to capture the breaker information. $d_2 = 1$ indicates that a breaker has opened and the

Table 14. Protection status of IED Relay R1

Time	R1				
t/s	Pickup	Trip	AR	52A	52B
0.139	0	0	0	0	1
1.004	1	0	0	0	1
1.937	1	1	0	0	1
2.007	1	1	0	1	0
2.010	0	0	0	1	0

R1: IED relay 1, Pickup: pickup time, Trip: trip time, AR: auto re-closer (disabled), 52A and 52B: breaker auxiliary contacts in which '01': close; '10': open.

Table 15. IED data in time Series (decision system)

Time	R1		R2		R5		R7		Decision
t/s	V_1	I_1	V_2	I_2	V_5	I_5	V_7	I_7	$d_1 d_2$
0.139	N	N	N	N	N	N	N	N	N_S
1.003	N	H	N	N	N	N	N	N	A_0
1.004	L	H	N	N	N	N	N	H	E_0
1.005	L	H	N	L	N	H	L	H	E_0
1.006	L	H	L	L	L	H	L	H	E_0
2.007	L	N	L	L	L	H	L	H	E_1
2.011	L	N	L	N	L	H	L	N	A_1
2.012	L	L	N	N	N	N	N	N	S_1

State classification: N_S = normal with no breaker operation, A_0 = abnormal with no breaker operation, A_1 = abnormal with breaker(s) operation, E_0 = emergency with no breaker operation, E_1 = emergency with breaker(s) operation, S_1 = safe with breaker(s) operation.

respective line has been disconnected whereas $d_2 = 0$ indicates that a breaker has closed and the respective line has been connected.

By observing the set of events in Table 15, it may be easily identified that the problem is actually within the supervised region of the IED1 relay. However, given that there are n number of events and m number of relays, this may become impractical in an actual problem. Additionally, the data set is given in a perfect pattern, which may not be always the case with the real time data received from the control system. The irregularity in data pattern makes it much harder for a human to handle a complex situation.

Table 16 shows the discernibility matrix for Table 15. Due to the lack of space, the columns of the discernibility matrix in the table have been simplified such that R1 is given as "1", R2 as "2" and etc. The discernibility function is calculated using an absorption law in each column of Table 16.

Table 16. Discernibility matrix for Table 15

Time	0.139	1.003	1.004	1.005	1.006	2.007	2.011	2.012
0.139	∅							
1.003	1	∅						
1.004	1,7	1,7	∅					
1.005	1,2,5,7	1,2,5,7	∅	∅				
1.006	1,2,5,7	1,2,5,7	∅	∅	∅			
2.007	1,2,5,7	1,2,5,7	1,2,5,7	1,2,5	1	∅		
2.011	1,2,5,7	1,2,5,7	1,2,5,7	1,2,5,7	1,2,7	2,7	∅	
2.012	1	1	1,7	1,2,5,7	1,2,5,7	1,2,5,7	1,2,5,7	∅

Notation 1,2,5,7: {R1, R2, R5, R7} and 1,7: {R1, R7}.

The final discernibility function is: -

$$f(B) = f(0.139) \cdot f(1.003) \cdot f(1.004) \cdot f(1.005) \cdot \ldots f(2.007) \ldots \cdot f(2.011)$$
$$= 1 \cdot 1 \cdot (1+7) \cdot (1+2+5) \cdot \ldots (2+7) \ldots \cdot (1+2+5+7)$$
$$= 1 \cdot (2+7) \Rightarrow \{\mathrm{R1}\} \cdot \{\mathrm{R2} + \mathrm{R7}\}$$

The example shows that the R1 and R2 or R7 relays are the main source of information to justify the outcome of interest. Table 15 can then be reduced to Tables 17 and 18. Tables 21 and 22 show the list of the concise rules extracted using the relative discernibility matrices derived in Tables 19 and 20. For simplicity, the discernibility matrices are constructed for the subspace $\{R_1, R_2)\}$ and $\{R_1, R_7\}$ respectively rather than considering individual voltage and current measurements which can typically yield more concise rules.

The event rules formed from both Tables 21 and 22 are combined and categorised into 4 different classes according to their outcomes. To avoid confusion, they are renamed to Rule 1 - 5 as follows, omitting the first event which represents the normal operation and is not of interest for fault classification and event extraction: -

RULES applied to a FAULT within ZONE 11

Rule 1: IF NOMINAL voltage **AND** HIGH current ON BUS L1
$\Longrightarrow$ **CLASSIFIED STATE: ABNORMAL**. System at High Alert. Zone

Table 17. Reduct:{R1 · R2}

Time	R1		R2		Dec.	Yes
t/s	V_1	I_1	V_2	I_2	$d_1 d_2$	
0.139	N	N	N	N	N_S	√
1.003	N	H	N	N	A_0	√
1.004	L	H	N	N	E_0	√
1.005	L	H	N	L	E_0	√
1.006	L	H	L	L	E_0	√
2.007	L	N	L	L	E_1	×
2.011	L	N	L	N	A_1	√
2.012	L	L	N	N	S_1	√

Table 18. Reduct:{R1 · R7}

Time	R1		R7		Dec.	Yes
t/s	V_1	I_1	V_7	I_7	$d_1 d_2$	
0.139	N	N	N	N	N_S	√
1.003	N	H	N	N	A_0	√
1.004	L	H	N	H	E_0	√
1.005	L	H	L	H	E_0	√
2.007	L	N	L	H	E_1	√
2.011	L	N	L	N	A_1	√
2.012	L	L	N	N	S_1	√

Table 19. RDM for Table 17

Index	0.139	1.003	1.004	1.005	1.006	2.007	2.011	2.012
0.139	∅	R_1	R_1	R_1, R_2	R_1, R_2	R_1, R_2	R_1, R_2	R_1
1.003	R_1	∅	R_1	R_1, R_2	R_1, R_2	R_1, R_2	R_1, R_2	R_1
1.004	R_1	R_1	∅	∅	∅	R_1, R_2	R_1, R_2	R_1
1.005	R_1, R_2	R_1, R_2	∅	∅	∅	R_1, R_2	R_1, R_2	R_1, R_2
1.006	R_1, R_2	R_1, R_2	∅	∅	∅	R_1	R_1, R_2	R_1, R_2
2.007	R_1, R_2	R_1, R_2	R_1, R_2	R_1, R_2	R_1	∅	R_2	R_1, R_2
2.011	R_1, R_2	R_1, R_2	R_1, R_2	R_1, R_2	R_1, R_2	R_2	∅	R_1, R_2
2.012	R_1	R_1	R_1	R_1, R_2	R_1, R_2	R_1, R_2	R_1, R_2	∅
f (B)	R_1	R_1	R_1	$R_1 + R_2$	R_1	$R_1 \cdot R_2$	R_2	R_1

11 experiences LOW Voltage and High Current. Monitor the operation of relays (any pickups or tripping) and generate alarms to operator.

Rule 2: IF LOW voltage **and** NOMINAL current ON BUS L2 **OR** BUS L7 **AND** BRK1 OPEN

$\Longrightarrow$ **CLASSIFIED STATE: ABNORMAL**. Protection at Zone 11 has responded. System recovering. Situation under control but still unsafe. Monitor voltage threshold and initiate a voltage regulator if necessary.

Rule 3: IF LOW voltage **and** HIGH current ON BUS L1 **AND** HIGH CURRENT ON BUS L7 **OR** LOW CURRENT ON BUS L2

$\Longrightarrow$ **CLASSIFIED STATE: EMERGENCY**. System is unstable and urgent action is required. Protection has not yet responded. Check the condition of relays and initiate back up protection if necessary.

Table 20. RDM for Table 18

Time	0.139	1.003	1.004	1.005	2.007	2.011	2.012
0.139	$\emptyset$	R_1	R_1, R_7	R_1, R_7	R_1, R_7	R_1, R_7	R_1
1.003	R_1	$\emptyset$	R_1, R_7	R_1, R_7	R_1, R_7	R_1, R_7	R_1
1.004	R_1, R_7	R_1, R_7	$\emptyset$	$\emptyset$	R_1, R_7	R_1, R_7	R_1, R_7
1.005	R_1, R_7	R_1, R_7	$\emptyset$	$\emptyset$	R_1	R_1, R_7	R_1, R_7
2.007	R_1, R_7	R_1, R_7	R_1, R_7	R_1	$\emptyset$	R_7	R_1, R_7
2.011	R_1, R_7	R_1, R_7	R_1, R_7	R_1, R_7	R_7	$\emptyset$	R_1, R_7
2.012	R_1	R_1	R_1, R_7	R_1, R_7	R_1, R_7	R_1, R_7	$\emptyset$
f (B)	R_1	R_1	$R_1 + R_7$	R_1	$R_1 \cdot R_7$	R_7	R_1

Table 21. Rules for Reduct {R1 · R2}

Time	**R1**		**R2**		**Dec.**	Yes
t/s	V_1	I_1	V_2	I_2	d_1d_2	
0.139	N	N	•	•	N_S	×
1.003	N	H	•	•	A_0	√
1.004	L	H	•	•	E_0	√
1.005	L	H	•	•	E_0	R
1.005	•	•	N	L	E_0	×
1.006	L	H	•	•	E_0	R
2.007	L	N	L	L	E_1	×
2.011	•	•	L	N	A_1	√
2.012	L	L	•	•	S_1	√

Table 22. Rules for Reduct {R1 · R7}

Time	**R1**		**R7**		**Dec.**	Yes
t/s	V_1	I_1	V_7	I_7	d_1d_2	
0.139	N	N	•	•	N_S	×
1.003	N	H	•	•	A_0	R
1.004	L	H	•	•	E_0	R
1.004	•	•	N	H	E_0	×
1.005	L	H	•	•	E_0	R
2.007	L	N	L	H	E_1	√
2.011	•	•	L	N	A_1	R
2.012	L	L	•	•	S_1	R

Rule 4: IF LOW voltage **and** NOMINAL current ON BUS L1 **AND** HIGH CURRENT ON BUS L7 **OR** LOW CURRENT ON BUS L2 **AND** BRK1 OPEN
$\Longrightarrow$ **CLASSIFIED STATE: EMERGENCY**. Protection at Zone 11 has responded. Situation not yet under control and still very unstable. Monitor voltage and current threshold at Bus L1 and L7. Fault location within Zone 11.

Rule 5: IF LOW voltage **and** LOW current on BUS L1 **AND** BRK1 OPEN
$\Longrightarrow$ **CLASSIFIED STATE: SAFE**. System within the safe margin. Generate a fault analysis report that identifies the fault type and the affected region. The condition of the protection is evaluated. Restoration procedure and maintenance records also generated accordingly.

Despite the tedious way of extracting rules, detailed rules for each fault scenario can be obtained relatively easy via extension simulation of the power network. The rules generated are for EACH scenario and they are stored in the knowledge base system. The inference engine then uses a lookup table to retrieve the mapping between the input values and the rule's consequent(s) for each scenario. This means that if the fault symptom matches the list of rules given (the antecedents of the extracted rules), a fault in Zone Z11 (Zone 1 supervised by the relay R1) is concluded. Different sets of decisions could also be fired based on the rule's consequent(s). A lookup table can be thought of as a matrix, which has as many columns as there are inputs, and as many rows as outcomes. The inference engine thus takes a set of inputs and matches this input pattern with the patterns in the matrix, stored in rows. The best match determines the outcome. The matching process could sum the matching bits in a row, or carry out a multiplication of the input as a column vector with the matrix. The largest element in the product column selects the outcome. In a time independent system, where the outcome depends only on the instantaneous state of the inputs, the case structure can be very conveniently expressed in this form of matrix. The advantage of using this method is that the rules can be induced easily. This saves significant time and cost when developing a knowledge base.

The two examples show that the RS-GA approach is capable of inducing the decision rules from a substation database, even though the data may not be complete because the database has been simplified for this demonstration purpose. A more complete decision system and substation topology has been used and tested (see the following section) and the algorithm is still able to work efficiently despite the presence of some levels of missing/defective data in the database. The rules may also look predictable for a small substation in Fig. 2. However, when considering a larger substation or a complex power network with a significant number of protection system(s), extracting rules manually may be time-consuming and require significant resources. As such, this hybrid method will be useful to power utilities for exploiting substation rules. Furthermore, relying on the switching actions for estimating a faulty section might not always be adequate when considering relay failures and the complexity of a power network. Therefore, we believe that voltage and current components should also be considered in a fault section estimation procedure.

The classification sometimes does not clearly justify the status alarm (see the event at 1.005s, 2.007s in Table 21 and event at 1.004s in Table 22). If the extracted rule does not comply with the state classification (Normal, Abnormal, Emergency and Safe) set earlier, it does not mean that the rule extraction is inaccurate, because the data set does not contain adequate information to classify the events. This limited knowledge of the data set can be noticed by comparing the unmatched rules with the original table. These rules should be verified and refined by experts to improve their coverage.

Tables 23 and 24 show the reducts and rules generated using a GA. The outcomes are those obtained in the example but with additional rules from relay R5 $\{V_5, I_5\}$.

Table 23. GA Reducts for Case 3.2

No	Reducts	Support	Length
1	$\{V_1, I_1\}$	100	1
2	$\{V_2, I_2\}$	100	1
3	$\{V_5, I_5\}$	100	1
4	$\{V_7, I_7\}$	100	1
5	$\{V_1, I_1\}, \{V_2, I_2\}$	100	2
6	$\{V_1, I_1\}, \{V_5, I_5\}$	100	2
7	$\{V_1, I_1\}, \{V_7, I_7\}$	100	2

Table 24. GA Rules generated from Table 23

No	List of rules	LSup	RSup	Acc	LCov	RCov	Zone	Yes
1	$\{V_1(N), I_1(N)\}$	1	1	1.0	0.125	1.000	N_S	×
2	$\{V_1(N), I_1(H)\}$	1	1	1.0	0.125	1.000	A_0	√
3	$\{V_1(L), I_1(H)\}$	3	3	1.0	0.375	1.000	E_0	√
4	$\{V_2(N), I_2(L)\}$	1	1	1.0	0.125	0.333	E_0	×
5	$\{V_1(L), I_1(N)\} \wedge \{V_2(L), I_2(L)\}$	1	1	1.0	0.125	1.000	E_1	×
6	$\{V_2(L), I_2(N)\}$	1	1	1.0	0.125	1.000	A_1	√
7	$\{V_1(L), I_1(L)\}$	1	1	1.0	0.125	1.000	S_1	√
8	$\{V_7(N), I_7(H)\}$	1	1	1.0	0.125	0.333	E_0	×
9	$\{V_7(L), I_7(N)\}$	1	1	1.0	0.125	1.000	A_1	√
10	$\{V_1(L), I_1(N)\} \wedge \{V_7(L), I_7(H)\}$	1	1	1.0	0.125	1.000	E_1	√
11	$\{V_5(N), I_5(H)\}$	1	1	1.0	0.125	0.333	E_0	×
12	$\{V_5(L), I_5(N)\}$	1	1	1.0	0.125	1.000	A_1	√
13	$\{V_1(L), I_1(N)\} \wedge \{V_5(L), I_5(H)\}$	1	1	1.0	0.125	1.000	E_1	√

Depending on the fitness function setting, the GA searches all the possible solutions to characterise the events. The GA settings used in this case study are: mutation rate = 0.01, crossover rate = 0.6, inversion rate = 0.05, number of generations = 500, population = 70. This is however not the optimum setting for rule generation since the large inversion rate (of 0.05) could lead to lower fitness than 0.01 [31]. The fitness function bias, the weighting between the subset cost and hitfraction, is set to 0.9 to reward shorter elements/rule length as we want to pick the shortest length reducts that satisfy the results. These settings are applied intentionally to demonstrate that GA can provide us with more flexibility to generate a number of rules subject to the parameter settings we provide. Such function is important in the substation rule extraction. It leads to a simple problem formulation that is able to produce global optimal solutions. The rules generated by the GA are generally large and its advantage becomes much more visible when few events are available and more knowledge is required to justify the entire outcomes.

4 Example II

This example demonstrates a more complex distribution substation that is comprised of various types of relay models.

4.1 Fault Scenarios on Defined Protection Zones

Figure 3 shows a typical double bus 132/11kV substation network model developed using PSCAD/EMTDC. Like the Case Study 3.1 in Example 1, a selection of fault scenarios was applied to the network and the operating responses of the relays, circuit breakers, voltage and current sensors were collected and stored in an event database.

The network model is protected by several types of relay models, each of which includes one or more of the protection functions listed in Table 25. For instance, the directional relays at R5 and R6 also include non-directional time graded earth fault elements to protect the 11kV busbar and provide backup for the 11kV feeders supervised by relays R1, R2, R3 and R4. Table 26 shows the trip data of IED R6 and R8 collected from the PSCAD/EMTDC simulation. Unlike the previous example, four protection zones are defined in Fig. 3 due to the size of the supervised network and IED devices. Breakers BRK5 and BRK7 are regarded as one breaker zone labelled as "BZ2". Similarly, BRK6 and BRK8 are regarded as one labelled as "BZ3". Zone 1 represents the protection zones of R1, R2, R3 and R4. Zone 2 includes the protection zones of R5, R7, R9 and R11 and Zone 3 covers the protection zones of R6, R8, R10 and R12. Zone 4 is the busbar protection zone (not considered in this scenario). Zone 25 indicates the regional Zone 2 supervised by the relay R5.

The bus-coupler BC is assumed closed prior to the fault. To prevent both transformers from tripping as the result of a fault on the 11kV terminal, R5 and R6 are set to look into their respective transformers in accordance with IEEE nomenclature 67. Both 132/11kV transformers are protected by differential unit protection, restricted earth fault protection and balanced earth fault protection [32]. The sensitive earth fault protection is not required since the neutral of the transformer is solidly earthed. The set of collected situations for this example are approximately 7,000 - 10,000 cases (1ms time tagged) and more than 300

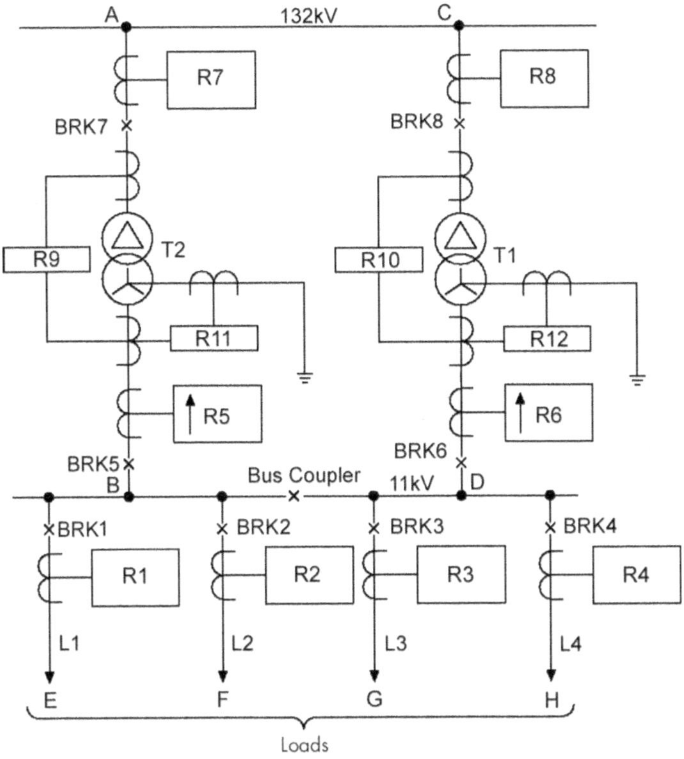

Fig. 3. 132/11kV substation model with a double bus arrangement

condition attributes (i.e. single phase and three phase power, voltage, current and phase angle as well as other protection trip signals and fault indictors) are available. However, only 35 attributes of three phase voltages and currents and respective protection trip signals are chosen.

The case covers a wide range of typical voltage and current situations that occur in each scenario. Due to its large size, only change of state data is presented.

Tables 27 and 28 lay out a list of recorded voltage and current patterns as well as the switching actions caused by the protection system(s) subject to various faults at different locations in the substation. Bx is the short form of the breaker BRKx in which x $= \{1, 2, 3, 4\}$. H1 indicates the current is flowing in the direction that would trigger one of the directional relays, i.e. R5 and R6.

Table 27 was converted into a discernibility matrix and the discernibility functions are derived from each column before the final discernibility function is calculated. A total of 24 reducts based on Relay Box measurements are generated and listed in Table 29. Tables 30 and 31 list the decision rules generated by the hybrid RS-GA model. The former considers a relay box that combines both the voltage and current measurements while the latter considers the two measurements separately. Combining the relay trip signals and breaker data from

Table 25. List of IED relay models in the 132/11kV double bus substation

IED Plate numbers	1,2,3,4	5,6	7,8	9,10	11,12
Instantaneous Overcurrent, 50	√	√	√	×	×
Instantaneous Earth Fault, 50N	√	√	√	×	×
Time delayed Overcurrent, 51	√	√	√	×	×
Time delayed Earth Fault, 51N	√	√	√	×	×
Balanced Earth Fault, 51G	×	×	√	×	×
Standby Earth Fault, 51NB	×	×	×	×	√
Directional Phase Overcurrent, 67	×	√	×	×	×
Directional Earth Fault, 67N	×	√	×	×	×
Auto-recloser, 79	√	×	×	×	×
Restricted Earth Fault, 87N	×	×	×	×	√
Transformer Differential, 87T	×	×	×	√	×

√: available, ×: not available.

Table 26. Protection status of IED Relay R6 and R8

Time	R6				R8		
t/s	67	50/51	52A	52B	50/51	52A	52B
0.139	0	0	0	1	0	0	1
1.039	1	1	0	1	0	0	1
1.119	1	1	1	0	0	1	0
1.133	0	1	1	0	0	1	0
1.139	0	0	1	0	0	1	0

Table 28, we can improve the coverage of the extracted rules. In Table 30, it can be seen that the rules 5, 8, 22 and 29 are not really useful because I_9, I_{10}, I_{11} and I_{12} are from the unit protections which are only important if there is a fault on the transformer. Rules 12, 19, 28 and 36 are also not good enough to be considered due to the lack of rule coverage. For instance, low voltage and current on the bus 7 or 8 as indicated by rule 12 and 28 may not necessarily indicate a fault on Zone Z38 or Z27 respectively. Disregarding the unit protection data, Rule 19 and Rule 36 are inconsistent as they are practically the same, but give different outcomes. Furthermore, high current and low voltage on branch 7 and 8 indicate that the fault is downstream but do not necessarily indicate the transformer at

Table 27. List of voltage and current patterns with defined fault zones in Fig. 3

R1		R2		R3		R4		R5		R6		R7		R8		R9	R10	R11	R12	ZONE
V_1	I_1	V_2	I_2	V_3	I_3	V_4	I_4	V_5	I_5	V_6	I_6	V_7	I_7	V_8	I_8	I_9	I_{10}	I_{11}	I_{12}	
L	H	L	L	L	L	L	L	L	H	L	H	N	H	N	H	L	L	L	L	Z11
L	L	L	L	L	L	L	L	L	H1	L	H	N	H	N	H	L	L	L	L	Z25
L	L	L	L	L	L	L	L	L	H	L	H1	N	H	N	H	L	L	L	L	Z36
L	L	L	L	L	L	L	L	L	L	L	L	L	H	L	L	L	L	L	L	Z27
L	L	L	L	L	L	L	L	L	L	L	L	L	L	L	H	L	L	L	L	Z38
L	L	L	L	L	L	L	L	L	H1	L	H	L	H	L	H	H	L	L	L	Z2T
L	L	L	L	L	L	L	L	L	H1	L	H	L	H	L	H	L	L	H	L	Z2T
L	L	L	L	L	L	L	L	L	H1	L	H	L	H	L	H	H	L	H	L	Z2T
L	L	L	L	L	L	L	L	L	H	L	H1	L	H	L	H	L	H	L	L	Z3T
L	L	L	L	L	L	L	L	L	H	L	H1	L	H	L	H	L	L	L	H	Z3T
L	L	L	L	L	L	L	L	L	H	L	H1	L	H	L	H	L	H	L	H	Z3T

Table 28. List of switching actions with defined fault zones in Fig. 3

R1	R2	R3	R4	R5	R6	R7	R8	R9	R10	R11	R12	B1	B2	B3	B4	BZ2	BZ3	ZONE
10	01	01	01	01	01	01	01	01	01	01	01	10	01	01	01	01	01	Z11
01	10	01	01	01	01	01	01	01	01	01	01	01	10	01	01	01	01	Z12
01	01	10	01	01	01	01	01	01	01	01	01	01	01	10	01	01	01	Z13
01	01	01	10	01	01	01	01	01	01	01	01	01	01	01	10	01	01	Z14
01	01	01	01	10	01	01	01	01	01	01	01	01	01	01	01	10	01	Z25
01	01	01	01	01	10	01	01	01	01	01	01	01	01	01	01	01	10	Z36
01	01	01	01	01	01	10	01	01	01	01	01	01	01	01	01	10	01	Z27
01	01	01	01	01	01	01	10	01	01	01	01	01	01	01	01	01	10	Z38
01	01	01	01	01	01	01	01	10	01	10	01	01	01	01	01	10	01	Z2T
01	01	01	01	01	01	01	01	10	01	01	01	01	01	01	01	10	01	Z2T
01	01	01	01	01	01	01	01	01	01	10	01	01	01	01	01	10	01	Z2T
01	01	01	01	01	01	01	01	01	10	01	10	01	01	01	01	01	10	Z3T
01	01	01	01	01	01	01	01	01	10	01	01	01	01	01	01	01	10	Z3T
01	01	01	01	01	01	01	01	01	01	01	10	01	01	01	01	01	10	Z3T

fault. Without the information about the switching actions of the relays and proper refinements, these rules 5, 8, 22 and 28 will not be good/fit enough for use. If the extracted rule does not comply with zone classification (decision) set earlier, the rule should thus be verified and refined by experts to improve its

Table 29. GA Reducts generated (based on Relay Box) for Case 4.1

No	Reducts	Support	Length
1	$\{V_1, I_1\}$	100	1
2	$\{V_5, I_5\} \cdot \{V_6, I_6\}$	100	2
3	$\{V_5, I_5\} \cdot \{V_8, I_8\}$	100	2
4	$\{V_5, I_5\} \cdot \{V_7, I_7\}$	100	2
5	$\{V_5, I_5\} \cdot I_9 \cdot I_{11}$	100	3
6	$\{V_1, I_1\} \cdot \{V_6, I_6\} \cdot \{V_7, I_7\}$	100	3
7	$\{V_1, I_1\} \cdot \{V_6, I_6\} \cdot \{V_8, I_8\}$	100	3
8	$\{V_1, I_1\} \cdot \{V_6, I_6\} \cdot I_9 \cdot I_{11}$	100	4
9	$\{V_6, I_6\} \cdot \{V_8, I_8\}$	100	2
10	$\{V_6, I_6\} \cdot \{V_7, I_7\}$	100	2
11	$\{V_6, I_6\} \cdot I_{10} \cdot I_{12}$	100	3
12	$\{V_1, I_1\} \cdot \{V_5, I_5\} \cdot \{V_8, I_8\}$	100	3
13	$\{V_1, I_1\} \cdot \{V_5, I_5\} \cdot \{V_7, I_7\}$	100	3
14	$\{V_1, I_1\} \cdot \{V_5, I_5\} \cdot I_{10} \cdot I_{12}$	100	4
15	$\{V_8, I_8\}$	100	1
16	$\{V_7, I_7\} \cdot I_9 \cdot I_{10} \cdot I_{11} \cdot I_{12}$	100	5
17	$\{V_7, I_7\}$	100	1
18	$\{V_8, I_8\} \cdot I_9 \cdot I_{10} \cdot I_{11} \cdot I_{12}$	100	5
19	$\{I_9\}$	100	1
20	$\{V_7, I_7\} \cdot \{V_8, I_8\} \cdot I_{10} \cdot I_{12}$	100	4
21	I_{11}	100	1
22	I_{10}	100	1
23	$\{V_7, I_7\} \cdot \{V_8, I_8\} \cdot I_9 \cdot I_{11}$	100	4
24	I_{12}	100	1

coverage. To form a more complete final set of rules for locating the fault section in the network, we combine the rules given in Table 30 or Table 31 with the switching actions of protection systems in Table 28 to improve the coverage and accuracy of the rules.

Voting results: Table 32 illustrates the results computed by the voting algorithm with the certainty coefficients computed for each decision class. In Case A,

Table 30. GA Rules generated from Table 29

No	List of rules	LSup	RSup	Acc	LCov	RCov	Zone	Yes
1	$\{V_1(L), I_1(H)\}$	1.0	1.0	1.0	0.091	1.000	Z11	√
2	$\{V_5(L), I_5(H)\} \wedge \{V_6(L), I_6(H)\}$	1.0	1.0	1.0	0.091	1.000	Z11	√
3	$\{V_5(L), I_5(H1)\} \wedge \{V_7(N), I_7(H)\}$	1.0	1.0	1.0	0.091	1.000	Z25	√
4	$\{V_5(L), I_5(H1)\} \wedge \{V_8(N), I_8(H)\}$	1.0	1.0	1.0	0.091	1.000	Z25	R
5	$\{V_5(L), I_5(H1)\} \wedge I_9(L) \wedge I_{11}(L)$	1.0	1.0	1.0	0.091	1.000	Z25	×
6	$\{V_1(L), I_1(L)\} \wedge \{V_6(L), I_6(H)\} \wedge \{V_7(N), I_7(H)\}$	1.0	1.0	1.0	0.091	1.000	Z25	√
7	$\{V_1(L), I_1(L)\} \wedge \{V_6(L), I_6(H)\} \wedge \{V_8(N), I_8(H)\}$	1.0	1.0	1.0	0.091	1.000	Z25	R
8	$\{V_1(L), I_1(L)\} \wedge \{V_6(L), I_6(H)\} \wedge I_9(L) \wedge I_{11}(L)$	1.0	1.0	1.0	0.091	1.000	Z25	×
9	$\{V_5(L), I_5(L)\} \wedge \{V_7(L), I_7(H)\}$	1.0	1.0	1.0	0.091	1.000	Z27	√
10	$\{V_6(L), I_6(L)\} \wedge \{V_7(L), I_7(H)\}$	1.0	1.0	1.0	0.091	1.000	Z27	R
11	$\{V_7(L), I_7(H)\} \wedge I_9(L) \wedge I_{10}(L) \wedge I_{11}(L) \wedge I_{12}(L)$	1.0	1.0	1.0	0.091	1.000	Z27	√
12	$\{V_8(L), I_8(L)\}$	1.0	1.0	1.0	0.091	1.000	Z27	×
13	$\{V_5(L), I_5(H1)\} \wedge \{V_7(L), I_7(H)\}$	3.0	3.0	1.0	0.373	1.000	Z2T	√
14	$\{V_5(L), I_5(H1)\} \wedge \{V_8(L), I_8(H)\}$	3.0	3.0	1.0	0.373	1.000	Z2T	R
15	$\{V_6(L), I_6(H)\} \wedge \{V_7(L), I_7(H)\}$	3.0	3.0	1.0	0.373	1.000	Z2T	√
16	$\{V_6(L), I_6(H)\} \wedge \{V_8(L), I_8(H)\}$	3.0	3.0	1.0	0.373	1.000	Z2T	R
17	$I_9(H)$	2.0	2.0	1.0	0.182	0.667	Z2T	√
18	$I_{11}(H)$	2.0	2.0	1.0	0.182	0.667	Z2T	√
19	$\{V_7(L), I_7(H)\} \wedge \{V_8(L), I_8(H)\} \wedge I_{10}(L) \wedge I_{12}(L)$	3.0	3.0	1.0	0.373	1.000	Z2T	×
20	$\{V_6(L), I_6(H1)\} \wedge \{V_8(N), I_8(H)\}$	1.0	1.0	1.0	0.091	1.000	Z36	√
21	$\{V_6(L), I_6(H1)\} \wedge \{V_7(N), I_7(H)\}$	1.0	1.0	1.0	0.091	1.000	Z36	R
22	$\{V_6(L), I_6(H1)\} \wedge I_{10}(L) \wedge I_{12}(L)$	1.0	1.0	1.0	0.091	1.000	Z36	×
23	$\{V_1(L), I_1(L)\} \wedge \{V_5(L), I_5(H)\} \wedge \{V_8(N), I_8(H)\}$	1.0	1.0	1.0	0.091	1.000	Z36	√
24	$\{V_1(L), I_1(L)\} \wedge \{V_5(L), I_5(H)\} \wedge \{V_7(N), I_7(H)\}$	1.0	1.0	1.0	0.091	1.000	Z36	R
25	$\{V_1(L), I_1(L)\} \wedge \{V_5(L), I_5(H)\} \wedge I_{10}(L) \wedge I_{12}(L)$	1.0	1.0	1.0	0.091	1.000	Z36	√
26	$\{V_6(L), I_6(L)\} \wedge \{V_8(L), I_8(H)\}$	1.0	1.0	1.0	0.091	1.000	Z38	√
27	$\{V_5(L), I_5(L)\} \wedge \{V_8(L), I_8(H)\}$	1.0	1.0	1.0	0.091	1.000	Z38	R
28	$\{V_7(L), I_7(L)\}$	1.0	1.0	1.0	0.091	1.000	Z38	×
29	$\{V_8(L), I_8(H)\} \wedge I_9(L) \wedge I_{10}(L) \wedge I_{11}(L) \wedge I_{12}(L)$	1.0	1.0	1.0	0.091	1.000	Z38	×
30	$\{V_5(L), I_5(H)\} \wedge \{V_8(L), I_8(H)\}$	3.0	3.0	1.0	0.373	1.000	Z3T	√
31	$\{V_5(L), I_5(H)\} \wedge \{V_7(L), I_7(H)\}$	3.0	3.0	1.0	0.373	1.000	Z3T	R
32	$\{V_6(L), I_6(H1)\} \wedge \{V_8(L), I_8(H)\}$	3.0	3.0	1.0	0.373	1.000	Z3T	√
33	$\{V_6(L), I_6(H1)\} \wedge \{V_7(L), I_7(H)\}$	3.0	3.0	1.0	0.373	1.000	Z3T	R
34	$I_{10}(H)$	2.0	2.0	1.0	0.182	0.667	Z3T	√
35	$I_{12}(H)$	2.0	2.0	1.0	0.182	0.667	Z3T	√
36	$\{V_7(L), I_7(H)\} \wedge \{V_8(L), I_8(H)\} \wedge I_9(L) \wedge I_{11}(L)$	3.0	3.0	1.0	0.373	1.000	Z3T	×

Table 31. GA Rules generated for Case 4.1 based on Individual Voltage and Current measurements to identify the faulty section

No	List of rules	LSup	RSup	Acc	LCov	RCov	Zone	Yes
1	$I_1(H)$	1	1	1	0.091	1.000	Z11	√
2	$I_5(H) \cdot I_6(H)$	1	1	1	0.091	1.000	Z11	√
3	$I_1(L) \cdot I_6(H) \cdot I_9(L) \cdot I_{11}(L)$	1	1	1	0.091	1.000	Z25	×
4	$I_1(L) \cdot I_6(H) \cdot V_7(N)$	1	1	1	0.091	1.000	Z25	√
5	$I_1(L) \cdot I_6(H) \cdot V_8(N)$	1	1	1	0.091	1.000	Z25	R
6	$I_5(H1) \cdot I_9(L) \cdot I_{11}(L)$	1	1	1	0.091	1.000	Z25	×
7	$I_5(H1) \cdot V_7(N)$	1	1	1	0.091	1.000	Z25	√
8	$I_5(H1) \cdot V_8(N)$	1	1	1	0.091	1.000	Z25	R
9	$I_5(L) \cdot I_7(H)$	1	1	1	0.091	1.000	Z27	√
10	$I_6(L) \cdot I_7(H)$	1	1	1	0.091	1.000	Z27	R
11	$I_8(L)$	1	1	1	0.091	1.000	Z27	×
12	$I_{11}(H)$	2	2	1	0.182	0.667	Z2T	√
13	$I_5(H1) \cdot V_7(L)$	3	3	1	0.273	1.000	Z2T	√
14	$I_5(H1) \cdot V_8(L)$	3	3	1	0.273	1.000	Z2T	R
15	$I_6(H) \cdot V_7(L)$	3	3	1	0.273	1.000	Z2T	√
16	$I_6(H) \cdot V_8(L)$	3	3	1	0.273	1.000	Z2T	R
17	$I_7(H) \cdot V_8(L) \cdot I_8(H) \cdot I_{10}(L) \cdot I_{12}(L)$	3	3	1	0.273	1.000	Z2T	√
18	$I_9(H)$	2	2	1	0.182	0.667	Z2T	√
19	$V_7(L) \cdot I_7(H) \cdot I_8(H) \cdot I_{10}(L) \cdot I_12(L)$	3	3	1	0.273	1.000	Z2T	×
20	$I_1(L) \cdot I_5(H) \cdot I_{10}(L) \cdot I_{12}(L)$	1	1	1	0.091	1.000	Z36	×
21	$I_1(L) \cdot I_5(H) \cdot V_7(N)$	1	1	1	0.091	1.000	Z36	√
22	$I_1(L) \cdot I_5(H) \cdot V_8(N)$	1	1	1	0.091	1.000	Z36	R
23	$I_6(H1) \cdot I_{10}(L) \cdot I_{12}(L)$	1	1	1	0.091	1.000	Z36	×
24	$I_6(H1) \cdot V_7(N)$	1	1	1	0.091	1.000	Z36	√
25	$I_6(H1) \cdot V_8(N)$	1	1	1	0.091	1.000	Z36	R
26	$I_5(L) \cdot I_8(H)$	1	1	1	0.091	1.000	Z38	√
27	$I_6(L) \cdot I_8(H)$	1	1	1	0.091	1.000	Z38	R
28	$I_7(L)$	1	1	1	0.091	1.000	Z38	×
29	$I_{10}(H)$	2	2	1	0.182	0.667	Z3T	√
30	$I_{12}(H)$	2	2	1	0.182	0.667	Z3T	√
31	$I_5(H) \cdot V_7(L)$	3	3	1	0.273	1.000	Z3T	√
32	$I_5(H) \cdot V_8(L)$	3	3	1	0.273	1.000	Z3T	R
33	$I_6(H1) \cdot V_7(L)$	3	3	1	0.273	1.000	Z3T	√
34	$I_6(H1) \cdot V_8(L)$	3	3	1	0.273	1.000	Z3T	R
35	$I_7(H) \cdot V_8(L) \cdot I_8(H) \cdot I_9(L) \cdot I_{11}(L)$	3	3	1	0.273	1.000	Z3T	×
36	$V_7(L) \cdot I_7(H) \cdot I_8(H) \cdot I_9(L) \cdot I_{11}(L)$	3	3	1	0.273	1.000	Z3T	×

Table 32. Accumulating the casted votes for all rules that fires

Index	Certainty	Fraction	Decimal
	certainty(x, (ZONE = Z11))	$^1/_{11}$	0.09
	certainty(x, (ZONE = Z25))	$^3/_{11}$	**0.27**
1a	certainty(x, (ZONE = Z27))	$^1/_{11}$	0.09
	certainty(x, (ZONE = Z36))	$^1/_{11}$	0.09
	certainty(x, (ZONE = Z38))	$^1/_{11}$	0.09
	certainty(x, (ZONE = Z2T))	$^3/_{11}$	**0.27**
	certainty(x, (ZONE = Z3T))	$^1/_{11}$	0.09
2a	certainty(x, (ZONE = Z25))	$^1/_2$	0.50
	certainty(x, (ZONE = Z2T))	$^1/_2$	0.50
	certainty(x, (ZONE = Z11))	$^1/_{15}$	0.07
	certainty(x, (ZONE = Z25))	$^4/_{15}$	**0.27**
1b	certainty(x, (ZONE = Z27))	$^2/_{15}$	0.13
	certainty(x, (ZONE = Z36))	$^1/_{15}$	0.07
	certainty(x, (ZONE = Z38))	$^2/_{15}$	0.13
	certainty(x, (ZONE = Z2T))	$^3/_{15}$	0.20
	certainty(x, (ZONE = Z3T))	$^2/_{15}$	0.13
2b	certainty(x, (ZONE = Z25))	$^3/_5$	**0.60**
	certainty(x, (ZONE = Z2T))	$^1/_5$	0.20
	certainty(x, (ZONE = Z3T))	$^1/_5$	0.20

assuming that only the rules presented with $V_5 = L$ or $I_5 = H1$ were fired, the voting algorithm for the support count index 1a (based on Table 30) concluded that the fault is in Zone 2. However, the algorithm is not able to distinguish specifically if the fault is in Zone 25 (fault on feeder 5) or Zone 2T (fault on Transformer T2).

The support count for the case $V_5 = L$ or $I_5 = H1$ with outcome Zone Z25 is 3, whereas the total support count for the case $V_5 = L$ or $I_5 = H1$ regardless any outcome is 11. The same procedure applies to Z2T in which the support count for the case $V_5 = L$ or $I_5 = H1$ with outcome Z2T is also 3. The support count index 2a (based on Table 31) also shows no marginal preferences for Zone 25 or Zone 2T. The reason for this conflict is because both fault locations give the similar effects on voltage and current on the feeder 5. To distinguish the two types of fault, extra information will be required.

Table 33. Classifier result using the 90% Training Set and 10% Test set

Training	Test Sets (10%)				Mean
Set (90%)	Split 1	Split 2	Split 3	Split 4	Accuracy
1	1.000	1.000	1.000	1.000	1.000
2	1.000	0.000	1.000	1.000	0.750
3	1.000	1.000	0.000	1.000	0.750
4	1.000	1.000	1.000	1.000	1.000
		Measure of Accuracy			**0.875**

Table 34. Classifier result using the 70% Training Set and 30% Test set

Training	Test Sets (30%)				Mean
Set (70%)	Split 1	Split 2	Split 3	Split 4	Accuracy
1	1.000	1.000	1.000	1.000	1.000
2	1.000	0.333	0.667	0.667	0.667
3	1.000	0.667	0.000	1.000	0.667
4	1.000	0.667	1.000	0.333	0.750
		Measure of Accuracy			**0.771**

Table 35. Classifier result using the 50% Training Set and 50% Test set

Training	Test Sets (50%)				Mean
Set (50%)	Split 1	Split 2	Split 3	Split 4	Accuracy
1	0.800	1.000	1.000	0.800	0.900
2	1.000	0.600	0.600	0.800	0.750
3	1.000	0.600	0.200	0.800	0.650
4	0.400	0.400	0.600	0.000	0.350
		Measure of Accuracy			**0.663**

Let us now assume that in Case B, the rules presented with $I_9 = L, V_5 = L$ or $I_5 = H1$ are fired. We have accumulated the cast votes for all rules that fire and divided them by the number of support count for all rules that fire, i.e 15 for support count index 1b (based on the rules in Table 30) and 5 for support count index 2b (based on the rules in Table 31). With more information given

Table 36. Protection status for Breaker BRK1 failure

Time	R1		R5		R6		Z11	BZ2	BZ3	R1 Failure
t/s	50/51	79	67	50/51	67	50/51	BRK1	BRK5-7	BRK6-8	Indicator
0.139	0	0	0	0	0	0	01	01	01	0
1.004	0	0	0	0	0	0	01	01	01	0
1.007	0	0	0	0	0	0	01	01	01	0
1.011	1	0	0	0	0	0	01	01	01	0
1.091	1	0	0	0	0	0	00	01	01	1
1.339	1	0	0	1	0	1	00	01	01	1
1.419	1	0	0	1	0	1	00	10	10	1
1.442	1	0	0	0	0	0	00	10	10	1
1.444	0	0	0	0	0	0	00	10	10	1

Table 37. IED data in time series for Breaker BRK1 failure

Time	R1		R2		R5		R6		R7		R8		BZ2	BZ3	S
t/s	V_1	I_1	V_2	I_2	V_5	I_5	V_6	I_6	V_7	I_7	V_8	I_8	BRK57	BRK68	a
0.139	N	N	N	N	N	N	N	N	N	N	N	N	01	01	
1.004	L	N	L	N	L	N	L	N	N	N	N	N	01	01	
1.005	L	H	L	L	L	N	L	N	N	N	N	N	01	01	
1.006	L	H	L	L	L	N	L	N	N	N	N	N	01	01	
1.007	L	H	L	L	L	H	L	H	N	N	N	N	01	01	
1.008	L	H	L	L	L	H	L	H	N	H	N	H	01	01	
1.419	L	H	L	L	L	H	L	H	N	H	N	H	10	10	
1.434	L	H	L	L	L	H	L	H	N	H	N	H	10	10	
1.437	L	H	L	L	L	H	L	H	N	H	N	H	10	10	
1.438	L	H	L	L	L	H	L	H	N	H	N	H	10	10	
1.439	L	H	L	L	L	H	L	H	N	H	N	H	10	10	
1.440	L	H	L	L	L	H	L	H	N	L	N	L	10	10	
1.442	L	H	L	L	L	L	L	L	N	L	N	L	10	10	
1.443	L	L	L	L	L	L	L	L	L	L	L	L	10	10	
1.444	L	L	L	L	L	L	L	L	L	L	L	L	10	10	
1.445	L	L	L	L	L	L	L	L	L	L	L	L	10	10	

BRK57: BRK5 and BRK7, BRK68: BRK6 and BRK8.

this time, the voting algorithm now indicates that the fault is in Zone 25 rather than in Zone 2T due to its higher support count in the given set of rules. This is in-line with expectations. However, the suggestion from the voting result should always be left, in the final analysis, to domain experts to decide the necessary action to be taken.

Table 38. GA Reducts (Relay Box) for Case 4.2 without Breaker data

No	Reducts	Support	Length
1	$\{V_1, I_1\}$	100	1
2	$\{V_2, I_2\}$	100	1
3	$\{V_5, I_5\}$	100	1
4	$\{V_6, I_6\}$	100	1
5	$\{V_7, I_7\}$	100	1
6	$\{V_8, I_8\}$	100	1
7	$\{V_1, I_1, V_5, I_5\}$	100	2
8	$\{V_1, I_1, V_6, I_6\}$	100	2
9	$\{V_1, I_1, V_7, I_7\}$	100	2
10	$\{V_1, I_1, V_8, I_8\}$	100	2
11	$\{V_2, I_2, V_5, I_5\}$	100	2
12	$\{V_2, I_2, V_6, I_6\}$	100	2
13	$\{V_2, I_2, V_7, I_7\}$	100	2
14	$\{V_2, I_2, V_8, I_8\}$	100	2
15	$\{V_5, I_5, V_7, I_7\}$	100	2
16	$\{V_5, I_5, V_8, I_8\}$	100	2
17	$\{V_6, I_6, V_7, I_7\}$	100	2
18	$\{V_6, I_6, V_8, I_8\}$	100	2

Table 39. GA Reducts (Relay Box) for Case 4.2 with Breaker data

No	Reducts	Support	Length
1	$\{V_1, I_1\}$, {BZ2}	100	2
2	$\{V_1, I_1\}$, {BZ3}	100	2
3	$\{V_2, I_2\}$, $\{V_7, I_7\}$, {BZ2}	100	3
4	$\{V_2, I_2\}$, $\{V_7, I_7\}$, {BZ3}	100	3
5	$\{V_2, I_2\}$, $\{V_8, I_8\}$, {BZ2}	100	3
6	$\{V_2, I_2\}$, $\{V_8, I_8\}$, {BZ3}	100	3

Table 40. GA Reducts (Individual Measurements) for Case 4.2 with Breaker data

Index No.	Number of reducts	Support	Length
1	I_1, V_2, {BZ2 OR BZ3}	100	3
2	I_1, V_3, {BZ2 OR BZ3}	100	3
3	I_1, V_4, {BZ2 OR BZ3}	100	3
4	I_1, V_5, {BZ2 OR BZ3}	100	3
5	I_1, V_6, {BZ2 OR BZ3}	100	3

Classifier's Performance: The original simulation data is randomly divided into three different training sets and test sets respectively with a partition of 90%, 70% and 50% of the data for training and 10%, 30% and 50% for testing. The procedure is repeated four times for four random splits of the data. This means that four different test sets were generated in each case and each of these was tested on every split of the training sets for a total of 4 runs. The splits are used to avoid results based on rules that were generated for a particular selection of events [33]. This makes the results more reliable and independent of one particular selection of events. **Note:** the classifier performance is assessed on the simulation data with relay box measurements, that is to use the combined voltage and current measurements rather than treating the two recorded sets of data separately.

The average accuracy of the classifier for the 10% test data is reasonably good, **87.5%** (see Table 33). The results have demonstrated that most of the rules generated from the training sets are able to classify the events in the test set. The inaccuracy occurred when the rules generated from the training sets 2 and 3 were used to classify the respective test sets 2 and 3. Only one event with the decision class Z3T was unclassified. The cause of this problem is due to the small data set used in this case study (only 11 events) and this leads to

Table 41. GA Rules generated from Table 38

No	List of rules	LSup	RSup	Acc	LCov	RCov	Zone	Yes
1	$\{V_1(N), I_1(N)\}$	1.0	1.0	1.0	0.063	1.000	N_S	×
2	$\{V_2(N), I_2(N)\}$	1.0	1.0	1.0	0.063	1.000	N_S	×
3	$\{V_5(N), I_5(N)\}$	1.0	1.0	1.0	0.063	1.000	N_S	×
4	$\{V_6(N), I_6(N)\}$	1.0	1.0	1.0	0.063	1.000	N_S	×
5	$\{V_1(L), I_1(N)\}$	1.0	1.0	1.0	0.063	1.000	A_0	√
6	$\{V_2(L), I_2(N)\}$	1.0	1.0	1.0	0.063	1.000	A_0	√
7	$\{V_1(L), I_1(H)\}, \{V_5(L), I_5(N)\}$	2.0	2.0	1.0	0.125	0.500	E_0	√
8	$\{V_1(L), I_1(H)\}, \{V_6(L), I_6(N)\}$	2.0	2.0	1.0	0.125	0.500	E_0	R
9	$\{V_1(L), I_1(H)\}, \{V_7(N), I_7(N)\}$	3.0	3.0	1.0	0.188	0.750	E_0	√
10	$\{V_1(L), I_1(H)\}, \{V_8(N), I_8(N)\}$	3.0	3.0	1.0	0.188	0.750	E_0	R
11	$\{V_2(L), I_2(L)\}, \{V_5(L), I_5(N)\}$	2.0	2.0	1.0	0.125	0.500	E_0	√
12	$\{V_2(L), I_2(L)\}, \{V_6(L), I_6(N)\}$	2.0	2.0	1.0	0.125	0.500	E_0	R
13	$\{V_2(L), I_2(L)\}, \{V_7(N), I_7(N)\}$	3.0	3.0	1.0	0.188	0.750	E_0	√
14	$\{V_2(L), I_2(L)\}, \{V_8(N), I_8(N)\}$	3.0	3.0	1.0	0.188	0.750	E_0	R
15	$\{V_5(L), I_5(H)\}, \{V_7(N), I_7(N)\}$	1.0	1.0	1.0	0.063	0.250	E_0	√
16	$\{V_5(L), I_5(H)\}, \{V_8(N), I_8(N)\}$	1.0	1.0	1.0	0.063	0.250	E_0	R
17	$\{V_6(L), I_6(H)\}, \{V_7(N), I_7(N)\}$	1.0	1.0	1.0	0.063	0.250	E_0	√
18	$\{V_6(L), I_6(H)\}, \{V_8(N), I_8(N)\}$	1.0	1.0	1.0	0.063	0.250	E_0	R
19a	$\{V_7(N), I_7(H)\}$	6.0	1.0	0.2	0.375	0.250	E_0	×
19b	$\{V_7(N), I_7(H)\}$	6.0	5.0	0.8	0.375	0.714	E_1	×
20a	$\{V_8(N), I_8(H)\}$	6.0	1.0	0.2	0.375	0.250	E_0	×
20b	$\{V_8(N), I_8(H)\}$	6.0	5.0	0.8	0.375	0.714	E_1	×
21	$\{V_1(L), I_1(H)\}, \{V_5(L), I_5(L)\}$	1.0	1.0	1.0	0.063	0.143	E_1	√
22	$\{V_1(L), I_1(H)\}, \{V_6(L), I_6(L)\}$	1.0	1.0	1.0	0.063	0.143	E_1	R
23	$\{V_7(N), I_7(L)\}$	2.0	2.0	1.0	0.125	0.286	E_1	×
24	$\{V_8(N), I_8(L)\}$	2.0	2.0	1.0	0.125	0.286	E_1	×
25	$\{V_1(L), I_1(L)\}$	3.0	3.0	1.0	0.188	1.000	S_1	√
26	$\{V_7(L), I_7(L)\}$	3.0	3.0	1.0	0.188	1.000	S_1	√
27	$\{V_8(L), I_8(L)\}$	3.0	3.0	1.0	0.188	1.000	S_1	R

higher classification errors since the 10% of the data we used for testing may be different than the 90% we used for training. As such, the rules generated do not cover the case in the 10% of the test data.

The same procedure is repeated again, this time with different training sets and split of data. For the training set that comprises of 70% of the data,leaving 30% for the test set, the accuracy of classification is approximately **77.1%** (see Table 34). In the final assessment, the accuracy of classification dropped, as expected, to approximately **66.3%** (see Table 35) when the smaller variety of

Table 42. Classifier result using the 90% Training Set and 10% Test set

Training	Test Sets (10%)				Mean
Set (90%)	Split 1	Split 2	Split 3	Split 4	Accuracy
1	1.000	1.000	1.000	1.000	1.000
2	1.000	1.000	1.000	1.000	1.000
3	1.000	1.000	0.750	1.000	0.938
4	1.000	1.000	1.000	1.000	1.000
		Measure of Accuracy			**0.984**

Table 43. Classifier result using the 70% Training Set and 30% Test set

Training	Test Sets (30%)				Mean
Set (70%)	Split 1	Split 2	Split 3	Split 4	Accuracy
1	0.909	1.000	1.000	1.000	0.977
2	1.000	1.000	1.000	1.000	1.000
3	1.000	1.000	0.909	1.000	0.977
4	1.000	1.000	1.000	1.000	1.000
		Measure of Accuracy			**0.989**

Table 44. Classifier result using the 50% Training Set and 50% Test set

Training	Test Sets (50%)				Mean
Set (50%)	Split 1	Split 2	Split 3	Split 4	Accuracy
1	0.941	1.000	0.941	1.000	0.971
2	1.000	1.000	1.000	1.000	1.000
3	0.706	0.765	0.471	0.765	0.676
4	1.000	1.000	1.000	1.000	1.000
		Measure of Accuracy			**0.912**

training set is used (only 50% of the data). The overall average accuracy for the three experiments is $(87.5 + 77.1 + 66.3)/3 = 76.97\%$. The reason of the lower accuracy in 30% and 50% test data cases is because the system has been trained on a smaller variety of events and therefore it may not recognise a larger variety of events for classifications. The larger size of the test set is used as it tends to have more events i.e. 3 events in the 30% partitions, but only 1 event in the 10%.

As the strength of the rule relies upon the large support basis that describes the number of events (which support each rule), the accuracy estimated from the 30% partition is therefore higher than the 10% partition. Another explanation is that the events are not equally distributed over the given decision classes when the data set is split. One decision class may dominate over the other decision classes. Due to lack of events in this analysis, the number of unclassified and/or misclassified events is more frequent. To overcome this problem, the events can be duplicated to ensure the same equal number of events are distributed over the decision classes in order to make the voting process more capable of detecting these events. This gives significant improvement in classification rate as can be seen in following Tables 42, 43 and 44.

The number of unclassified and/or misclassified events in power system should always be treated with high priority as it indicates the failure of the classifier and hence it needs to be verified by experts. The results showed that the overall classification rates are still good even though a small data set has been used.

4.2 Fault Data in Time Series

When a circuit breaker is tripped, the trip event and the disturbance must always be reviewed to verify the operating performance of the protection system. Despite the fact that the breaker failure in a substation is a rare event, it is still an important issue for substation data analysis. This case study was carried out to evaluate the data patterns in Fig. 3 when the overcurrent relay R1 or breaker BRK1 failed to operate. A single phase A to earth fault was applied at 1.0s to the load bus L1. The failure of breaker BRK1 at 1.091s has caused the upstream relays R5 and R6 to trip at 1.339s. Both IED relays, which serve as a backup to the downstream protection, saw the fault at the busbar. They both tripped the breakers BRK5 and BRK6 and inter-tripped the respective breakers BRK7 and BRK8 simultaneously. The trip category for the breaker BRK1 failure is given in Table 36 with inclusion of the fault indicator data while Table 37 listed out the voltage and current measurements captured before and after the breaker BRK1 failed. R9, R10, R11, R12 are excluded from Table 37 since these unit protection relays do not contribute to this fault analysis.

Table 38 shows a set of 18 reducts computed using the RS-GA model. In the parameter setting, the size of the keep list is set to 10. With the same setting, if we enlarge the size of the keep list by a factor of 10 times, the number of results generated would be 36. If we are interested only in shorter length of reducts, we can reward the short length reducts and filter out the longer reducts. Table 39 considers GA reducts in the relay box with breaker data, while Table 40 considers the GA reducts in individual voltage and current measurement and breaker data. The inclusion of breaker data reduces the number of rules generated but increases the restriction of using other relays for fault diagnosis. In other words, it decreases the coverage.

Table 41 listed the decision rules derived from the reduct sets in Table 38. The first four rules that represent the steady state normal operation, N_S are ignored since they do not play an essential part in the decision making/support. Some of the listed rules are redundant labelled as 'R' due to the similar patterns captured by different IED relays. As previously noted, they can be useful when one or more of these IED relays in operation are not working or in service. Therefore, these redundant rules can be used as the 'reserved rules' in the knowledge based system. Two inconsistencies were identified in Rules 19 and 20. The rules are not able to distinguish the outcome of E_0 and E_1. Rules 23 and 24 are discarded because they are not good enough to justify the emergency situation due to the lack of information. As for the smaller system proposed in the previous section, the idea is to extract, analyse and evaluate the decision rules for each simulated scenario before adding them into the knowledge based system. Each fault scenario has different pattern of rules. If the symptom matches the list of all given rules for a particular fault as in this case, for instance, then it can be concluded that the fault is at the load bus L1 and the breaker BRK1 has malfunctioned. The drawback of this system however is that each time the network topology changes, the knowledge base has to be updated/revised. An automatic update of the knowledge base can be achieved relatively easily by linking the simulation results performed on all topologies in which that the network can possibly be configured. However, this will greatly increase the size of knowledge base. The solution is to develop a generic knowledge base that could reuse some of the knowledge which are common to all the scenarios (See the following Sect. 5: Future Work).

Classifier's Performance: At **98.4%**, the average accuracy of the classifier for the 10% test data is good (see Table 42). The results demonstrated that most of the rules generated from the training sets are able to classify the events in the test set. The inaccuracy occurred when the rules generated from the training set 3 were used to classify the test set 3. Only one event with the decision class E_1 was left unclassified as the classifier was not able to distinguish it from the events with the decision class E_0. The cause of this problem is likely due to the small dataset used in this study (only 35 events).

The same procedure is repeated again, this time with different training sets and splits of data. For the training set that consists of 70% of the data with 30% for the test set, the accuracy of classification is approximately **98.9%** (see Table 43). In the final assessment, the accuracy of classification dropped, as expected, to approximately **91.2%** (see Table 44) when the smaller variety of training set is used (only 50% of the data). The overall average accuracy for all the three experiments is $(98.4 + 98.9 + 91.2)/3 = 96.17\%$. From the classifier results, we can see that only one event was actually unclassified in the 10% of data for testing whereas in the 30% and 50% partition, there are two events and six events left unclassified, respectively. The classification rates for the three tests in Tables 42, 43 and 44 perform better than the previous example simply because the events have been duplicated before the splits to ensure equal distribution of decision classes. This verifies that the classification rate in the first example is largely caused by unequal distribution of decision classes as the result of the small set of data used, rather than any misclassification of rules extracted in the training sets.

5 Future Work

There are many kinds of primary substations and each of them may have different configurations, operational rationale etc. It is therefore necessary to build a knowledge extraction system suitable for every kind of primary substation. In addition to that, increased penetration of distributed generation (DG) within a distribution network will alter the direction of the conventional power flows in the network and frequently trigger changes of network topology (disconnection and reconnection of the DGs). To overcome this challenge, a direct way is to use multiple sets of training database that consist of a set of rules that suits particular network configuration in the knowledge based system. However, such an approach will result a very large knowledge based system that may be costly to build and difficult to maintain. In order to resolve this problem it is necessary to share generic knowledge to reduce the cost of building each knowledge based system. This generic knowledge base is composed of general rules that operate the power system with the minimum human interaction. It should be applicable to most primary substations overlooking a portion of a distribution network and is easily adaptable even when the configurations of primary substations are changed. Building a generic knowledge base requires the understanding of the problem solving process of an expert, identifying the suitable method used, and characteristics of the problem solving knowledge.

In the event of a fault, at least one piece of equipment/asset in parallel could be overloaded as the result of losing one or more feeders. The current carrying capacity of the overloaded feeder will be twice higher than its previous nominal value after the event. In the case involving the tripping of one transformer, it shows that the system current returns to normal after the faulty feeder has been isolated. However, it does not indicate that the current state is in the full load or overload condition, meaning that if one transformer is isolated from the system, the other could have temporarily taken or exceeded the full-load capacity. Therefore, the future work is to develop a generic knowledge base system that is partially independent of network topology and is able to inform responsible engineers about the status of other transformers after the system is restored.

6 Conclusion

This chapter suggests the use of a hybrid RS-GA method to process and extract implicit knowledge from operational data derived from relays and circuit breakers. The proposed technique simplifies the rule generation process, reduces the rule maintenance costs, the outage response time and resources required to develop a rule-based diagnostic system. Two given examples have demonstrated how knowledge can be induced from data sets and the promise for practical application. The approach is also able to extract rules from time series data. The advantage of integrating RST and GA approach is that the former can effectively reduce a large quantity of data and generate efficient rules whereas the latter provides us flexibility to generate a number of rules subject to the parameter settings we provide, hence produce global optimal solutions. More importantly, when the population of nearly optimal (but different) individuals was computed, new objects/events can also be added dynamically. This is because the genetic algorithm is able to adapt itself to this new situation after a few generations [27]. The methodology is more attractive than some techniques e.g. Bayesian approach and Dempster Shafer theory, because no assumption about the independence of the attributes is necessary nor is any background knowledge about the data required [34].

Classic expert systems were developed to handle a specific task for a *fixed* network topology. Increased penetration of distributed generation (DGs) will alter the distribution network topology from time to time and could complicate the fault diagnosis task. Classic rules developed in the conventional knowledge based systems are no longer efficient enough to supervise active networks. An automated rule induction system is therefore essential for extracting rules from various simulated topologies in a distribution system influenced by the connection/disconnection of DGs. Future work is to develop generic rules applicable to all the simulated topologies that are identified and stored in one common database while other rules unique to specific topologies are kept separately in other databases.

References

1. McDonald, J.: Overwhelmed by Alarms: Blackout Puts Filtering and Suppression Technologies in the Spotlight. Electricity Today (8) (2003)
2. Hor, C., Shafiu, A., Crossley, P., Dunand, F.: Modeling a Substation in a Distribution Network: Real Time Data Generation for Knowledge Extraction. In: IEEE PES Summer Meeting, Chicago, Illinois, USA, July 21st-25th (2002)
3. Hor, C., Kangvansaichol, K., Crossley, P., Shafiu, A.: Relays Models for Protection Studies. In: IEEE Power Tech, Bologna, Italy, June 23rd-26th (2000)
4. Hor, C., Crossley, P.: Knowledge extraction from intelligent electronic devices. In: Peters, J.F., Skowron, A. (eds.) Transactions on Rough Sets III. LNCS, vol. 3400, pp. 82–111. Springer, Heidelberg (2005)
5. Hor, C., Crossley, P.: Extracting Knowledge from Substations for Decision Support. IEEE Transactions on Power Delivery 20(2), 595–602 (2005)
6. Hor, C., Crossley, P.: Unsupervised Event Extraction within Substations Using Rough Classification. IEEE Transactions on Power Delivery 21(4), 1809–1816 (2006)
7. Hor, C., Crossley, P., Watson, S.: Building Knowledge for Substation based Decision Support using Rough Sets. IEEE Transactions on Power Delivery 22(3), 1372–1379 (2007)
8. Ackerman, W.J.: Substation automation and the EMS. In: IEEE Transmission and Distribution Conference, USA, April 11-16, vol. 1(1), pp. 274–279 (1999)
9. McDonald, J.: Substation integration and automation. In: Grigsby, L.L. (ed.) Electric Power Substations Engineering. CRC Press, Boca Raton (2007)
10. McDonald, J.: Substation Automation Basics - The Next Generation. Electric Energy T& D Magazine (May-June 2007)

11. Brunner, C., Kern, T., Kruimer, B., Schimmel, G., Schwarz, K.: IEC 61850 based digital communication as interface to the primary equipment Evaluation of system architecture and real time behavior. CIGRE 2004, Study committee B3 (2004)
12. CIGRE Study Committee B5, The automation of New and Existing Substations: why and how, International Council on Large Electric Systems, 21 rue d'Artois, 75008 Paris, France (August 2003)
13. Brand, K.-P., Lohmann, V., Wimmer, W.: Substation Automation Handbook. Utility Automation Consulting Lohmann (2003), `http://www.uac.ch` ISBN 3-85758-951-5
14. Komorowski, J., Pawlak, Z., Polkowski, L., Skowron, A.: Rough Sets: A Tutorial. In: Pal, S.K., Skowron, A. (eds.) Rough Fuzzy Hybridization – A New Trend in Decision Making. Springer, Singapore (1999)
15. Komorowski, J., Pawlak, Z., Polkowski, L., Skowron, A.: A Rough Set Perspective on Data and Knowledge. In: Klsgen, W., Zytkow, J. (eds.) Handbook of Data Mining and Knowledge Discovery. Oxford University Press, Oxford (2000)
16. Hang, X., Dai, H.: An optimal strategy for extracting probabilistic rules by combining rough sets and genetic algorithm. In: Grieser, G., Tanaka, Y., Yamamoto, A. (eds.) DS 2003. LNCS, vol. 2843, pp. 153–165. Springer, Heidelberg (2003)
17. Zhai, L.Y., Khoo, L.P., Fok, S.C.: Feature extraction using rough set theory and genetic algorithms–an application for the simplification of product quality evaluation. Computers and Industrial Engineering 43(3), 661–676 (2002)
18. Khoo, L.P., Zhai, L.Y.: A prototype genetic algorithm enhanced rough set based rule induction system. Computers in Industry 46(1), 95–106 (2001)
19. Pawlak, Z., Busse, J., Slowinkis, R., Ziarko, W.: Rough Sets. Artificial Intelligence Emerging Technologies magazine, Communications of the ACM 38, 89–95 (1995)
20. Aldridge, C.: A Theory of Empirical Spatial Knowledge Supporting Rough Set based Knowledge Discovery in Geographic Databases. PhD thesis, Department of Information Science, University of Otago, Dunedin, New Zealand (1999)
21. Mollestad, T., Skowron, A.: A Rough Set Framework for Data mining of propositional default rules. In: Proceeding of the 9th International Symposium on Methodologies for Intelligent Systems, Zakopane, Poland (1996)
22. Pawlak, Z.: Rough Set approach to Knowledge based Decision Support. In: Proceeding of the 14th European Conference on Operational Research, Jerusalem, Israel (1995)
23. Skowron, A., Rauszer, C.: The discernibility matrices and functions in information systems. In: Intelligent Decision Support - Handbook of Applications and Advances of the Rough Set Theory, pp. 331–362. Kluwer Academic Publishers, Dordrecht (1992)
24. Hvidsten, T.: Fault Diagnosis in Rotating Machinery using Rough Set Theory and ROSETTA. Technical report, Department of Computer Science and Information Science, Norwegian University of Science and Technology, Trondheim, Norway (1999)
25. IEEE Tutorial on Modern Heuristic Optimisation Techniques with Applications to Power Systems, IEEE PES Tutorial 02TP160
26. Jensen, R., Shen, Q.: Rough set based feature selection: A review. In: Hassanien, A.E., Suraj, Z., Slezak, D., Lingras, P. (eds.) Rough Computing: Theories, Technologies and Applications. Idea Group Inc (IGI) Publisher, Europe (2007)
27. Wroblewski, J.: Finding minimal reducts using Genetic Algorithm. In: Second Annual Joint Conference on Information Sciences, pp. 186–189 (1995)

28. Øhrn, A.: Discernibility and Rough Sets in Medicine: Tools and Applications. PhD thesis, Department of Computer Science and Information Science, Norwegian University of Science and Technology (NTNU), Trondheim, Norway (2000)
29. Roed, G.: Knowledge Extraction from Process Data: A Rough Set Approach to Data Mining on Time Series. Master thesis, Department of Computer and Information Science, Norwegian University of Science and Technology, Trondheim, Norway (1999)
30. Gross, G., Bose, A., DeMarco, C., Pai, M., Thorp, J., Varaiya, P.: Real Time Security Monitoring and Control of Power Systems, Technical report, The Consortium for Electric Reliability Technology Solutions (CERTS) Grid of the Future White Paper (1999)
31. Hill, S., O'Riordan, C.: Inversion Revisited - Analysing an Inversion Operator Using Problem Generators. In: Proceeding of the Bird of a Feather Workshops, Genetic and Evolutionary Computation Conference (GECCO), Chigaco, pp. 34–41 (2003)
32. East Midlands Electricity, Long Term Development Statement for East Midlands Electricity. Summary Information, East Midlands Electricity (2003)
33. Bjanger, M.: Vibration A nalysis in Rotating Machinery using Rough Set Theory and ROSETTA. Technical report, Department of Computer Science and Information Science, Norwegian University of Science and Technology, Trondheim, Norway (1999)
34. Aasheim, Ø., Solheim, H.: Rough Sets as a Framework for Data Mining. Technical report, Knowledge Systems Group, Faculty of Computer Systems and Telematics, The Norwegian University of Science and Technology, Trondheim, Norway (1996)

Index

Author Index